Stochastic Processes and Filtering Theory

This is Volume 64 in
MATHEMATICS IN SCIENCE AND ENGINEERING
A series of monographs and textbooks
Edited by RICHARD BELLMAN, *University of Southern California*

The complete listing of books in this series is available from the Publisher
upon request.

STOCHASTIC PROCESSES AND FILTERING THEORY

Andrew H. Jazwinski

Analytical Mechanics Associates, Inc.
Seabrook, Maryland

1970

ACADEMIC PRESS   New York and London

ACADEMIC PRESS, INC.
111 Fifth Avenue, New York, New York 10003

United Kingdom Edition published by
ACADEMIC PRESS, INC. (LONDON) LTD.
Berkeley Square House, London W1X 6BA

LIBRARY OF CONGRESS CATALOG CARD NUMBER: 71-84150
AMS 1968 SUBJECT CLASSIFICATIONS 6050, 6075, 6520, 9360

Second Printing, 1972

PRINTED IN THE UNITED STATES OF AMERICA

TO ANITA

Preface

This book presents a unified treatment of linear and nonlinear filtering theory for engineers, with sufficient emphasis on applications to enable the reader to use the theory. The need for this book is twofold. First, although linear estimation theory is relatively well known, it is largely scattered in the journal literature and has not been collected in a single source. Second, available literature on the continuous nonlinear theory is quite esoteric and controversial, and thus inaccessible to engineers uninitiated in measure theory and stochastic differential equations. Furthermore, it is not clear from the available literature whether the nonlinear theory can be applied to practical engineering problems. In attempting to fill the stated needs, I have retained as much mathematical rigor as I felt was consistent with my prime objective—to explain the theory to engineers. Thus I have avoided measure theory in this book by using mean square convergence and calculus rather than with probability one convergence, on the premise that everyone knows how to average. As a result, I only require of the reader background in advanced calculus, theory of ordinary differential equations, and matrix analysis. A review of probability and stochastic processes, on an engineering level, is included in the first chapters of the book. Since mathematical rigor has not been completely sacrificed, this book should also be of interest to the applied mathematician. Of particular interest might be the relatively complete treatment of mean square calculus and the chapter on stochastic differential equations. The book is relatively complete and self-contained, and is thus suitable for either a full two-semester graduate course in filtering theory, linear and nonlinear, or a single-semester graduate course in linear filtering theory.

The theme of the book is the probabilistic or Bayesian approach. Dynamical systems are modeled by finite-dimensional Markov processes; outputs of stochastic difference and differential equations. The fundamental entity in our approach to filtering is the conditional probability density function of the state of our dynamical system, conditioned on the available measurements made on the system. This approach seems essential for a development of the nonlinear theory. We depart from the main theme on occasion to rederive the results using more classical statistical methods. Nonlinear filtering results are derived first, and these are then specialized to linear systems. The treatment of linear filtering includes filter stability and model error sensitivity. We emphasize the application of linear filters to nonlinear problems, including the important problem of filter divergence and a variety of model error compensation techniques designed to prevent the divergence. The last chapter deals with the development of approximate nonlinear filters and presents concrete applications in nonlinear problems. The performance of these nonlinear filters is critically analyzed.

This book is an outgrowth of notes for a series of graduate lectures and seminars given at the NASA Goddard Space Flight Center, the NASA Electronics Research Center, and MIT in 1966. In the course of these seminars, which dealt primarily with nonlinear filtering theory, I found that the engineers and students generally had an insufficient background in probability and stochastic processes to understand filtering theory. I therefore devoted several weeks to a review of these subjects. It is because I believe deficiency in these subjects is the rule rather than the exception that I have included a review of probability theory and stochastic processes in the book.

ANDREW H. JAZWINSKI

Seabrook, Maryland
September 1969

Acknowledgments

It is with a great deal of pleasure that I acknowledge the active support and encouragement of Mr. R. K. Squires and the Special Projects Branch of the NASA Goddard Space Flight Center in the preparation of this book. My thanks also go to the management of Analytical Mechanics Associates, Inc. for their support, without which the preparation of this book would not have been possible. I am indebted to Mrs. A. Bailie and Mr. N. Levine of Analytical Mechanics Associates for most of the numerical results appearing in this book. Their excellent and often bugless programming sometimes amazed me. I wish to thank Prof. H. J. Kushner of Brown University for his critical review of the manuscript and several helpful suggestions. My thanks also go to Dr. W. M. Wonham of the NASA Electronics Research Center, Mr. W. R. Hahn of the U. S. Naval Ordnance Laboratory, and Messrs. Gentry Lee and W. E. Wagner of the Martin Company for reading the manuscript and offering their useful critiques. I am also indebted to Mrs. A. Michalowski for her outstanding editorial assistance, and to Mrs. K. Brice for her excellent typing. The efforts and cooperation of the Academic Press staff is also gratefully acknowledged. Last, but not least, I thank my wife Anita for her encouragement and understanding during the preparation of this book.

My own research, which is contained in this book, was supported by the NASA Goddard Space Flight Center under Contracts NAS 5-9298 and NAS 5-11048.

Contents

4. Stochastic Differential Equations

5. Introduction to Filtering Theory

6. Nonlinear Filtering Theory

7. Linear Filtering Theory

8. Applications of Linear Theory

9. Approximate Nonlinear Filters

1

Introduction

1. INTRODUCTION

Physical systems are designed and built to perform certain defined functions. Submarines, aircraft, and spacecraft must navigate in their respective environments to accomplish their objectives, whereas an electric power system network must meet the load demands. In order to determine whether a system is performing properly, and ultimately to control the system performance, the engineer must know what the system is "doing" at any instant of time. In other words, the engineer must know the *state* of his system. In navigation, the state consists of position and velocity of the craft in question; in an ac electric power system, the state may be taken as voltages and phase angles at network nodes. Physical systems are often subject to random disturbances, so that the system state may itself be random. In order to determine the state of his system, the engineer builds a measuring device (e.g., sextant, radar, voltmeter) and takes *measurements* or *observations* on his system. These measurements are generally contaminated with noise caused by the electronic and mechanical components of the measuring device.

The problem of determining the state of a system from noisy measurements is called *estimation*, or *filtering*, and is the main subject of this book. It is of central importance in engineering, since state estimates are required in the monitoring, and for the control of systems. Furthermore, a large class of (system) identification problems can be regarded as problems of filtering.

This book takes the state-space approach to filtering. The physical (dynamical) system is modeled by a finite-dimensional Markov process, the output of a stochastic (random) differential or difference equation. The quasi-steady state estimation in an ac electric power system network is an example of a static problem, which is a special case of the dynamical model system representation used in this book. We take the Bayesian, or probabilistic, point of view in filtering. That is, we look upon the conditional probability density function of the state, given the measurements, as the fundamental entity in the problem. This conditional density embodies all the information about the state of the system which is contained in the available measurements, and all estimates of the state can be constructed from this density.

Our approach enables us to give a rather rigorous and relatively complete and self-contained treatment of nonlinear, as well as linear, filtering theory. Yet we avoid the measure-theoretic aspects of probability theory, thus making our work accessible to engineers. In fact, this book is principally intended for engineers. Because mathematical rigor has not been completely sacrificed, our work should also be of interest to the applied mathematician.

2. SCOPE AND OBJECTIVES

Following the fundamental work of Kalman and Bucy [3, 4][1] in linear filtering theory, a host of papers and reports appeared, formally deriving their linear filtering algorithm via "least squares," "maximum likelihood," and other classical, statistical methods.[2] These statistical methods were also formally applied to the nonlinear estimation problem. Using linearization of one sort or another, Kalman-like filtering algorithms were developed and applied to nonlinear problems. This statistical work received its impetus from the aerospace dollar. Work was duplicated, triplicated; everyone derived his own Kalman filter, perhaps partly because of a lack of understanding of Kalman's original work.

---

[1] Numbers in brackets refer to references or bibliography at the end of each chapter.

[2] A sketch of the developments in filtering theory, including the earlier work of Wiener, may be found in Chapter 5, especially in Sections 5.1 and 5.3.[3]

[3] Section 5.1 refers to Section 1 of Chapter 5. We use the following convention. In referring to a section of another chapter, the section number is preceded by the chapter number, as in Section 5.1. When referring to a section of the same chapter, only the section number is used. Thus if we refer to Section 1 here, we mean Section 1 of the present chapter.

In the meantime, such important aspects of the linear theory as filter stability, error sensitivity, and the problem of modeling, some of which were treated in Kalman's original work, went largely ignored. It is a truism, however, that the step from theory to practice is a large and difficult one, and that the aerospace community contributed much to the reduction of the theory to practice. The statistical work served to bring linear filtering within the grasp of the practicing engineer, and led to the applications[4] of the theory in satellite orbit determination, submarine and aircraft navigation, and space flight, including the Ranger, Mariner, and Apollo missions. On the other hand, the statistical methods tend to mask the fundamental probabilistic structure of the filtering problem. This structure consists of a stochastic process evolving in time, whose probability law is to be determined from observations of another, related, stochastic process. It becomes increasingly important to exploit this probabilistic structure when dealing with the nonlinear filtering problem.

While Kalman and Bucy were formulating the linear theory in the United States, Stratonovich [9] was pioneering in the development of the probabilistic approach to nonlinear filtering in Russia. The work of Stratonovich was not immediately known in the West, however, and Kushner [5, 6] and also Wonham [10] independently developed the continuous nonlinear theory here. Ho and Lee [1] and Jazwinski [2], respectively, applied the probabilistic approach to discrete and continuous-discrete[5] nonlinear problems. Much of the subsequent theoretical work in nonlinear filtering was done by Kushner.

Most of the work in nonlinear filtering is very theoretical, involving such hitherto obscure and difficult subjects as "stochastic differential equations" and the "Itô calculus," which require a fair facility with measure theory for understanding. An engineer interested in applications is virtually lost. Furthermore, it is not clear from the literature whether the nonlinear theory can be fruitfully applied to any significant practical problem. A student or engineer interested in learning the linear theory also has a difficult task before him. There is a large body of journal literature on the subject which thus far has not been collected in a single source.

Our objective in writing this book is twofold. First, it is to present the current state-of-the-art in nonlinear filtering theory in a form accessible to engineers, yet without completely sacrificing mathematical rigor. This includes the development of approximate nonlinear filters suitable

---

[4] See Chapter 8 for aerospace applications.

[5] Problems with continuous dynamics, but with discrete observations.

for applications, and an analysis of their efficacy in nonlinear problems. Second, it is to gather together, in a unified presentation, many of the results in linear filtering theory. This includes such important properties of linear filters as stability and model error sensitivity. We emphasize the applications of the linear theory to nonlinear problems, and treat the current problem area of filter divergence.

We have attempted to make our book relatively complete and self-contained. To this end, we have included essentially all the material on probability theory and stochastic processes we need for our presentation of filtering theory. This includes a thorough treatment of stochastic differential equations. We avoid measure theory by using mean square convergence rather than probability one convergence. The background we require of the reader consists essentially of advanced calculus, theory of ordinary differential equations, and matrix analysis, plus some mathematical sophistication. Aside from mean square calculus and stochastic differential equations, which we develop fully in this book, we use probability theory and stochastic processes on an engineering level comparable to that in Parzen [8] or Papoulis [7].

The unifying theme in the book is the probabilistic or Bayesian approach. We develop the nonlinear theory first, and obtain linear filtering algorithms as special cases of the nonlinear results. We do, however, include many of the more popular derivations of linear filters in examples. Thus a reader interested only in the linear theory and its applications can avoid much of the mathematics, as we explain in Section 3. Our book is thus suitable as a text for a full two-semester graduate course in filtering theory (linear and nonlinear), or for a one-semester graduate course in linear filtering theory. We include a number of examples that illustrate the concepts. These can be supplemented by the instructor or the reader himself.

3. A GUIDED TOUR

We describe here the contents of our book and indicate how readers, with varying interests and mathematical backgrounds, might approach it.

Background material in probability theory and stochastic processes is given in Chapters 2 and 3. Most of this material is on the engineering level comparable to Parzen [8] or Papoulis [7], with the following exceptions. Certain properties of moments and important inequalities are developed in Section 2.4. That same section also contains an introduction to (random) vectors and matrices. The concepts of convergence of random sequences are defined in Section 3.3, and mean square calculus is developed in Section 3.4. These are required for the study of stochastic differential equations and stochastic calculus in Chapter 4, which in turn are required for the development of the continuous nonlinear filter in Chapter 6. Section 3.9 is devoted to a study of stochastic difference equations, which is prerequisite for discrete filtering.

The reader familiar with probability theory and stochastic processes on the engineering level might, with the exceptions just described, use Chapters 2 and 3 for reference or review. We would advise the review of Section 2.5 on conditional probabilities and expectations, Section 2.6 on Gaussian random variables, and Section 3.8 on white noise. We might add that our treatment of probability theory and stochastic processes is restricted to continuous random variables and stochastic processes with a continuous state space. Furthermore, we deal only with Markov processes. These are the processes generated by stochastic difference and differential equations; the models we use for dynamical systems.

Chapter 4 is devoted to a study of stochastic differential equations. The theme here is the modeling of a certain stochastic process as the solution of a (stochastic) differential equation. This leads to definition of the Itô stochastic integral and proof of existence and uniqueness of solutions of stochastic differential equations. It is here that we use the mean square calculus developed in Section 3.4. This is followed by a formal derivation of Itô's lemma, which is a fundamental tool in continuous nonlinear filter theory. We next introduce another stochastic integral, that of Stratonovich, and show its relationship with the Itô integral. We then deal, insofar as our mathematical tools permit us, with the important problem of modeling a physical process by a stochastic differential equation. We use the Itô integral throughout the book, and our justification for so doing is given in Section 4.10. Kolmogorov's equation is derived in Section 4.9. It describes the evolution of the unconditioned probability density function of our process and thus is a fundamental tool in continuous nonlinear filtering theory.

Chapter 5 begins the study of the main subject of our book by introducing and defining the problems of filtering, prediction, and smoothing. Our probabilistic approach is described, with emphasis on optimality and optimality criteria in estimation. We also describe the statistical approach to filtering and its probabilistic interpretations.

Chapter 6 presents the (mathematical) solutions to nonlinear filtering problems. We develop equations of evolution for the conditional probability density function, for moments, and for the mode, in continuous, continuous-discrete, and discrete filtering and prediction problems. We also develop the concept of limited memory filtering, with a view toward applications in problems with erroneous system models.

The nonlinear theory is specialized to linear problems in Chapter 7, and linear filters and predictors are developed. More popular derivations of the linear filters, using nonprobabilistic methods, are included in examples. We then introduce the concepts of information, observability, and controllability, and use these in a study of error bounds and stability of the linear filters. The emphasis here is on the discrete filter, since this has not been explicitly treated in the literature. We then study the sensitivity of filter performance to errors in the statistical and dynamical system model. Finally, the linear limited memory filter is developed.

The last two chapters deal with applications.

Chapter 8 develops the applications of linear filter theory, particularly to nonlinear problems. Here we emphasize the discrete and continuous-discrete problems, and go into considerable detail to assist the reader in getting a firm grasp on the theory, by applying the filter to a concrete example problem. We then present actual applications of linear filters in orbit determination and reentry. The balance of the chapter is devoted to the problem of filter divergence due to model errors, and techniques found useful in preventing divergence. We present simulations of two such techniques: the adaptive and limited memory filters.

Chapter 9 is devoted to the development, analysis, and simulations of approximate nonlinear filters. We critically assess the state-of-the-art in nonlinear filtering, and suggest some areas for future research.

Readers who are primarily interested in linear filtering, or those who either do not have the mathematical background required for the non-linear theory and/or prefer to study the linear theory first, might proceed as follows. Find any of a number of derivations of the linear filtering algorithm via nonprobabilistic methods in the examples of Chapter 7 (Examples 7.1–7.4 for the discrete filter; Examples 7.9–7.13 for the continuous filter). Then proceed to a study of the linear theory in the remainder of Chapter 7, and its applications in Chapter 8. This route, of course, still requires an engineering background in probability and stochastic processes.

REFERENCES

1. Y. C. Ho and R. C. K. Lee, A Bayesian Approach to Problems in Stochastic Estimation and Control, *IEEE Trans. on Automatic Control* **9**, 333–339 (1964).

2. A. H. Jazwinski, Nonlinear Filtering with Discrete Observations, AIAA 3rd Aerospace Sciences Meeting, New York, Paper No. 66–38, January 1966.

3. R. E. Kalman, A New Approach to Linear Filtering and Prediction Problems, *Trans. ASME, Ser. D: J. Basic Eng.* **82**, 35–45 (1960).

4. R. E. Kalman and R. S. Bucy, New Results in Linear Filtering and Prediction Theory, *Trans. ASME, Ser. D: J. Basic Eng.* **83**, 95–108 (1961).

5. H. J. Kushner, On the Dynamical Equations of Conditional Probability Density Functions, with Applications to Optimal Stochastic Control Theory, *J. Math. Anal. Appl.* **8**, 332–344 (1964).

6. H. J. Kushner, On the Differential Equations Satisfied by Conditional Probability Densities of Markov Processes, *SIAM J. Control* **2**, 106–119 (1964).

7. A. Papoulis, "Probability, Random Variables, and Stochastic Processes." McGraw-Hill, New York, 1965.

8. E. Parzen, "Stochastic Processes." Holden-Day, San Francisco, California, 1962.

9. R. L. Stratonovich, Conditional Markov Processes, *Theor. Probability Appl.* **5**, 156–178 (1960).

10. W. M. Wonham, Stochastic Problems in Optimal Control, *IEEE Intern. Conv. Record* **11**, 114–124 (1963).

2

Probability Theory and Random Variables

1. INTRODUCTION

This chapter offers some background material in probability theory for those readers not already familiar with this subject. The level of presentation corresponds to that in Parzen [6] or Papoulis [5], that is, it is nonmeasure theoretic. Our development is by no means complete. Only those topics are included which are required in the subsequent chapters. Although the presentation is somewhat terse (this is not a text on probability theory), a number of examples are included for the purpose of illustrating various concepts. Readers desiring a more complete and detailed account of the theory are referred to the books by Parzen and Papoulis, which have already been cited. Those readers with a knowledge of measure theory might consult Loeve [4]. For the benefit of those who are familiar with measure theory, we note some of the relationships with that subject in footnotes.

Before proceeding to the subject proper, we delineate and define probability theory and the role it plays in "problem solving," which is, after all, our ultimate objective. We have in mind a physical phenomenon (*experiment*) which is random. That is to say, the outcome of any one occurrence (*trial*) of the phenomenon (*experiment*) cannot be predicted, but is rather governed by chance. For concreteness, consider the rolling of a fair die. The problem is to determine the number of times an even number will come up in a thousand rolls.

We proceed as follows: (1) There are six equally likely possible outcomes (*events*) on a single roll of a fair die: the six sides. Therefore, the probability of any one side coming up on a single roll is $\frac{1}{6}$. (2) Since the six possible events are mutually exclusive, that is, the occurrence of one excludes the others, and three of the possible events result in the event *an even number comes up*, we conclude that the probability of an even number on a single roll is $\frac{1}{6} + \frac{1}{6} + \frac{1}{6} = \frac{1}{2}$. (3) We therefore predict that an even number will come up about 500 times in 1000 rolls.

It is seen that the solution of the foregoing problem proceeds in three distinct steps. These are

1. determination of (*prior*) probabilities of certain events,

2. operation on these probabilities with certain *rules* to obtain probabilities of other events, and

3. making physical predictions on the basis of the probabilities obtained in Step 2.

Now probability theory deals *only* with Step 2. It consists of certain *rules* derived from *axioms* by deduction. The axioms are, of course, consistent with our intuitive and experimental notions of probabilities. These rules tell us how to operate on the probabilities of certain events to obtain probabilities of other events. Steps 1 and 3 belong to the science of statistics and will not concern us here.

It is worth noting that there are generally three ways of assigning prior probabilities (Step 1). These are

(a) *relative frequency:* conducting a large number of experiments and computing the ratios N_A/N, where N_A is the number of times event A occurs and N is the total number of experiments;

(b) *classical method:* prior probabilities are found *a priori* without experimentation by counting all the equally likely alternatives;

(c) assigning prior probabilities as a *measure of belief,* based on the precept that subjective knowledge is better than no knowledge at all.

In our example of the die experiment, we used method (b) to assign prior probabilities. Method (a) is widely used in practice, especially in experiments that are conceptually too difficult to apply the classical method (b). We shall, in fact, use relative frequency notions to motivate some of our definitions whose rigorous treatment is beyond the scope of this book.

2. PROBABILITY AXIOMS

The underlying concept in probability theory is that of the **probability space**, which we denote by Ω, with **elements** denoted by ω. Ω is the basic sample space, and its elements ω are samples or experimental outcomes. Certain subsets of Ω (collections of outcomes) are called **events**. The class of events will be defined more precisely later.

Referring to the problem discussed in the introductory section, the rolling of a die, we might take $\Omega = \{1, 2, 3, 4, 5, 6\}$, where the element $1 \leqslant \omega \leqslant 6$ is the experimental outcome *the number ω turns up when the die is rolled*. $\Lambda_1 = \{2\} \subset \Omega$ is the event *2 turns up when the die is rolled*, and $\Lambda_2 = \{2, 4, 6\} \subset \Omega$ is the event *an even number turns up when the die is rolled*. If on a particular trial of this experiment the outcome is $\omega = 2$, then both events Λ_1 and Λ_2 occur on that trial. It is clear, that in order to proceed with a probabilistic analysis, all events have to have probabilities assigned to them. In the die experiment, all the ω sets (subsets of Ω) are events and can have assigned probabilities.

Recall the following set theory notation. If A and B are two sets, then $A \cup B$ is their *union*, $A \cap B$ is their *intersection*. $A - B$ is the *complement* of B with respect to A. $\varnothing$ denotes the *empty set*. If $A \cap B = \varnothing$, that is, the sets A and B are *disjoint*, then the events A and B are said to be **mutually exclusive**.

We assign probabilities to events via a **probability function** $\Pr\{\cdot\}$ defined on the class of events. That is, to each event Λ we assign a number $\Pr\{\Lambda\}$, called the *probability of Λ*. The probability function satisfies the following axioms:

(i) $\Pr\{\Lambda\} \geqslant 0$;

(ii) $\Pr\{\Omega\} = 1$;

(iii) if $\Lambda_i \cap \Lambda_j = \varnothing$, $i \neq j$, $i, j = 1, \dots, n$, then
$\Pr\{\Lambda_1 \cup \Lambda_2 \cup \cdots \cup \Lambda_n\} = \Pr\{\Lambda_1\} + \cdots + \Pr\{\Lambda_n\}$;

(iv) if $\Lambda_i \cap \Lambda_j = \varnothing$, $i \neq j$, $i, j = 1, \dots, n, \dots$, then
$\Pr\{\Lambda_1 \cup \cdots \cup \Lambda_n \cup \cdots\} = \Pr\{\Lambda_1\} + \cdots + \Pr\{\Lambda_n\} + \cdots$.[1]

Probability theory is based on these axioms. It consists of rules obtained from these axioms by deduction. These axioms, of course, conform with the relative frequency interpretation of probabilities.

---

[1] As is evident from the axioms, $\Pr\{\cdot\}$ is a completely additive, nonnegative set function that maps ω sets into the positive real numbers, normalized so that $\Pr\{\Omega\} = 1$. The class of events will be a class of measurable sets, and the probability function is then nothing more than a probability measure.

From the probability axioms, it is, for example, easy to show that

$$\Pr\{\varnothing\} = 0, \tag{2.1}$$

$$\Pr\{\Lambda\} = 1 - \Pr\{\Omega - \Lambda\} \leqslant 1, \tag{2.2}$$

$$\Pr\{\Lambda_1 \cup \Lambda_2\} = \Pr\{\Lambda_1\} + \Pr\{\Lambda_2\} - \Pr\{\Lambda_1 \cap \Lambda_2\}$$
$$\leqslant \Pr\{\Lambda_1\} + \Pr\{\Lambda_2\}, \tag{2.3}$$

$$\text{if} \quad \Lambda_1 \subset \Lambda_2, \quad \text{then} \quad \Pr\{\Lambda_2\} = \Pr\{\Lambda_1\} + \Pr\{\Lambda_2 \cap (\Omega - \Lambda_1)\} \geqslant \Pr\{\Lambda_1\}. \tag{2.4}$$

PROOF: (2.1): Since $\Omega \cap \varnothing = \varnothing$ and $\Omega = \Omega \cup \varnothing$, we have, from Axiom (iii),

$$\Pr\{\Omega\} = \Pr\{\Omega \cup \varnothing\} = \Pr\{\Omega\} + \Pr\{\varnothing\},$$

and then (2.1) follows by Axiom (ii).

(2.2): Since $(\Omega - \Lambda) \cap \Lambda = \varnothing$ and $(\Omega - \Lambda) \cup \Lambda = \Omega$, we have from Axiom (iii),

$$\Pr\{\Omega\} = \Pr\{(\Omega - \Lambda) \cup \Lambda\} = \Pr\{\Omega - \Lambda\} + \Pr\{\Lambda\}.$$

Now using Axiom (ii),

$$\Pr\{\Lambda\} = 1 - \Pr\{\Omega - \Lambda\} \leqslant 1,$$

where the last inequality follows from Axiom (i).

(2.3): The sets $\Lambda_1 \cup \Lambda_2$ and Λ_2 can be written as unions of two disjoint sets:

$$\Lambda_1 \cup \Lambda_2 = \Lambda_1 \cup [(\Omega - \Lambda_1) \cap \Lambda_2],$$
$$\Lambda_2 = (\Lambda_1 \cap \Lambda_2) \cup [(\Omega - \Lambda_1) \cap \Lambda_2].$$

Then by Axiom (iii),

$$\Pr\{\Lambda_1 \cup \Lambda_2\} = \Pr\{\Lambda_1\} + \Pr\{(\Omega - \Lambda_1) \cap \Lambda_2\},$$
$$\Pr\{\Lambda_2\} = \Pr\{\Lambda_1 \cap \Lambda_2\} + \Pr\{(\Omega - \Lambda_1) \cap \Lambda_2\}.$$

Eliminating $\Pr\{(\Omega - \Lambda_1) \cap \Lambda_2\}$ from these last two equations, we get (2.3). The inequality in (2.3), of course, follows from Axiom (i). (2.4): Since $\Lambda_1 \subset \Lambda_2$,

$$\Lambda_2 = \Lambda_1 \cup [(\Omega - \Lambda_1) \cap \Lambda_2],$$

which is the union of two disjoint sets. Equation (2.4) then follows by Axioms (iii) and (i). ■

We have yet to define the class of events on which the probability function is defined. In the die experiment, we can define $\Pr\{\,\cdot\,\}$ on the class of all the subsets of Ω. That is, probabilities can be assigned to all the ω sets; all ω sets are events. This is unfortunately not true in general.[2] Now in defining the class of events, we want the set operations (unions, intersections, complements) performed on events to yield sets that are also events. A class of sets having these properties is called a Borel field. More precisely, a class F of ω sets is called a **Borel field** if

(1) $\Omega \in F$;

(2)   if $\Lambda \in F$, then $\Omega - \Lambda \in F$;

(3)   if $\Lambda_1, \Lambda_2, ..., \Lambda_n \in F$, then $\bigcup_1^n \Lambda_i \in F$ and $\bigcap_1^n \Lambda_i \in F$;

(4)   if $\Lambda_1, \Lambda_2, ... \in F$, then $\bigcup_1^\infty \Lambda_i \in F$ and $\bigcap_1^\infty \Lambda_i \in F$.

We state without proof that, given a class F_o of ω sets, there is a unique Borel field of ω sets $B(F_o)$ with the properties

(1) $F_o \subset B(F_o)$;

(2)   if F_1 is a Borel field of ω sets and if $F_o \subset F_1$, then $B(F_o) \subset F_1$.

$B(F_o)$ is the *smallest* Borel field of ω sets which contains all the sets of F_o. It is called the **Borel field generated by** F_o. If, for example, $\Omega = \{real\ numbers\}$, we can take $F_o = \{$intervals of the form $\omega \leqslant x_1$, for all $x_1\}$. Then $B(F_o)$ contains all the open and closed intervals, points, and all countable unions and intersections of intervals and points, that is, all the sets on the real line which are of interest in applications. If $\Omega = E^2$ (the plane), B consists of rectangles and all countable unions and intersections of rectangles. For $\Omega = E^n$ (Euclidean n-space), B is similarly defined.

The **class of events** is a Borel field B and the probability function is defined on B. We suppose that the probability space Ω, Borel field B, and probability function $\Pr\{\,\cdot\,\}$ have been defined. The triplet $(\Omega, B, \Pr)$ is called an **experiment**. It will be understood that all sets referred to are Borel sets, that is, sets belonging to B.

Example 2.1.   Consider again the experiment of rolling a die (once). We can define a probability space $\Omega_1 = \{1, 2, 3, 4, 5, 6\}$. We can take the set of all subsets of Ω_1 as our Borel field B_1. However, if we are only interested in betting on the events $\{even\}$ or $\{odd\}$, we can take the

---

[2] All ω sets may not be measurable. It may be impossible to assign probabilities to all ω sets and satisfy probability Axiom (iv).

field $B_2 = \{\varnothing, \{1, 3, 5\}, \{2, 4, 6\}, \Omega_1\}$. It is easily seen how to define a probability function on B_1 and B_2. Note, however, that

$$A = \{\varnothing, \{1, 3, 5\}, \{2, 4, 6\}, \Omega_1, \{1\}\}$$

is not a field, since $\{1\} \cup \{2, 4, 6\} = \{1, 2, 4, 6\} \notin A$, even though $\{1, 2, 4, 6\} \subset \Omega_1$ and $\Omega_1 \in A$. ▲

Even the definition of a probability space for a given physical experiment is not unique. In our die experiment, we could define Ω_2 as the set of all real numbers. The probability of any Ω_2 subset is then $\frac{1}{6}$ the number of the points $1, 2, ..., 6$ which it contains. Thus it is clear that the experiment (Ω, B, Pr) must be specified. It is not determined by the physical experiment.

We note that $\mathrm{Pr}\{A\} = 0$ does not imply that $A = \varnothing$. For instance, in Ω_2 of Example 2.1, $\mathrm{Pr}\{(7, 8)\} = 0$, but that open interval is obviously not empty. Similarly, $\mathrm{Pr}\{A\} = 1$ does not imply that $A = \Omega$. If P is a property that defines a Borel set and $\mathrm{Pr}\{P\} = 1$, we say that the property P is true **with probability 1** (abbreviated as wp 1). The equivalent terminology **almost everywhere** or **for almost all** ω is also used. $\mathrm{Pr}\{P\} = 1$ means that the probability that property P does not hold is zero.

3. RANDOM VARIABLES

A real finite-valued function $\mathrm{x}(\cdot)$ defined on Ω is called a (real) **random variable** if, for every real number x, the inequality

$$\mathrm{x}(\omega) \leqslant x \tag{2.5}$$

defines an ω set whose probability is defined.[3] This means that the ω set defined by inequality (2.5) is an event. We use lower case roman letters to denote random variables. Lower case italic letters denote **sample values** or **realizations** of the random variable. Thus x is a random variable and x is its realization. That is, for some $\omega \in \Omega$, $\mathrm{x}(\omega) = x$. We often use the abbreviation rv for random variable.

In view of the definition of a random variable, the function

$$F_\mathrm{x}(x) \triangleq \mathrm{Pr}\{\mathrm{x}(\omega) \leqslant x\} \tag{2.6}$$

is defined for all real x and is called the **distribution function** of the

---

[3] A random variable is a Borel measurable function.

random variable x. The subscript x on $F_x(x)$ serves to identify the random variable. In view of (2.4), $F_x(x)$ is monotone, nondecreasing

$$F_x(x_1) \leqslant F_x(x_2) \qquad \text{if} \qquad x_1 < x_2 . \qquad (2.7)$$

It is easy to see that

$$\lim_{x \to -\infty} F_x(x) = 0 \qquad (2.8)$$

and

$$\lim_{x \to \infty} F_x(x) = 1. \qquad (2.9)$$

The distribution function completely describes the properties of an rv. That is, given $F_x(x)$, we can compute probabilities that the rv x takes on values in any set on the real line which is of interest. For example, if $x_1 < x_2$, then

$$\{x(\omega) \leqslant x_2\} = \{x(\omega) \leqslant x_1\} \cup \{x_1 < x(\omega) \leqslant x_2\},$$

where the two ω sets on the right-hand side are disjoint. Therefore, by probability axiom (iii) and the definition of the distribution function, we have

$$\Pr\{x_1 < x(\omega) \leqslant x_2\} = F_x(x_2) - F_x(x_1). \qquad (2.10)$$

Now let $x_1 = x - \epsilon$ and $x_2 = x$ in (2.10). Then

$$\Pr\{x - \epsilon < x(\omega) \leqslant x\} = F_x(x) - F_x(x - \epsilon).$$

Letting $\epsilon \to 0$,

$$\Pr\{x(\omega) = x\} = F_x(x) - F_x(x^-). \qquad (2.11)$$

So if $F_x(\cdot)$ is discontinuous at x, its jump equals $\Pr\{x(\omega) = x\}$. Now

$$\{x_1 \leqslant x(\omega) \leqslant x_2\} = \{x(\omega) = x_1\} \cup \{x_1 < x(\omega) \leqslant x_2\},$$

so that

$$\Pr\{x_1 \leqslant x(\omega) \leqslant x_2\} = F_x(x_1) - F_x(x_1^-) + F_x(x_2) - F_x(x_1)$$

$$= F_x(x_2) - F_x(x_1^-). \qquad (2.12)$$

Similarly,

$$\Pr\{x(\omega) < x_1\} = F_x(x_1^-), \qquad (2.13)$$

$$\Pr\{x(\omega) \geqslant x_1\} = 1 - F_x(x_1^-), \qquad (2.14)$$

$$\Pr\{x(\omega) > x_1\} = 1 - F_x(x_1), \qquad (2.15)$$

and so on. The reader may compute the probabilities of other sets.

A random variable x is called **discrete** if there exists a **mass function** $m_x(\cdot)$ such that

$$F_x(x) = \sum_{\substack{\xi \leqslant x \\ m_x(\xi) > 0}} m_x(\xi). \tag{2.16}$$

It follows from (2.16) and (2.11) that

$$F_x(x) - F_x(x^-) = m_x(x) = \Pr\{x(\omega) = x\} \tag{2.17}$$

and in view of (2.16) and (2.9)

$$\sum_{\substack{\xi \\ m_x(\xi) > 0}} m_x(\xi) = 1. \tag{2.18}$$

Essentially, a discrete rv can assume only a countable number of values (realizations).

Example 2.2. Consider the experiment of tossing a coin. $\Omega = \{h, t\}$ where h is *heads* and t is *tails*. $B = \{\varnothing, h, t, \Omega\}$. $\Pr\{h\} = p$, $\Pr\{t\} = q$, $p + q = 1$. Define the rv x by $x(h) = 1$, $x(t) = 0$. x is a so-called *zero-one* (discrete) random variable associated with the event h. We determine the distribution function $F_x(x)$. If $x \geqslant 1$, then

$$\{x(\omega) \leqslant x\} = \{x(\omega) = 0\} \cup \{x(\omega) = 1\} = \Omega,$$

so that

$$F_x(x) = \Pr\{x(\omega) \leqslant x\} = 1 \qquad (x \geqslant 1).$$

If $0 \leqslant x < 1$, then

$$\{x(\omega) \leqslant x\} = \{x(\omega) = 0\} = t,$$

so that

$$F_x(x) = q \qquad (0 \leqslant x < 1).$$

If $x < 0$, then $\{x(\omega) \leqslant x\} = \varnothing$, so that $F_x(x) = 0$. Combining these results we have

$$F_x(x) = \begin{cases} 1, & x \geqslant 1 \\ q, & 0 \leqslant x < 1 \\ 0, & x < 0. \end{cases}$$

It is easy to see that there exists a mass function that is given by

$$m_x(x) = \begin{cases} q, & x = 0 \\ p, & x = 1 \\ 0, & \text{otherwise.} \end{cases}$$

The distribution and mass functions are displayed in Figs. 2.1 and 2.2, respectively. ▲

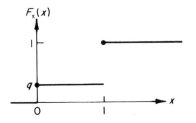

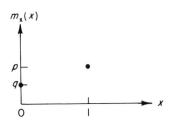

Fig 2.1. Distribution function of a discrete random variable.

Fig. 2.2. Probability mass function.

A random variable x is called **continuous** if there exists a **density function** $p_x(\cdot)$ such that

$$F_x(x) = \int_{-\infty}^{x} p_x(\xi) \, d\xi, \qquad -\infty < x < \infty.^4 \qquad (2.19)$$

The density function exists if the distribution function is absolutely continuous, that is, the number of points at which $F_x(\cdot)$ is not differentiable is countable. It is immediately evident from (2.19) that $F_x(x)$ is then continuous at all x, and that

$$p_x(x) = \frac{d}{dx} F_x(x) \qquad (2.20)$$

at all x at which the derivative exists.

Since $F_x(\cdot)$ is continuous, it follows from (2.11) that

$$\Pr\{x(\omega) = x\} = 0. \qquad (2.21)$$

Since $F_x(\cdot)$ is monotonic, and in view of (2.19),

$$p_x(x) \geqslant 0. \qquad (2.22)$$

Also, in view of (2.19) and (2.9),

$$\int_{-\infty}^{\infty} p_x(\xi) \, d\xi = 1. \qquad (2.23)$$

---

[4] In general, we assume that this integral and, until further notice, all other integrals in this book are *Riemann* integrals. This will lead us into some subtle mathematical difficulties that need not, however, concern an engineer. We shall explain these difficulties as they arise.

A function $p_x(\cdot)$ satisfying (2.22) and (2.23) is a density function. That is, it defines a distribution function which in turn determines an experiment $(\Omega, B, \Pr)$ and a random variable (see Papoulis [5, p. 99]).

Now in view of (2.19), for $x_1 < x_2$,

$$F_x(x_2) - F_x(x_1) = \int_{x_1}^{x_2} p_x(\xi)\, d\xi.$$

Therefore, using (2.10),

$$\Pr\{x_1 \leqslant x(\omega) \leqslant x_2\} = \int_{x_1}^{x_2} p_x(\xi)\, d\xi,$$

and, for Δx sufficiently small,

$$\Pr\{x \leqslant x(\omega) \leqslant x + \Delta x\} \simeq p_x(x)\, \Delta x. \tag{2.24}$$

Thus we have a complete analogy between a probability density and a mass density, if we imagine the mass distributed continuously along the real line.

Example 2.3. A telephone call occurs at random in the time interval $[0, T]$. Let $\Omega = [0, T]$; $\Pr\{t_1 \leqslant \omega \leqslant t_2\} = (t_2 - t_1)/T$; $t_1, t_2 \in [0, T]$. Define the rv x by $x(\omega) = \omega$. We determine its distribution and density functions. If $x > T$, then

$$\{x(\omega) \leqslant x\} = \Omega, \qquad F_x(x) = 1.$$

If $0 \leqslant x \leqslant T$, then

$$\{x(\omega) \leqslant x\} = \{0 \leqslant \omega \leqslant x\}, \qquad F_x(x) = x/T.$$

If $x < 0$, then

$$\{x(\omega) \leqslant x\} = \varnothing, \qquad F_x(x) = 0.$$

Differentiating this distribution function, we get

$$p_x(x) = \begin{cases} 0, & x < 0 \\ 1/T, & 0 < x < T \\ 0, & x > T. \end{cases}$$

The distribution function is not differentiable at 0 and T, but it is easy to see that the density function may be arbitrarily defined at these two points without affecting the experiment. These distribution and density functions are displayed in Figs. 2.3 and 2.4. ▲

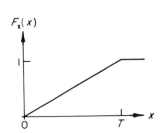

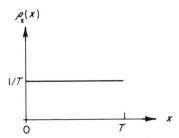

Fig. 2.3. Uniform distribution function.   Fig. 2.4. Uniform density function.

The random variable of this example is called a **uniformly distributed** (continuous) random variable (uniformly distributed on the interval $[0, T]$). Its density function could have been written down immediately from the probability function by observing that every interval of equal length in $[0, T]$ is equally probable, and intervals outside $[0, T]$ are impossible. In engineering applications, it is common to characterize an rv by directly specifying its density function.

Consider the "random variable" $x(\omega) = c$ for all $\omega \in \Omega$. This rv is the constant c, independent of the experimental outcomes. For $x < c$,

$$\Pr\{x(\omega) \leqslant x\} = F_x(x) = 0,$$

and, for $x \geqslant c$,

$$\Pr\{x(\omega) \leqslant x\} = F_x(x) = 1.$$

Thus,

$$F_x(x) = H(x - c), \tag{2.25}$$

where $H(\cdot)$ is the Heaviside unit step function. Proceeding formally, we have

$$p_x(x) = \frac{d}{dx} F_x(x) = \delta(x - c), \tag{2.26}$$

where $\delta(\cdot)$ is the Dirac delta function. Thus we see that constants may be formally regarded as random variables, with density functions that are Dirac delta functions.

In view of the fact that the density (mass) function determines the distribution function, the density (mass) function completely characterizes a continuous (discrete) random variable. In this book, we will be concerned only with continuous random variables. We henceforth assume that all our continuous rv's possess density functions with all the required smoothness properties. Thus, instead of specifying an experiment $(\Omega, B, \Pr)$ and defining the rv x on Ω, we shall equivalently specify a random variable with a given probability density function. Properties of discrete rv's can be written by analogy with continuous rv's by replacing the density function by the mass function and integrals by sums. We note, in passing, that there are random variables that are of the so-called mixed type. That is, their distribution function cannot be written as in (2.16) or in (2.19), but rather as a combination of integrals and sums. We shall not encounter such rv's in this book.

The **expectation** (*mean, average, first moment*) of a (continuous) random variable x is defined (when it exists) by

$$\mathscr{E}\{x\} \triangleq \int_{-\infty}^{\infty} x p_x(x)\, dx.^5 \tag{2.27}$$

We say that $\mathscr{E}\{x\}$ *exists* if $\mathscr{E}\{|\,x\,|\} < \infty$, that is, if the integral in (2.27) is absolutely convergent. Therefore, $\mathscr{E}\{x\}$ exists if and only if $\mathscr{E}\{|\,x\,|\} < \infty$, and then

$$|\,\mathscr{E}\{x\}\,| \leqslant \mathscr{E}\{|\,x\,|\}. \tag{2.28}$$

In writing the integral in (2.27) we shall generally, in the future, omit the limits of integration. It will be understood that, in the absence of limits, the integral is over the whole real line.

A *fixed* (nonrandom) real function $f(\cdot)$ of the random variable x is also a random variable,[6] say $y = f(x)$. Its expectation may be computed in two ways:

$$\mathscr{E}\{y\} \triangleq \int y p_y(y)\, dy = \int f(x)\, p_x(x)\, dx.^7 \tag{2.29}$$

Let a and b be fixed constants and $y = a$. Then

$$\mathscr{E}\{a\} = a \tag{2.30}$$

in view of (2.23). Now let $y = ag_1(x) + bg_2(x)$. Then

$$\mathscr{E}\{ag_1(x) + bg_2(x)\} = a\mathscr{E}\{g_1(x)\} + b\mathscr{E}\{g_2(x)\}. \tag{2.31}$$

[5] The expectation can also be written in terms of the *Stieltjes* integral $\int_{-\infty}^{\infty} x\, dF_x(x)$. A definition of the expectation in terms of this latter integral is more general than ours. In view of (2.20) and our assumption of the existence and smoothness of the density function, it is equivalent to our definition.

[6] Under the mild condition that $\{x : f(x) \leqslant y\} \in B$ for every y. This ensures that $\{y[x(\omega)] \leqslant y\} \in B$ (is an event). This means that $f(\cdot)$ is a *Borel* measurable (*Baire*) function. It is assumed that all functions in this book are Baire functions.

[7] For the last integral to exist as a Riemann integral, $f(\cdot)$ must be more restricted, e.g., Riemann integrable over every interval. That the two evaluations of $\mathscr{E}\{y\}$ are equal follows from the properties of the integral. The proof, however, is beyond the scope of this book.

Now

$$\mathscr{E}\{x^n\} \triangleq \int x^n p_x(x)\, dx \tag{2.32}$$

defines the **nth moment** of x. $\mathscr{E}\{x^2\}$ is called the **mean square** value of x.

$$\mathscr{E}\{(x - \mathscr{E}\{x\})^n\} \triangleq \int (x - \mathscr{E}\{x\})^n p_x(x)\, dx \tag{2.33}$$

defines the **nth moment of x about the mean (nth central moment)**. The second moment about the mean is called the **variance** of x:

$$\mathrm{var}\{x\} = \mathscr{E}\{(x - \mathscr{E}\{x\})^2\} = \mathscr{E}\{x^2\} - \mathscr{E}^2\{x\}. \tag{2.34}$$

Thus the variance is the mean square minus the square mean. Finally, the **standard deviation** of x is defined by

$$\sigma\{x\} \triangleq (\mathrm{var}\{x\})^{1/2}. \tag{2.35}$$

It is a measure of the *dispersion* about the mean in the samples of x. Note the analogy between the variance and the mass moment of inertia about the center of mass.

Example 2.4. Referring to the random variable of Example 2.3, it is easy to compute the mean and variance to be $m = \mathscr{E}\{x\} = T/2$, $\sigma^2 = T^2/12$. ▲

Example 2.5. It was noted that the variance of a random variable is a measure of the dispersion in its samples. We show here that, if c is a fixed constant and if $\mathscr{E}\{(x - c)^2\} = 0$, then x = c wp 1. Suppose the assertion is not true. Then there exists an $\epsilon > 0$ such that

$$\Pr\{|\,x - c\,| > \epsilon\} = \int_{|x-c|>\epsilon} p_x(x)\, dx > 0.$$

But then

$$\mathscr{E}\{(x - c)^2\} = \int_{-\infty}^{\infty} (x - c)^2\, p_x(x)\, dx \geqslant \int_{|x-c|>\epsilon} (x - c)^2\, p_x(x)\, dx$$

$$\geqslant \epsilon^2 \int_{|x-c|>\epsilon} p_x(x)\, dx > 0.$$

In particular, if $\mathscr{E}\{x^2\} = 0$, then x = 0 wp 1. ▲

A random variable may also be specified in terms of its **characteristic function** defined by

$$\varphi_x(u) \triangleq \mathscr{E}\{\exp(iux)\}, \qquad i^2 = -1. \tag{2.36}$$

The importance of this function lies in the fact that one can obtain all the moments of the rv x by differentiating its characteristic function. It is easy to show that

$$\mathscr{E}\{x^n\} = (1/i^n)\frac{d^n}{du^n}\varphi_x(0) \tag{2.37}$$

for all n, provided of course that $\varphi_x(u)$ is analytic. The characteristic function always exists since

$$|\mathscr{E}\{\exp(iux)\}| \leqslant \mathscr{E}\{|\exp(iux)|\} = 1.$$

From its definition, we see that the characteristic function is just the Fourier transform of the density function

$$\varphi_x(u) = \int_{-\infty}^{\infty} \exp(iux)\, p_x(x)\, dx. \tag{2.38}$$

Therefore, if the characteristic function is absolutely integrable, the inverse Fourier transform

$$p_x(x) = (1/2\pi)\int_{-\infty}^{\infty} \exp(-iux)\,\varphi_x(u)\, du \tag{2.39}$$

gives the density function.

Of central importance in this book will be random variables that are **Gaussian**, or equivalently, **normally distributed**. A random variable is Gaussian if its density function is given by

$$p_x(x) = (1/2\pi\sigma^2)^{1/2} \exp\left[-\frac{1}{2}\left(\frac{x-m}{\sigma}\right)^2\right], \tag{2.40}$$

where m and σ^2 are (constant) parameters. An easy computation shows that

$$m = \mathscr{E}\{x\}, \qquad \sigma^2 = \mathrm{var}\{x\}. \tag{2.41}$$

Note that these two parameters characterize the Gaussian density. If the rv x is Gaussian with mean m and variance σ^2, we sometimes write

$$x \sim N(m, \sigma^2). \tag{2.42}$$

The normal density function has the familiar bell shape. The normal distribution function is, of course, given by

$$F_{\mathbf{x}}(x) = (1/2\pi\sigma^2)^{1/2} \int_{-\infty}^{x} \exp\left[-\frac{1}{2}\left(\frac{\xi - m}{\sigma}\right)^2\right] d\xi. \qquad (2.43)$$

The importance of Gaussian random variables will become apparent in subsequent chapters. Readers interested in other density functions that are important in applications may consult Papoulis [5, Chapter 4], for example. The normal density and distribution functions (with $m = 0$ and $\sigma^2 = 1$) are plotted in Figs. 2.5 and 2.6.

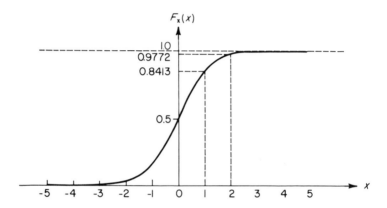

FIG. 2.5.   Normal distribution function.

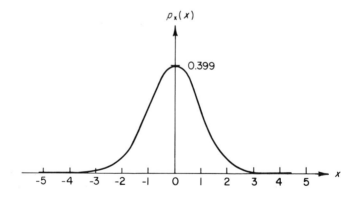

FIG. 2.6.   Normal density function.

Example 2.6. Let the rv x be normally distributed with mean m and variance σ^2. Compute its characteristic function

$$\varphi_x(u) = (1/2\pi\sigma^2)^{1/2} \int_{-\infty}^{\infty} \exp\left[iux - \frac{1}{2}\left(\frac{x-m}{\sigma}\right)^2\right] dx.$$

Letting $y = (x - m)/\sigma$,

$$\varphi_x(u) = (1/2\pi)^{1/2} \exp(ium) \int_{-\infty}^{\infty} \exp(iu\sigma y - \tfrac{1}{2}y^2)\, dy$$

$$= (1/2\pi)^{1/2} \exp(ium - \tfrac{1}{2}u^2\sigma^2) \int_{-\infty}^{\infty} \exp[-\tfrac{1}{2}(y - iu\sigma)^2]\, dy$$

$$= \exp(ium - \tfrac{1}{2}u^2\sigma^2). \tag{2.44}$$

Let us compute all the central moments of x (moments about the mean). In view of the symmetry about the mean of the Gaussian density function (2.40), it is evident that all the odd central moments are zero. Since the central moments of x are the same as the moments of $x - m$ (from definition), we compute the characteristic function of $x - m$ and use (2.37). From the definition of the characteristic function (2.36) and (2.44),

$$\varphi_{x-m}(u) = \exp(-ium)\, \varphi_x(u)$$

$$= \exp(-\tfrac{1}{2}u^2\sigma^2). \tag{2.45}$$

We may note from (2.45) that the central moments of x will depend only on its variance σ^2. It is easy to establish by induction that

$$\varphi^{(n+2)}(u) = -(n+1)\,\sigma^2\varphi^{(n)}(u) - u\sigma^2\varphi^{(n+1)}(u), \qquad n = 0, 1,..., \tag{2.46}$$

with $\varphi^{(0)}(u) = \varphi(u)$, $\varphi^{(1)}(u) = -u\sigma^2\varphi(u)$. We have omitted the subscript $x - m$ on φ. The superscript denotes the derivative

$$\varphi^{(n)}(u) = (d^n/du^n)\,\varphi(u).$$

Evaluating (2.46) at $u = 0$,

$$\varphi^{(n+2)}(0) = -(n+1)\,\sigma^2\varphi^{(n)}(0). \tag{2.47}$$

Consider (2.47) as a difference equation with initial condition $\varphi^{(1)}(0) = 0$. It immediately follows that

$$\varphi^{(n)}(0) = 0, \qquad \text{all odd} \quad n \geqslant 1.$$

With initial condition $\varphi^{(0)}(0) = 1$, (2.47) has the solution

$$\varphi^{(n)}(0) = (i^n)\, 1 \cdot 3 \cdot 5 \cdots (n-1)\, \sigma^n, \qquad \text{all even} \quad n \geqslant 2.$$

Therefore, in view of (2.37),

$$\mathscr{E}\{(x - m)^n\} = \begin{cases} 0, & \text{all odd } n \geqslant 1 \\ 1 \cdot 3 \cdot 5 \cdots (n-1)\,\sigma^n, & \text{all even } n \geqslant 2. \end{cases} \qquad (2.48)$$

We see that for any $\sigma > 0$, no matter how small, the even central moments grow without bound $(n \to \infty)$. ▲

4. JOINTLY DISTRIBUTED RANDOM VARIABLES

The (continuous) random variables $x_1, x_2, ..., x_n$ are said to be **jointly distributed** if they are defined on the same probability space. They may be characterized by their **joint distribution function**

$$F_{x_1,...,x_n}(x_1, ..., x_n) \triangleq \Pr\{x_1(\omega) \leqslant x_1, ..., x_n(\omega) \leqslant x_n\}, \qquad (2.49)$$

where

$$\{x_1(\omega) \leqslant x_1, ..., x_n(\omega) \leqslant x_n\} = \{x_1(\omega) \leqslant x_1\} \cap \cdots \cap \{x_n(\omega) \leqslant x_n\},$$

or by their **joint density function**

$$F_{x_1,...,x_n}(x_1, ..., x_n) = \int_{-\infty}^{x_1} \cdots \int_{-\infty}^{x_n} p_{x_1,...,x_n}(\xi_1, ..., \xi_n)\, d\xi_1, ..., d\xi_n. \qquad (2.50)$$

It follows from (2.50) that

$$p_{x_1,...,x_n}(x_1, ..., x_n) = \frac{\partial^n}{\partial x_1, ..., \partial x_n} F_{x_1,...,x_n}(x_1, ..., x_n) \qquad (2.51)$$

at all $(x_1, ..., x_n)$ at which the foregoing derivative exists. The properties of the joint distribution and density functions follow by analogy with the single variable case of the preceding section and will not be developed here.

Example 2.7. Let x and y be jointly distributed rv's with joint distribution function $F_{x,y}(x, y)$ and joint density $p_{x,y}(x, y)$. Convince yourself from the definitions and reasoning analogous to that used in Eqs. (2.7)–(2.15) and (2.21)–(2.24) that

$$F_{x,y}(-\infty, y) = F_{x,y}(x, -\infty) = F_{x,y}(-\infty, -\infty) = 0,$$

$$F_{x,y}(\infty, \infty) = 1,$$

$$F_{x,y}(x_2, y) - F_{x,y}(x_1, y) = \Pr\{x_1 < x(\omega) \leqslant x_2, y(\omega) \leqslant y\} \geqslant 0,$$

$(x_2 > x_1)$, and similarly in y. Also,

$$p_{\mathrm{x,y}}(x, y) \geqslant 0,$$

$$\Pr\{x < \mathrm{x}(\omega) \leqslant x + \varDelta x, y < \mathrm{y}(\omega) \leqslant y + \varDelta y\} \simeq p_{\mathrm{x,y}}(x, y)\, \varDelta x\, \varDelta y,$$

$$\iint p_{\mathrm{x,y}}(x, y)\, dx\, dy = 1. \quad \blacktriangle$$

We are often interested only in *some* of the rv's $x_1, \ldots, x_n$; in $x_1, \ldots, x_m$, $m < n$, say. Their joint distribution function is

$$F_{\mathrm{x}_1,\ldots,\mathrm{x}_m}(x_1, \ldots, x_m) = F_{\mathrm{x}_1,\ldots,\mathrm{x}_n}(x_1, \ldots, x_m, \infty, \ldots, \infty), \qquad (2.52)$$

as is evident from (2.49). The distribution in (2.52) is sometimes called **marginal**. Differentiating (2.52) and using (2.50) and (2.51), we obtain the **marginal density function**

$$p_{\mathrm{x}_1,\ldots,\mathrm{x}_m}(x_1, \ldots, x_m) = \int_{-\infty}^{\infty} \cdots \int p_{\mathrm{x}_1,\ldots,\mathrm{x}_n}(x_1, \ldots, x_n)\, dx_{m+1} \cdots dx_n. \quad (2.53)$$

In view of this and the definition (2.27), the **expectation** of x_k, $1 \leqslant k \leqslant n$, is given by

$$\mathscr{E}\{\mathrm{x}_k\} = \int \cdots \int x_k\, p_{\mathrm{x}_1,\ldots,\mathrm{x}_n}(x_1, \ldots, x_n)\, dx_1, \ldots, dx_n. \quad (2.54)$$

Example 2.8. Let x and y be jointly **uniformly** distributed on $[0, T] \times [0, T]$, that is,

$$p_{\mathrm{x,y}}(x, y) = \begin{cases} 1/T^2 & \text{on} \quad [0, T] \times [0, T] \\ 0 & \text{elsewhere.} \end{cases}$$

The marginal densities are

$$p_{\mathrm{x}}(x) = \int_{-\infty}^{\infty} p_{\mathrm{x,y}}(x, y)\, dy = \int_0^T (1/T^2)\, dy = 1/T \qquad (x \in [0, T]),$$

$$p_{\mathrm{x}}(x) = 0 \qquad (x \notin [0, T]),$$

and, similarly,

$$p_{\mathrm{y}}(y) = \begin{cases} 1/T, & y \in [0, T] \\ 0, & \text{elsewhere.} \end{cases}$$

Therefore,

$$p_{\mathrm{x,y}}(x, y) = p_{\mathrm{x}}(x)\, p_{\mathrm{y}}(y);$$

x and y are *independent*. Compute $\Pr\{x \leqslant T/2, y \leqslant T/2\} = 1/4$. Compute $\mathscr{E}\{x\} = \mathscr{E}\{y\} = T/2$. ▲

In general, if f is a fixed, real function of the jointly distributed random variables $x_1, ..., x_n$, say $y = f(x_1, ..., x_n)$, then

$$\mathscr{E}\{y\} \triangleq \int y p_y(y) \, dy = \int \cdots \int f(x_1, ..., x_n) \, p_{x_1, ..., x_n}(x_1, ..., x_n) \, dx_1 \cdots dx_n \quad (2.55)$$

when it exists. The expectation of x_k has already been given in (2.54). The **second moment** of x_k is

$$\mathscr{E}\{x_k^2\} = \int \cdots \int x_k^2 p_{x_1, ..., x_n}(x_1, ..., x_n) \, dx_1 \cdots dx_n . \quad (2.56)$$

Higher moments and central moments are written in an obvious way. In addition, joint moments of the form

$$\mathscr{E}\{x_k^\alpha x_l^\beta\}$$

and joint central moments

$$\mathscr{E}\{(x_k - \mathscr{E}\{x_k\})^\alpha (x_l - \mathscr{E}\{x_l\})^\beta\}$$

can be defined. Of particular importance is the **covariance** of x_k and x_l which is defined by

$$\text{cov}\{x_k, x_l\} \triangleq \mathscr{E}\{(x_k - \mathscr{E}\{x_k\})(x_l - \mathscr{E}\{x_l\})\}. \quad (2.57)$$

We have that

$$\text{cov}\{x_k, x_l\} = \mathscr{E}\{x_k x_l\} - \mathscr{E}\{x_k\} \mathscr{E}\{x_l\}. \quad (2.58)$$

Note that

$$\text{cov}\{x_k, x_k\} = \text{var}\{x_k\}.$$

The ratio

$$\frac{\text{cov}\{x_k, x_l\}}{\sigma\{x_k\} \sigma\{x_l\}} = \rho(x_k, x_l) \quad (2.59)$$

defines the **correlation coefficient** of x_k and x_l. It is, of course, assumed in (2.59) that x_k and x_l are nondegenerate (have positive variances) and that their variances are finite. See Example 2.5.

Having defined various moments and joint moments, it is important to know under what conditions these moments exist. We now develop several useful inequalities and use these to establish conditions for the existence of moments.

Let x and y be jointly distributed rv's. We first note that the expectation operator $\mathscr{E}$ is *linear*. That is, if a and b are fixed constants, then

$$\mathscr{E}\{ax + by\} = a\mathscr{E}\{x\} + b\mathscr{E}\{y\}. \qquad (2.60)$$

This fact was already used [in (2.58)] and, of course, follows from like properties of the integral that defines the expectation. See also (2.31).

Now consider

$$\mathscr{E}\{(a \mid x \mid - \mid y \mid)^2\} = a^2\mathscr{E}\{\mid x \mid^2\} - 2a\mathscr{E}\{\mid xy \mid\} + \mathscr{E}\{\mid y \mid^2\} \geqslant 0$$

for all a. This is a quadratic in a which is nonnegative for all a. As a result, it cannot have two real, distinct roots, so its discriminant cannot be positive; that is,

$$\mathscr{E}^2\{\mid xy \mid\} - \mathscr{E}\{\mid x \mid^2\}\,\mathscr{E}\{\mid y \mid^2\} \leqslant 0.$$

This is just *Schwarz's* inequality ($\mathscr{E}$ operator is an integral operator)

$$\mathscr{E}^2\{\mid xy \mid\} \leqslant \mathscr{E}\{\mid x \mid^2\}\,\mathscr{E}\{\mid y \mid^2\}. \qquad (2.61)$$

The inequality

$$\mathscr{E}\{\mid x + y \mid\} \leqslant \mathscr{E}\{\mid x \mid\} + \mathscr{E}\{\mid y \mid\} \qquad (2.62)$$

follows from the triangle inequality for real numbers (just take expectations of both sides). Using the triangle inequality, it is also easy to verify that

$$\mathscr{E}\{\mid x + y \mid^2\} \leqslant 2\mathscr{E}\{\mid x \mid^2\} + 2\mathscr{E}\{\mid y \mid^2\}. \qquad (2.63)$$

Now we can easily prove the following theorems.

Theorem 2.1. *If* $\mathscr{E}\{\mid x \mid^2\} < \infty$, *then* $\mathscr{E}\{\mid x \mid\} < \infty$.

PROOF: Taking the square root in Schwarz's inequality and setting $y \equiv 1$, we have

$$\mathscr{E}\{\mid x \mid\} \leqslant \mathscr{E}^{1/2}\{\mid x \mid^2\} < \infty. \quad \blacksquare$$

Thus the existence of the second moment implies the existence of the first moment. In fact, the following theorem can be proved (see Loeve [4, p. 155]).

Theorem 2.2. *If* $\mathscr{E}\{\mid x \mid^n\} < \infty$, *then* $\mathscr{E}\{\mid x \mid^m\} < \infty$ *for all* $1 \leqslant m \leqslant n$.

The existence of the nth moment guarantees the existence of all lower-order moments.

Theorem 2.3. *If* $\mathscr{E}\{|x|^2\} < \infty$ *and* $\mathscr{E}\{|y|^2\} < \infty,$ *then* (i) $\mathscr{E}\{|x + y|^2\} < \infty$ *and* (ii) $\mathscr{E}\{|xy|\} < \infty.$

PROOF: (i) follows with the aid of (2.63), and (ii) via Schwarz's inequality. ∎

The following theorem can also be proved (see Loeve [4, p. 155]).

Theorem 2.4. *If* $\mathscr{E}\{|x|^n\} < \infty$ *and* $\mathscr{E}\{|y|^n\} < \infty,$ *then* $\mathscr{E}\{|x + y|^n\} < \infty.$

Example 2.9. Let x and y be random variables with *finite second moments.* That is, $\mathscr{E}\{|x|^2\} < \infty$ and $\mathscr{E}\{|y|^2\} < \infty.$ By Theorem 2.1, $\mathscr{E}\{x\}$ and $\mathscr{E}\{y\}$ exist. Therefore, var{x} and var{y} exist. By Theorem 2.3, $\mathscr{E}\{xy\}$ exists; therefore cov{x, y} also exists. ▲

Random variables with finite second moments constitute an important class of rv's. We shall have much more to say about them in Chapter 3.

Two jointly distributed random variables x_1 and x_2 are said to be **independent** if any of the following *equivalent* conditions is satisfied:

$$F_{x_1,x_2}(x_1, x_2) = F_{x_1}(x_1) F_{x_2}(x_2), \quad \text{all} \quad x_1, x_2, \quad (2.64)$$

$$p_{x_1,x_2}(x_1, x_2) = p_{x_1}(x_1) p_{x_2}(x_2), \quad \text{all} \quad x_1, x_2, \quad (2.65)$$

$$\mathscr{E}\{f(x_1) g(x_2)\} = \mathscr{E}\{f(x_1)\} \mathscr{E}\{g(x_2)\}, \quad (2.66)$$

for all fixed functions $f(\cdot), g(\cdot)$, provided these expectations exist. This concept generalizes to more than two random variables. We say that $x_1, ..., x_n$ are **mutually independent** if

$$p_{x_1,...,x_n}(x_1, ..., x_n) = p_{x_1}(x_1) \cdots p_{x_n}(x_n) \quad (2.67)$$

for all $x_1, ..., x_n$. Conditions (2.64) and (2.66) generalize in an obvious way. Also, $x_1, ..., x_n$ are said to be **jointly independent** of $y_1, ..., y_m$ if

$$p_{x_1,...,x_n,y_1,...,y_m}(x_1, ..., x_n, y_1, ..., y_m) = p_{x_1,...,x_n}(x_1, ..., x_n) p_{y_1,...,y_m}(y_1, ..., y_m)$$

$$(2.68)$$

for all $x_1, ..., x_n, y_1, ..., y_m$.

We say that the two jointly distributed random variables x_k and x_l are **uncorrelated** if their second moments are finite and if

$$\text{cov}\{x_k, x_l\} = 0. \quad (2.69)$$

Their correlation coefficient (2.59) is then zero.

Theorem 2.5. *Let the jointly distributed random variables* x_k *and* x_l *be independent. Then they are also uncorrelated.*

PROOF:

$$\text{cov}\{x_k, x_l\} = \iint x_k x_l p_{x_k, x_l}(x_k, x_l)\, dx_k\, dx_l - \mathscr{E}\{x_k\}\, \mathscr{E}\{x_l\}$$

$$= \iint x_k x_l p_{x_k}(x_k)\, p_{x_l}(x_l)\, dx_k\, dx_l - \mathscr{E}\{x_k\}\, \mathscr{E}\{x_l\}$$

$$= \int x_k p_{x_k}(x_k)\, dx_k \cdot \int x_l p_{x_l}(x_l)\, dx_l - \mathscr{E}\{x_k\}\, \mathscr{E}\{x_l\}$$

$$= \mathscr{E}\{x_k\} \cdot \mathscr{E}\{x_l\} - \mathscr{E}\{x_k\}\, \mathscr{E}\{x_l\} = 0. \quad \blacksquare$$

Note that the converse to Theorem 2.5 is, in general, not true (see Theorem 2.6, however). Lack of correlation means that (2.66) holds for functions $f(x) = g(x) = x$, whereas independence requires (2.66) to hold for *all* functions f and g.

Example 2.10. Consider the two random variables y_1 and y_2 defined by

$$y_1 = \sin 2\pi x, \qquad y_2 = \cos 2\pi x,$$

where x is uniformly distributed on $[0, 1]$. It is easy to compute $\mathscr{E}\{y_1\} = \mathscr{E}\{y_2\} = \mathscr{E}\{y_1 y_2\} = 0$. Therefore, $\text{cov}\{y_1, y_2\} = 0$, so that y_1 and y_2 are uncorrelated. However, $y_1{}^2 + y_2{}^2 = 1$, so y_1 and y_2 are clearly not independent. ▲

Two jointly distributed rv's x_k and x_l are said to be **orthogonal** if their second moments are finite and if

$$\mathscr{E}\{x_k x_l\} = 0. \tag{2.70}$$

Note that if x_k and x_l are uncorrelated, then $x_k - \mathscr{E}\{x_k\}$ and $x_l - \mathscr{E}\{x_l\}$ are orthogonal. Also, if x_k and x_l are uncorrelated and either $\mathscr{E}\{x_k\}$ or $\mathscr{E}\{x_l\}$ is zero, then they are orthogonal.

Example 2.11. Let x and y be uncorrelated. Then the variance of their sum is the sum of their variances. ▲

Example 2.12. Let x and y be rv's with joint density function

$$p_{x,y}(x, y) = (1/2\pi) \exp[-(1/2)(x^2 + y^2)].$$

Let the reader show that x and y are independent, uncorrelated, orthogonal and are both *identically* distributed (they are both Gaussian). See (2.40). ▲

Example 2.13. We note here the properties of the correlation coefficient (2.59). If, in the derivation of Schwarz's inequality (2.61), we replace | x | by x − $\mathscr{E}\{x\}$ and | y | by y − $\mathscr{E}\{y\}$, we obtain

$$\mathscr{E}^2\{(x - \mathscr{E}\{x\})(y - \mathscr{E}\{y\})\} \leqslant \mathscr{E}\{(x - \mathscr{E}\{x\})^2\}\,\mathscr{E}\{(y - \mathscr{E}\{y\})^2\}.$$

Thus $cov^2\{x, y\} \leqslant var\{x\}\,var\{y\}$, and we immediately have that $0 \leqslant |\rho(x, y)| \leqslant 1$. Obviously $\rho(x, y) = 0$ if and only if x and y are uncorrelated. We now show that $|\rho| = 1$ if and only if $y = ax + b$, where a and b are fixed constants.[8] Suppose $y = ax + b$. Then

$$\mathscr{E}\{y\} = a\mathscr{E}\{x\} + b,$$

$$var\{y\} = a^2\,var\{x\},$$

$$cov\{x, y\} = a\,var\{x\},$$

so that $\rho = \pm 1$. Now suppose $|\rho| = 1$. Then $cov^2\{x, y\} = var\{x\}\,var\{y\}$. Let

$$a = \frac{cov\{x, y\}}{var\{x\}},$$

$$b = \mathscr{E}\{y\} - \frac{cov\{x, y\}\,\mathscr{E}\{x\}}{var\{x\}},$$

and use the result in Example 2.5. ▲

The **characteristic function** of the jointly distributed rv's $x_1, ..., x_n$ is defined by

$$\varphi_{x_1,...,x_n}(u_1,..., u_n) \triangleq \mathscr{E}\left\{\exp\left(i \sum_{j=1}^{n} u_j x_j\right)\right\}. \tag{2.71}$$

The **marginal** characteristic function of $x_1, ..., x_m$, $m < n$, is clearly

$$\varphi_{x_1,...,x_n}(u_1,..., u_m, 0,..., 0). \tag{2.72}$$

If $x_1, ..., x_n$ are *independent*, then [see (2.66) or (2.67)]

$$\varphi_{x_1,...,x_n}(u_1,..., u_n) = \varphi_{x_1}(u_1) \cdots \varphi_{x_n}(u_n). \tag{2.73}$$

---

[8] We can only show that $y = ax + b$ wp 1. We often do not make a distinction between identities and wp 1 equalities. For engineering purposes, they are the same.

The converse can also be shown to be true. Moments and joint moments of the rv's $x_1, ..., x_n$ can be obtained from their joint characteristic function by analogy with the single variable case.

The random variables x_1 and x_2 are said to be **Gaussian** or **jointly normally distributed** if their joint density function is given by

$$p_{x_1, x_2}(x_1, x_2) = \frac{1}{2\pi\sigma\{x_1\}\,\sigma\{x_2\}(1 - \rho^2)^{1/2}} \exp\left\{-\frac{1}{2(1 - \rho^2)}\left[\left(\frac{x_1 - \mathscr{E}\{x_1\}}{\sigma\{x_1\}}\right)^2\right.\right.$$
$$\left.\left. - 2\rho\left(\frac{x_1 - \mathscr{E}\{x_1\}}{\sigma\{x_1\}}\right)\left(\frac{x_2 - \mathscr{E}\{x_2\}}{\sigma\{x_2\}}\right) + \left(\frac{x_2 - \mathscr{E}\{x_2\}}{\sigma\{x_2\}}\right)^2\right]\right\}, \quad (2.74)$$

where $\rho(x_1, x_2)$ is the correlation coefficient. We see that the joint Gaussian density in (2.74) is undefined if $|\rho| = 1$. See Example 2.13. In that case, however, x_1 completely determines x_2 and *vice versa*, and it is sufficient to specify the density of just one of these rv's. Also compare (2.74) with Example 2.12. Now x_2 may be integrated out in (2.74) to show that the marginal density of x_1 is Gaussian with mean $\mathscr{E}\{x_1\}$ and variance $\sigma^2\{x_1\}$. See also Theorem 2.12. We may note that if x_1 and x_2 are uncorrelated ($\rho = 0$), then (2.74) reduces to

$$p_{x_1, x_2}(x_1, x_2) = p_{x_1}(x_1)\,p_{x_2}(x_2), \quad (2.75)$$

where the two densities on the right-hand side are Gaussian [see (2.40)]. Thus, two Gaussian rv's that are uncorrelated are also independent. Compare this with Theorem 2.5, the comments following it, and Example 2.10.

In dealing with jointly distributed random variables, it will be very convenient to arrange these rv's in a vector and take advantage of the tools of vector-matrix algebra. Let $x_1, ..., x_n$ be jointly distributed rv's. We introduce the **random vector** x:

$$x = \begin{bmatrix} x_1 \\ x_2 \\ \vdots \\ x_n \end{bmatrix}, \qquad x^T = [x_1, ..., x_n]. \quad (2.76)$$

A vector will always be a column vector. Superscript T denotes the vector (matrix) transpose. Thus x^T is a row vector. We make no notational distinction between a random vector x and a (scalar) random variable x. This is because many of our equations can stand equally well with x interpreted as a scalar or as a vector. For example,

$$p_x(x)$$

can be interpreted as the density function of the scalar rv x or as the joint density of $x_1, ..., x_n$ which we could write in longhand (if we cared to) as

$$p_{x_1,...,x_n}(x_1, ..., x_n).$$

Also,

$$\int p_x(x)\, dx = 1$$

could mean (for a scalar rv x) the integral over the whole real line of the density of the scalar rv x. If we interpret dx as the volume element in E^n (n-dimensional Euclidean space), then the preceding integral can be interpreted as the integral over E^n:

$$\int \cdots \int p_{x_1,...,x_n}(x_1, ..., x_n)\, dx_1 \cdots dx_n = 1.$$

The reason why a notational distinction is not important is, of course, the fact that joint density functions are scalar functions. In any event, when we write x we shall either state whether it is a scalar or a vector, if it does matter, or else that fact will be obvious from the context.

If x is a random vector, its **expectation** is the vector

$$\mathscr{E}\{x\} = \begin{bmatrix} \mathscr{E}\{x_1\} \\ \vdots \\ \mathscr{E}\{x_n\} \end{bmatrix}, \qquad \mathscr{E}\{x^T\} = [\mathscr{E}\{x_1\}, ..., \mathscr{E}\{x_n\}]. \tag{2.77}$$

$\mathscr{E}\{x\}$ is often called the **mean vector** (of x). If $A = [a_{ij}]$ is a **random matrix** (matrix whose elements are random variables), then, similarly,

$$\mathscr{E}\{A\} = [\mathscr{E}\{a_{ij}\}]. \tag{2.78}$$

The square ($n \times n$) matrix of elements $\text{cov}\{x_i, x_j\}$ is called the **covariance matrix** of the random vector x and is often designated by P_x:

$$P_x = [\text{cov}\{x_i, x_j\}] = \mathscr{E}\{(x - \mathscr{E}\{x\})(x - \mathscr{E}\{x\})^T\}$$

$$= \begin{bmatrix} \text{var}\{x_1\} & \text{cov}\{x_1, x_2\} & \cdots & \text{cov}\{x_1, x_n\} \\ \text{cov}\{x_1, x_2\} & \text{var}\{x_2\} & \cdots & \text{cov}\{x_2, x_n\} \\ \vdots & \vdots & \vdots & \vdots \\ \text{cov}\{x_1, x_n\} & \text{cov}\{x_2, x_n\} & \cdots & \text{var}\{x_n\} \end{bmatrix}. \tag{2.79}$$

The random matrix $(x - \mathscr{E}\{x\})(x - \mathscr{E}\{x\})^T$ is, we point out, the outer product of the random vector $(x - \mathscr{E}\{x\})$ with itself. We note that the covariance matrix is *symmetric*. It can be shown that it is *positive semi-definite*.[9] This follows immediately, for the case $n = 2$, from the inequality in Example 2.13. Other inequalities can be similarly developed to prove this fact for arbitrary n. Note that, in case the elements of x are (pairwise) uncorrelated, the covariance matrix P_x is *diagonal*. We note that higher-order joint moments (of x_i, x_j) cannot be expressed in matrix notation. They would be three-dimensional arrays (tensors).

Using the vector-matrix notation just developed, we write the (joint) density function of n **jointly normally distributed** random variables $x_1, ..., x_n$:

$$p_x(x) = [(2\pi)^{n/2} \mid P_x \mid^{1/2}]^{-1} \exp[-\tfrac{1}{2}(x - \mathscr{E}\{x\})^T P_x^{-1}(x - \mathscr{E}\{x\})]. \quad (2.80)$$

$\mid \cdot \mid$ denotes the determinant, $\mid P_x \mid = \det P_x$, and superscript -1 denotes the matrix inverse. Compare this with (2.74) for $n = 2$. Note that the parameters $\mathscr{E}\{x\}$ and P_x completely characterize the multivariate Gaussian density function. If the random vector x is Gaussian with mean vector $\mathscr{E}\{x\}$ and covariance matrix P_x, we sometimes write

$$x \sim N(\mathscr{E}\{x\}, P_x). \quad (2.81)$$

We note that the Gaussian density is defined only if the covariance matrix P_x is positive definite. Analogously to the $n = 2$ case and Example 2.13, if P_x is only semidefinite, say of rank $(n - 1)$, there exists a wp 1 linear relationship between the x_i. One of these rv's can therefore be eliminated.

We have already noted that, if the elements of x are uncorrelated, then P is diagonal. But then, as it is easy to see, Eq. (2.80) reduces to

$$p_x(x) = p_{x_1}(x_1) \cdots p_{x_n}(x_n). \quad (2.82)$$

We therefore have (compare with Theorem 2.5):

Theorem 2.6. *If the jointly normally distributed random variables* $x_1, ..., x_n$ *are (pairwise) uncorrelated, then they are independent.*

Example 2.14. Using vector-matrix notation, we rewrite the joint characteristic function in (2.71) as

$$\varphi_x(u) = \mathscr{E}\{\exp(iu^T x)\}, \quad (2.83)$$

---

[9] The matrix A is said to be *positive definite (positive semidefinite)* iff $x^T A x > 0$ ($x^T A x \geqslant 0$) for all vectors $x \neq 0$. If A is positive definite (positive semidefinite), we often write $A > 0$ ($A \geqslant 0$).

where

$$x^T = [x_1, ..., x_n], \qquad u^T = [u_1, ..., u_n].$$

It is not very difficult to verify that if $x_1, ..., x_n$ are jointly *normally* distributed, then

$$\varphi_x(u) = \exp(iu^T \mathscr{E}\{x\} - \tfrac{1}{2}u^T P_x u), \tag{2.84}$$

where P_x is the covariance matrix of (2.79). Note that the characteristic function exists even if P_x is semidefinite. ▲

Let the random n-vector y be a (vector-valued) function of the random n-vector x, say $y = f(x)$. If we know the (joint) density function $p_x(x)$, we can determine $p_y(y)$ in terms of it.

Theorem 2.7. *Let* x, y *be random n-vectors with* $y = f(x)$. *Suppose* f^{-1} *exists and both* f *and* f^{-1} *are continuously differentiable. Then*

$$p_y(y) = p_x[f^{-1}(y)] \left\| \frac{\partial f^{-1}(y)}{\partial y} \right\|, \tag{2.85}$$

where $\| \partial f^{-1}(y)/\partial y \| > 0$ *is the absolute value of the Jacobian determinant.*

PROOF:　For an arbitrary set $S \subset E^n$,

$$\Pr\{y(\omega) \in S\} = \Pr\{f[x(\omega)] \in S\} = \Pr\{x(\omega) \in f^{-1}(S)\}.$$

But

$$\Pr\{y(\omega) \in S\} = \int_S p_y(y) \, dy$$

and

$$\Pr\{x(\omega) \in f^{-1}(S)\} = \int_{f^{-1}(S)} p_x(x) \, dx$$

$$= \int_S p_x[f^{-1}(y)] \left\| \frac{\partial f^{-1}(y)}{\partial y} \right\| dy,$$

where the last integral follows by well-known rules for transformation of integrals. Therefore,

$$\int_S \left[p_y(y) - p_x[f^{-1}(y)] \left\| \frac{\partial f^{-1}(y)}{\partial y} \right\| \right] dy = 0 \tag{2.86}$$

for arbitrary S. Suppose (2.85) is not true. Then there is some $\bar{y} \in E^n$ at which the integrand in (2.86) is positive (say). But then, by (assumed) continuity, there is some region S containing $\bar{y}$ in which the integrand in (2.86) is positive. This contradicts (2.86). ∎

The density $p_y(y)$ in (2.85) is sometimes called a **derived density**, since it is derived from the density function of x. The assumptions we made on the function f in Theorem 2.7 can be relaxed to obtain more general relationships for the derived density. See Papoulis [5, pp. 126, 201] for more general results in the cases of $n = 1$ and $n = 2$. Numerous special cases are also treated in Papoulis [5, Chapters 5 and 7]. Theorem 2.7 will have important applications in our work in filtering theory.

Example 2.15. Let $y_1 = x_1 + x_2$. We wish to determine the density of y_1 from the densities of x_1 and x_2. Now this transformation is a projection, and Theorem 2.7 is not directly applicable. Let us adjoin to this the transformation $y_2 = x_2$ (say) and consider

$$y_1 = x_1 + x_2, \qquad y_2 = x_2.$$

In vector-matrix notation,

$$y = Ax,$$

where

$$A = \begin{bmatrix} 1 & 1 \\ 0 & 1 \end{bmatrix}, \qquad A^{-1} = \begin{bmatrix} 1 & -1 \\ 0 & 1 \end{bmatrix}, \qquad x^T = [x_1, x_2], \qquad y^T = [y_1, y_2].$$

Now Theorem 2.7 is applicable, and we have

$$p_{y_1, y_2}(y_1, y_2) = p_{x_1, x_2}(y_1 - y_2, y_2). \tag{2.87}$$

We get the (marginal) density of y_1 by integrating y_2 out in (2.87); that is,

$$p_{y_1}(y_1) = \int p_{x_1, x_2}(y_1 - y_2, y_2)\, dy_2. \tag{2.88}$$

In case x_1 and x_2 are *independent*, Eq. (2.88) reduces to

$$p_{y_1}(y_1) = \int p_{x_1}(y_1 - y_2)\, p_{x_2}(y_2)\, dy_2, \tag{2.89}$$

which is a convolution of the densities of x_1 and x_2. If x_1 and x_2 are independent,

$$\mathscr{E}\{\exp[iu(x_1 + x_2)]\} = \mathscr{E}\{\exp(iux_1) \cdot \exp(iux_2)\}$$
$$= \mathscr{E}\{\exp(iux_1)\}\, \mathscr{E}\{\exp(iux_2)\},$$

so that

$$\varphi_{x_1+x_2}(u) = \varphi_{x_1}(u)\,\varphi_{x_2}(u), \qquad (2.90)$$

which is another reason for the importance of characteristic functions. ▲

5. CONDITIONAL PROBABILITIES AND EXPECTATIONS

A rigorous treatment of this subject unfortunately requires rather advanced mathematical tools, and, as a result, we must limit ourselves to a formal treatment. We shall attempt to motivate our definitions via the relative frequency notion of probability. A mathematical treatment of this subject may be found in Loeve [4, Chapter VII] or Doob [3, Chapter I]. Papoulis [5, 2.3, 4.3, 6.2] presents numerous examples and certain cases not treated here.

Given two events A and B, we define the **conditional probability function** $\Pr\{A \mid B\}$ of event A *given* the event B by

$$\Pr\{A \mid B\} \triangleq \frac{\Pr\{A \cap B\}}{\Pr\{B\}} \qquad (\Pr\{B\} > 0). \qquad (2.91)$$

$\Pr\{A \mid B\}$ is undefined if $\Pr\{B\} = 0$. Loosely speaking, $\Pr\{A \mid B\}$ is the probability of event A, having already observed the occurrence of event B. Thus it is clearly unimportant that $\Pr\{A \mid B\}$ is undefined if $\Pr\{B\} = 0$.

Example 2.16. Consider the die experiment (Example 2.1) with $\Omega = \{1, 2, 3, 4, 5, 6\}$, Borel field F the class of all subsets of Ω, $\Pr\{i\} = 1/6$, $1 \leqslant i \leqslant 6$. Let the event $B = \{2, 4, 6\} = \{even\}$. Having observed the event *even*, we must revise our probability function. Clearly now $\Pr\{1 \mid B\} = \Pr\{3 \mid B\} = \Pr\{5 \mid B\} = 0$, $\Pr\{2 \mid B\} = \Pr\{4 \mid B\} = \Pr\{6 \mid B\} = \frac{1}{3}$. ▲

From the experiment $(\Omega, F, \Pr\{\cdot\})$, we can construct a new experiment $(\Omega, F, \Pr\{\cdot \mid B\})$. It is easy to show that $\Pr\{\cdot \mid B\}$, as defined in (2.91), satisfies the probability axioms of Section 2.

Our definition of the conditional probability function coincides with the intuitive relative frequency notion of probability. For example, consider an experiment repeated N times, N very large. Let N_A denote the number of occurrences of event A, N_B the number of occurrences of event B, and N_{AB} the number of occurrences of both A and B. Then, clearly,

$$\Pr\{A\} = N_A/N, \qquad \Pr\{B\} = N_B/N,$$

$$\Pr\{A \cap B\} = N_{AB}/N, \qquad \Pr\{A \mid B\} = N_{AB}/N_B ,$$

so that

$$\Pr\{A \mid B\} = N_{AB}/N \cdot N/N_B = \Pr\{A \cap B\}/\Pr\{B\}$$

which coincides with our definition. Now in case A and B are independent [see (2.64)], $\Pr\{A \cap B\} = \Pr\{A\}\Pr\{B\}$, so that

$$\Pr\{A \mid B\} = \Pr\{A\}, \tag{2.92}$$

which adds intuitive content to our definition of independence.

Let x and y be jointly distributed continuous random variables. Given the joint distribution and density of x and y, we seek the distribution and density of x *given* a realization y of y. We shall specify the events A and B in (2.91) in terms of conditions on our rv's, and then proceed formally to define the desired conditional distribution and density functions by a limiting process. Let $A = \{x(\omega) \leqslant x\}$. From (2.91),

$$\Pr\{x(\omega) \leqslant x \mid B\} = \frac{\Pr\{x(\omega) \leqslant x, B\}}{\Pr\{B\}}. \tag{2.93}$$

This is just the distribution function defined on the experiment $(\Omega, F, \Pr\{\cdot \mid B\})$ and, as can easily be shown, has properties (2.7), (2.8), and (2.9). Thus

$$F_{x|B}(x \mid B) = \frac{\Pr\{x(\omega) \leqslant x, B\}}{\Pr\{B\}}. \tag{2.94}$$

Now let $B = \{y < y(\omega) \leqslant y + \Delta y\}$ and assume that $p_y(y) > 0$ for the y in question. Then [see (2.10)]

$$\Pr\{B\} = F_y(y + \Delta y) - F_y(y)$$

and (2.94) becomes

$$F_{x|\{y<y(\omega)\leqslant y+\Delta y\}}(x \mid y < y(\omega) \leqslant y + \Delta y)$$

$$= \frac{\Pr\{x(\omega) \leqslant x, y < y(\omega) \leqslant y + \Delta y\}}{F_y(y + \Delta y) - F_y(y)}. \tag{2.95}$$

Now the numerator in (2.95) can be written as

$$F_{x,y}(x, y + \Delta y) - F_{x,y}(x, y)$$

(see Example 2.7). Dividing the numerator and denominator in (2.95) by Δy, that equation now becomes

$$F_{x|\{y<y(\omega)\leqslant y+\Delta y\}}(x \mid y < y(\omega) \leqslant y + \Delta y)$$
$$= \frac{[F_{x,y}(x, y + \Delta y) - F_{x,y}(x, y)]/\Delta y}{[F_y(y + \Delta y) - F_y(y)]/\Delta y}. \tag{2.96}$$

We formally define $F_{x|y}(x \mid y)$, the distribution x *given* $\{y(\omega) = y\}$, as the $\Delta y = 0$ limit of (2.96), which is

$$F_{x|y}(x \mid y) = \frac{\partial F_{x,y}(x, y)/\partial y}{dF_y(y)/dy} = \frac{\int_{-\infty}^{x} p_{x,y}(\xi, y)\, d\xi}{p_y(y)}. \tag{2.97}$$

The last equality follows from (2.50) (differentiate). Note also that $p_y(y)$ is the marginal density

$$p_y(y) = \int p_{x,y}(x, y)\, dx. \tag{2.98}$$

Differentiating (2.97) with respect to x, we get

$$p_{x|y}(x \mid y) = \frac{p_{x,y}(x, y)}{p_y(y)}. \tag{2.99}$$

Note the analogy with (2.91).

Motivated by the preceding formal derivation, we *define* the **conditional density function** $p_{x|y}(x \mid y)$ of x *given* $\{y(\omega) = y\}$ for all x and y such that $p_y(y) > 0$ by

$$p_{x|y}(x \mid y) \triangleq \frac{p_{x,y}(x, y)}{p_y(y)} = \frac{p_{x,y}(x, y)}{\int p_{x,y}(x, y)\, dx}.^{10} \tag{2.100}$$

The qualification that $p_y(y) > 0$ is no restriction, since, for any observed realization y of y, $p_y(y) > 0$ wp 1. Reversing the roles of x and y in (2.100),

$$p_{x,y}(x, y) = p_{y|x}(y \mid x)\, p_x(x), \tag{2.101}$$

---

[10] We note in passing that, if the rv's x and y are jointly discrete, then the conditional probability mass function is defined by

$$m_{x|y}(x \mid y) = \frac{m_{x,y}(x, y)}{m_y(y)},$$

and the conditional expectation is then

$$\mathscr{E}\{x \mid y\} = \sum_x x m_{x|y}(x \mid y).$$

and, substituting this in (2.100), we have

$$p_{x|y}(x \mid y) = \frac{p_{y|x}(y \mid x) \, p_x(x)}{p_y(y)}. \tag{2.102}$$

This is known as **Bayes' rule** or **Bayes' theorem**. Now our formal derivation of (2.99) needs only slight modifications in case x and y are jointly distributed random vectors. *Therefore, in the definition* (2.100), *as well as in* (2.101) *and* (2.102), x *and* y *are, in general, random vectors.* We write (2.100) out in "longhand" in the case where x is a random n-vector and y is a random m-vector:

$$p_{x_1,\ldots,x_n|y_1,\ldots,y_m}(x_1,\ldots,x_n \mid y_1,\ldots,y_m)$$

$$= \frac{p_{x_1,\ldots,x_n,y_1,\ldots,y_m}(x_1,\ldots,x_n,y_1,\ldots,y_m)}{p_{y_1,\ldots,y_m}(y_1,\ldots,y_m)}$$

$$= \frac{p_{x_1,\ldots,x_n,y_1,\ldots,y_m}(x_1,\ldots,x_n,y_1,\ldots,y_m)}{\int \cdots \int p_{x_1,\ldots,x_n,y_1,\ldots,y_m}(\xi_1,\ldots,\xi_n,y_1,\ldots,y_m)\,d\xi_1\cdots d\xi_n}. \tag{2.100}$$

The **conditional expectation** of the random variable (vector) x *given* the random variable (vector) y (given $\{y(\omega) = y\}$) is defined [by analogy with (2.27)] by

$$\mathscr{E}\{x \mid y\} \triangleq \int x p_{x|y}(x \mid y) \, dx. \tag{2.103}$$

If x is a random n-vector, then

$$\mathscr{E}\{x \mid y\} = [\mathscr{E}\{x_1 \mid y\},\ldots,\mathscr{E}\{x_n \mid y\}]^{\mathsf{T}}. \tag{2.104}$$

Compare this with (2.77) and (2.54). Note that, in view of (2.100), $\mathscr{E}\{x \mid y\}$ exists if $\mathscr{E}\{x\}$ exists.

We note that $p_{x|y}(x \mid y)$ depends on the realizations of y and, as a result, is *itself* a random variable. It is a function of the random variable (vector) y [see (2.107) below]. Similarly, $\mathscr{E}\{x \mid y\}$ is also a random variable (vector), being a function of y [see (2.110) below].

The formal properties of conditional density functions and conditional expectations are summarized in the following theorems.

Theorem 2.8 (Conditional Density Function). *Let* x *and* y *be jointly distributed random variables (vectors). Then*

$$p_{x|y}(x \mid y) \geqslant 0, \tag{2.105}$$

$$\int p_{x|y}(x \mid y) \, dx = 1, \tag{2.106}$$

$$p_x(x) = \mathscr{E}\{p_{x|y}(x \mid y)\} = \int p_{x|y}(x \mid y)\, p_y(y)\, dy, \qquad (2.107)$$

$$p_{x|y}(x \mid y) = p_x(x) \qquad \text{if } x \text{ and } y \text{ are independent}. \qquad (2.108)$$

These properties follow immediately from (2.100) and (2.68). The proof is left to the reader.

Theorem 2.9 (Conditional Expectations). *Let* x, y, z *be jointly distributed random variables (vectors); c, d fixed constants; g($\cdot$) a scalar-valued function. Assume that $\mathscr{E}\{x\}$, $\mathscr{E}\{z\}$, $\mathscr{E}\{g(y)\,x\}$ exist. Then*

$$\mathscr{E}\{x \mid y\} = \mathscr{E}\{x\} \qquad \text{if } x \text{ and } y \text{ are independent}, \qquad (2.109)$$

$$\mathscr{E}\{x\} = \mathscr{E}\{\mathscr{E}\{x \mid y\}\}, \qquad (2.110)$$

$$\mathscr{E}\{g(y)\,x \mid y\} = g(y)\,\mathscr{E}\{x \mid y\}, \qquad (2.111)$$

$$\mathscr{E}\{g(y)\,x\} = \mathscr{E}\{g(y)\,\mathscr{E}\{x \mid y\}\}, \qquad (2.112)$$

$$\mathscr{E}\{c \mid y\} = c, \qquad (2.113)$$

$$\mathscr{E}\{g(y) \mid y\} = g(y), \qquad (2.114)$$

$$\mathscr{E}\{cx + dz \mid y\} = c\mathscr{E}\{x \mid y\} + d\mathscr{E}\{z \mid y\}. \qquad (2.115)$$

PROOF: These properties follow easily from (2.100), (2.103), and Theorem 2.8. We point out that (2.112) follows from (2.111) with the help of (2.110). As for (2.115),

$$\mathscr{E}\{cx + dz \mid y\} = \int (cx + dz)\, p_{x,z|y}(x, z \mid y)\, dx\, dz,$$

from whence (2.115) follows. The rest of the proof is left to the reader. ∎

We comment that (2.110) means, roughly speaking, that to compute the average of x we may compute its average for each y and then average over the y's. In view of (2.115), the conditional expectation operator $\mathscr{E}\{\cdot \mid y\}$ is linear. It should be obvious that Schwarz's inequality (2.61), as well as inequalities (2.62) and (2.63), hold for conditional expectations. As a result, so do Theorems 2.1 and 2.3.

Example 2.17. The conditional mean (vector) was noted in (2.104). The *conditional covariance matrix* is appropriately [compare with (2.79)] defined by

$$P_{x|y} \triangleq \mathscr{E}\{(x - \mathscr{E}\{x \mid y\})(x - \mathscr{E}\{x \mid y\})^{\mathrm{T}} \mid y\}. \qquad (2.116)$$

Clearly, $P_{x|y}$ is a *random matrix* (matrix whose elements are rv's). What about $\mathscr{E}\{P_{x|y}\}$? We now show that

$$P_x = \mathscr{E}\{P_{x|y}\} + P_{\mathscr{E}\{x|y\}}, \tag{2.117}$$

where P_x is the unconditional covariance matrix defined in (2.79) and

$$P_{\mathscr{E}\{x|y\}} = \mathscr{E}\{(\mathscr{E}\{x \mid y\} - \mathscr{E}\{x\})(\mathscr{E}\{x \mid y\} - \mathscr{E}\{x\})^{\mathrm{T}}\} \tag{2.118}$$

is the covariance matrix of the conditional mean.

Using (2.110), we first write

$$P_x = \mathscr{E}\{\mathscr{E}\{(x - \mathscr{E}\{x\})(x - \mathscr{E}\{x\})^{\mathrm{T}} \mid y\}\} \tag{2.119}$$

and next develop the inner expectation. That expectation can be written as

$$\mathscr{E}\{[(x - \mathscr{E}\{x \mid y\}) + (\mathscr{E}\{x \mid y\} - \mathscr{E}\{x\})]$$
$$\cdot [(x - \mathscr{E}\{x \mid y\}) + (\mathscr{E}\{x \mid y\} - \mathscr{E}\{x\})]^{\mathrm{T}} \mid y\}. \tag{2.120}$$

Now $\mathscr{E}\{x \mid y\} - \mathscr{E}\{x\}$ is a function of y. Using (2.111), we therefore compute the cross terms in (2.120) to be zero. As a result, (2.120) reduces to

$$\mathscr{E}\{(x - \mathscr{E}\{x \mid y\})(x - \mathscr{E}\{x \mid y\})^{\mathrm{T}} \mid y\} + (\mathscr{E}\{x \mid y\} - \mathscr{E}\{x\})(\mathscr{E}\{x \mid y\} - \mathscr{E}\{x\})^{\mathrm{T}}, \tag{2.121}$$

where, on the last term, we used (2.114). Taking the expectation of (2.121), we have the desired result (2.117). ▲

Example 2.18. The *conditional characteristic function* of the random vector x given the random vector y is defined by

$$\varphi_{x|y}(u \mid y) = \mathscr{E}\{\exp(iu^{\mathrm{T}}x) \mid y\} \tag{2.122}$$

[compare with (2.83)]. Conditional moments can be obtained from the conditional characteristic function in the same way moments are obtained from the unconditional characteristic function. In view of (2.110),

$$\varphi_x(u) = \mathscr{E}\{\varphi_{x|y}(u \mid y)\}. \quad ▲ \tag{2.123}$$

Example 2.19. The joint Gaussian density of rv's x_1 and x_2 is given in (2.74). The exponent in (2.74) can be rewritten as

$$-\frac{1}{2}\left(\frac{x_1 - \mathscr{E}\{x_1\}}{\sigma\{x_1\}}\right)^2 - \frac{1}{2}\left[\frac{x_2 - \mathscr{E}\{x_2\} - \rho(\sigma\{x_2\}/\sigma\{x_1\})(x_1 - \mathscr{E}\{x_1\})}{(1 - \rho^2)^{1/2}\,\sigma\{x_2\}}\right]^2.$$

Therefore, we can write the joint Gaussian density as

$$p_{x_1,x_2}(x_1, x_2) = [(2\pi)^{1/2} \sigma\{x_1\}]^{-1} \exp\left[-\frac{1}{2}\left(\frac{x_1 - \mathscr{E}\{x_1\}}{\sigma\{x_1\}}\right)^2\right]$$

$$\cdot [(2\pi)^{1/2} \sigma\{x_2\}(1 - \rho^2)^{1/2}]^{-1}$$

$$\cdot \exp\left[-\frac{1}{2}\left(\frac{x_2 - \mathscr{E}\{x_2\} - \rho(\sigma\{x_2\}/\sigma\{x_1\})(x_1 - \mathscr{E}\{x_1\})}{\sigma\{x_2\}(1 - \rho^2)^{1/2}}\right)^2\right]. \quad (2.124)$$

But the first factor in the foregoing is the *Gaussian* density of x_1 [compare with (2.40)]. The second factor is also a Gaussian density with mean

$$\mathscr{E}\{x_2\} + \rho[\sigma\{x_2\}/\sigma\{x_1\}](x_1 - \mathscr{E}\{x_1\}) \quad (2.125)$$

and variance

$$\sigma^2\{x_2\}(1 - \rho^2). \quad (2.126)$$

Comparing (2.124) with (2.100), we conclude that the second factor in (2.124) is the conditional density

$$p_{x_2|x_1}(x_2 \mid x_1)$$

and that this density is Gaussian with parameters given in (2.125) and (2.126). Therefore, we have that, if x_1 and x_2 are jointly normally distributed, then the conditional distribution of x_2 given x_1 is also normal with

$$\mathscr{E}\{x_2 \mid x_1\} = \mathscr{E}\{x_2\} + \text{cov}\{x_2, x_1\} \text{var}^{-1}\{x_1\}(x_1 - \mathscr{E}\{x_1\}), \quad (2.127)$$

$$\mathscr{E}\{(x_2 - \mathscr{E}\{x_2 \mid x_1\})^2 \mid x_1\} = \text{var}\{x_2\} - \text{cov}\{x_2, x_1\} \text{var}^{-1}\{x_1\} \text{cov}\{x_1, x_2\}. \quad (2.128)$$

In loose notation, we sometimes say that the "random variable" $x_2 \mid x_1$ is Gaussian. ▲

6. PROPERTIES OF GAUSSIAN RANDOM VARIABLES

The density function of a scalar Gaussian rv was given in (2.40) and its characteristic function in (2.44). It was noted that the density function is characterized by two parameters, the mean and variance of the random variable. It was also noted that the Gaussian density function is symmetric, and, as a result, all the odd central moments are zero. All the even central moments were expressed in terms of the variance in Example 2.6. If the rv's x_1 and x_2 are jointly normal, then, as noted below (2.74), x_1 and x_2 are marginally normal. Furthermore, as shown in Example 2.19, the conditional rv $x_2 \mid x_1$ is also normal. According to Theorem 2.6, if x_1 and x_2 are uncorrelated, then they are independent. These properties of Gaussian random variables, their generalizations, and additional properties to be developed here, account for their importance in mathematics. Their engineering importance will become apparent in subsequent chapters.

Let the random $(n + m)$-vector $z^T = [x^T, y^T]$, where x is an n-vector and y is an m-vector, be $N \sim (m_z, P_z)$. That is, its density and characteristic functions [see (2.80) and (2.84)] are given by

$$p_z(z) = [(2\pi)^{(n+m)/2} \mid P_z \mid^{1/2}]^{-1} \exp[-\tfrac{1}{2}(z - m_z)^T P_z^{-1}(z - m_z)], \quad (2.129)$$

$$\varphi_z(u) = \exp(iu^T m_z - \tfrac{1}{2}u^T P_z u), \quad (2.130)$$

where

$$m_z{}^T = \mathcal{E}\{z^T\} = [\mathcal{E}\{x^T\}, \quad \mathcal{E}\{y^T\}] = [m_x{}^T, \quad m_y{}^T], \quad (2.131)$$

$$P_z = \mathcal{E}\{(z - m_z)(z - m_z)^T\}$$

$$= \begin{bmatrix} \mathcal{E}\{(x - m_x)(x - m_x)^T\}, & \mathcal{E}\{(x - m_x)(y - m_y)^T\} \\ \mathcal{E}\{(y - m_y)(x - m_x)^T\}, & \mathcal{E}\{(y - m_y)(y - m_y)^T\} \end{bmatrix}$$

$$= \begin{bmatrix} P_x & P_{xy} \\ P_{yx} & P_y \end{bmatrix}, \quad (2.132)$$

$$\mid P_z \mid = \mid P_x \mid \mid P_y - P_{yx}P_x^{-1}P_{xy} \mid = \mid P_y \mid \mid P_x - P_{xy}P_y^{-1}P_{yx} \mid.^{11} \quad (2.133)$$

Now assume that the elements of x are uncorrelated with the elements of y. That is, $P_{xy} = P_{yx}^T = 0$. Then we can write the joint characteristic function in (2.130) as

$$\varphi_z(u) = \varphi_{x,y}(u_1, u_2)$$

$$= \exp\left[i[u_1{}^T, u_2{}^T]\begin{bmatrix} m_x \\ m_y \end{bmatrix} - \tfrac{1}{2}[u_1{}^T, u_2{}^T]\begin{bmatrix} P_x & 0 \\ 0 & P_y \end{bmatrix}\begin{bmatrix} u_1 \\ u_2 \end{bmatrix}\right]$$

$$= \exp(iu_1{}^T m_x - \tfrac{1}{2}u_1{}^T P_x u_1)\exp(iu_2{}^T m_y - \tfrac{1}{2}u_2{}^T P_y u_2). \quad (2.134)$$

But these two exponential functions are characteristic functions of normal random vectors, one $N \sim (m_x, P_x)$, the other $N \sim (m_y, P_y)$. In view of (2.131) and (2.132), these are the characteristic functions of x and y, respectively. Therefore, we have proved an extension to Theorem 2.6.

---

[11] See Bellman [2, p. 103]. P_z is assumed positive definite here.

Theorem 2.10. *If the Gaussian random vectors* x *and* y *are uncorrelated, they are independent. Furthermore,* x *and* y *are marginally Gaussian (with appropriate parameters).*

Theorem 2.11. *Let the p-vector* $z \sim N(m_z, P_z)$. *Let* $w = Cz + a$, *where C is a* $q \times p$ *constant matrix, a is a constant q-vector, and* w *is a random q-vector. Then* $w \sim N(Cm_z + a, CP_zC^T)$.

PROOF: Write the characteristic function of w. By definition,

$$\varphi_w(u) = \mathscr{E}\{\exp(iu^Tw)\} = \exp(iu^Ta)\,\mathscr{E}\{\exp(iu^TCz)\}$$
$$= \exp(iu^Ta)\,\varphi_z(C^Tu).$$

But z is Gaussian, so that

$$\varphi_z(C^Tu) = \exp(iu^TCm_z - \tfrac{1}{2}u^TCP_zC^Tu).$$

Therefore,

$$\varphi_w(u) = \exp[iu^T(Cm_z + a) - \tfrac{1}{2}u^TCP_zC^Tu].$$

But this is the characteristic function of a Gaussian random vector with mean $Cm_z + a$ and covariance matrix CP_zC^T. ■

The meaning of Theorem 2.11 is essentially that linear operations on Gaussian random vectors produce Gaussian random vectors.

Theorem 2.12. *If the random vectors* x *and* y *are jointly normally distributed [according to (2.129)], then* y *(say) is also normally distributed (with appropriate parameters).*

PROOF: Let $w^T = [w_1^T, w_2^T]$, and consider the transformation

$$w_1 = x - P_{xy}P_y^{-1}y, \qquad w_2 = y. \tag{2.135}$$

In view of Theorem 2.11, w is normally distributed. An easy computation shows that w_1 and w_2 are uncorrelated. In fact,

$$\mathscr{E}\{(w_1 - \mathscr{E}\{w_1\})(w_1 - \mathscr{E}\{w_1\})^T\}$$
$$= \mathscr{E}\{[(x - m_x) - P_{xy}P_y^{-1}(y - m_y)][(x - m_x) - P_{xy}P_y^{-1}(y - m_y)]^T\}$$
$$= P_x - P_{xy}P_y^{-1}P_{yx},$$

so that

$$\mathscr{E}\{(w - \mathscr{E}\{w\})(w - \mathscr{E}\{w\})^T\} = \begin{bmatrix} P_x - P_{xy}P_y^{-1}P_{yx}, & 0 \\ 0, & P_y \end{bmatrix}. \tag{2.136}$$

Therefore, by Theorem 2.10, w_1 and w_2 are normal. But $w_2 = y$, and so y is normal. ∎

Theorem 2.13. *Let* x *and* y *be jointly normally distributed* [*according to* (2.129)]. *Then the conditional density of* x *given* y *is normal with mean*

$$m_x + P_{xy}P_y^{-1}(y - m_y) \qquad (2.137)$$

and covariance matrix

$$P_x - P_{xy}P_y^{-1}P_{yx}. \qquad (2.138)$$

(*That is,* x | y *is normal with these parameters.*)

PROOF: We have seen in the proof of Theorem 2.12 that w_1 and w_2 in (2.135) are normal and independent. As a result, their joint density (also normal) is the product of their individual densities

$$w_1 \sim N(m_x - P_{xy}P_y^{-1}m_y, P_x - P_{xy}P_y^{-1}P_{yx}), \qquad (2.139)$$

$$w_2 \sim N(m_y, P_y), \qquad (2.140)$$

as is evident from (2.135) and (2.136). That is,

$$p_{w_1,w_2}(w_1, w_2) = [(2\pi)^{n/2} \mid P_x - P_{xy}P_y^{-1}P_{yx} \mid^{1/2}]^{-1}$$

$$\cdot \exp\{-\tfrac{1}{2}(w_1 - m_x + P_{xy}P_y^{-1}m_y)^T (P_x - P_{xy}P_y^{-1}P_{yx})^{-1}$$

$$\cdot (w_1 - m_x + P_{xy}P_y^{-1}m_y)\}$$

$$\cdot [(2\pi)^{m/2} \mid P_y \mid^{1/2}]^{-1} \exp\{-\tfrac{1}{2}(w_2 - m_y)^T P_y^{-1}(w_2 - m_y)\}. \qquad (2.141)$$

Now the Jacobian of the transformation (2.135) is 1, and so, according to Theorem 2.7, we obtain the density $p_{x,y}(x, y)$ by eliminating w_1 and w_2, in (2.141), in favor of x and y via (2.135). As a result, we get

$$p_{x,y}(x, y) = [(2\pi)^{n/2} \mid P_x - P_{xy}P_y^{-1}P_{yx} \mid^{1/2}]^{-1}$$

$$\cdot \exp\{-\tfrac{1}{2}[x - m_x - P_{xy}P_y^{-1}(y - m_y)]^T [P_x - P_{xy}P_y^{-1}P_{yx}]^{-1}$$

$$\cdot [x - m_x - P_{xy}P_y^{-1}(y - m_y)]\}$$

$$\cdot [(2\pi)^{m/2} \mid P_y \mid^{1/2}]^{-1} \exp\{-\tfrac{1}{2}(y - m_y)^T P_y^{-1}(y - m_y)\}. \qquad (2.142)$$

[Note that, in view of (2.133), the coefficient in (2.142) agrees with that in (2.129). In fact, we have just indirectly *proved* (2.133) !] Now the second density in (2.142) is the marginal density of y (Theorem 2.12). Therefore, a comparison with (2.100) (definition of conditional density function) reveals that the first density is $p_{x|y}(x \mid y)$. ∎

In the proofs of Theorems 2.12 and 2.13, it was assumed that the covariance matrix P_z is positive definite. It can be shown that (2.137) and (2.138) are valid even if P_z is singular. See Anderson [1, Chapter 2] for the singular case.

REFERENCES

1. T. W. Anderson, "Introduction to Multivariate Statistical Analysis." Wiley, New York, 1958.
2. R. Bellman, "Introduction to Matrix Analysis." McGraw-Hill, New York, 1960.
3. J. L. Doob, "Stochastic Processes." Wiley, New York, 1953.
4. M. Loeve, "Probability Theory." Van Nostrand, Princeton, New Jersey, 1963.
5. A. Papoulis, "Probability, Random Variables, and Stochastic Processes." McGraw-Hill, New York, 1965.
6. E. Parzen, "Modern Probability Theory and Its Applications." Wiley, New York, 1960.

3

Stochastic Processes

1. INTRODUCTION

A scalar (vector) **stochastic process** $\{x_t\,, t \in T\}$ is a family of random variables (vectors) indexed by the parameter set T.[1] The words *stochastic* and *random* are used synonymously here. The parameter t will refer to *time* in our applications. If the random variables (vectors) x_t are discrete, we say that the stochastic process has a *discrete state space*. If they are continuous, the process is said to have a *continuous state space*. The parameter set T may also be discrete ($T = \{1, 2,..., n\}$, $T = \{1, 2,...\}$) or continuous ($T = [0, 1]$, $T = \{t : t \geqslant 0\}$). If the parameter set is discrete, the stochastic process is a *discrete parameter process*; if it is continuous, we say that the stochastic process is a *continuous parameter process*. A diagram of the classification we have just described appears in Table 3.1.

[1] In Chapter 2, we made a notational distinction between random variables and their realizations. Lower-case roman letters were used for the former, and lower-case italic letters for the latter. This, we felt, was helpful to a reader not acquainted with the theory. The reader should have enough practice by now to make this distinction unnecessary. Whether we are talking about an rv or its sample value should be clear from the context. Henceforth, therefore, we shall use lower-case italic characters for random variables (and vectors) as well as their realizations. Furthermore, when there is no possible confusion, we shall drop the cumbersome subscripts on distribution and density functions. The argument will identify the rv under discussion.

TABLE 3.1

CLASSIFICATION OF STOCHASTIC PROCESSES

State space	Continuous	Random sequence	(Stochastic) process Random function
	Discrete	Discrete parameter chain	Continuous parameter chain
		Discrete	Continuous
		Parameter set	

Except for isolated examples, we shall be concerned here *only* with continuous state space, discrete parameter and continuous parameter stochastic processes. We shall refer to continuous state space, discrete parameter processes as **random sequences**. A continuous state space, continuous parameter process will often be called a **stochastic process** or simply a **process**, when no confusion is likely to arise. Such a process will sometimes be called a **random function**.

Note that a stochastic process is really a function of two variables, the *time parameter t* and the *probability parameter ω*. The notation $\{x_t, t \in T\}$ is shorthand for $\{x_t(\omega), t \in T, \omega \in \Omega\}$. For each t, $x_t(\cdot)$ is a random variable (vector). For each ω, $x.(\omega)$ is a **realization** of the process: a sample function, if the parameter set is continuous; and a sample sequence, if it is discrete. $x_t(\omega)$ is sometimes written $x(t, \omega)$, although we shall not use this notation here. The inclusion of the argument t in parentheses [as in $y(t)$] will be reserved for fixed (nonrandom) functions.

Example 3.1 (Random Walk). Suppose that, at time instants 0, 1, 2,..., we toss a coin. If heads (h) comes up, we take a step $+\Delta x$ forward, and if tails (t) comes up, we take a $-\Delta x$ step backward ($\Delta x \geqslant 0$). Assume that we execute the steps instantaneously. Let $\Omega = \{h, t\}$. Let $\Pr\{h\} = p$, $\Pr\{t\} = q = 1 - p$ ($p = q = \frac{1}{2}$ for a *fair* coin). For each time instant, define the random variable

$$w_n(\omega) = \begin{cases} +\Delta x, & \omega = h \\ -\Delta x, & \omega = t, \end{cases} \tag{3.1}$$

so that

$$\Pr\{w_n(\omega) = +\Delta x\} = p, \qquad \Pr\{w_n(\omega) = -\Delta x\} = q. \qquad (3.2)$$

Let x_n denote our position at instant n, but before the execution of a step at that instant. Assuming we start at the origin, $x_0 = 0$, our position at n is given by

$$x_n = \sum_{i=0}^{n-1} w_i, \qquad n = 1, 2, \dots . \qquad (3.3)$$

It is clear that $\{x_n, n = 0, 1, \dots\}$ is a **discrete parameter chain**. For each n, $x_n(\cdot)$ is a discrete rv given by the sum of rv's in (3.3). Each realization of the chain is a sequence of real numbers of the form $\pm k \Delta x$, $k = 0, 1, \dots$. If we observe the process in continuous time, each realization is a staircase-type function. One realization of the chain is shown in Fig. 3.1. The stochastic process we have generated is called a (simple)

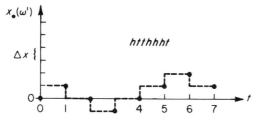

FIG. 3.1. Random walk.

random walk. Since we have defined the random walk as a function of the rv's w_i, we need not concern ourselves about the probability space associated with the walk. It is interesting to note that Ω', the probability space on which the random walk is defined, is the space of sequences of the form

$$\omega' = htthhht \cdots.$$

Now the statistics (probability law) of the w_n could be different. For example, the w_n could be specified to be continuous, independent, each identically distributed with density function $p(w)$. In that case, the process $\{x_n, n = 0, 1, \dots\}$ generated by (3.3) would be a **random sequence**. For each n, $x_n(\cdot)$ would be a continuous rv. Realizations would be real sequences. Note that (3.3) is the solution of the *difference equation*

$$x_{n+1} = x_n + w_n; \qquad n = 0, 1, \dots; \qquad x_0 = 0. \qquad (3.4)$$

This is a *random* difference equation [see Section 9]. Its solution is a random sequence. ▲

Example 3.2. Consider the emission of particles by a radiating source. The emission can be considered to be continuous on a macroscopic scale. Let x_t be the number of particles emitted in the time interval $[0, t]$ (the *population* at time t). Then $\{x_t, t \geqslant 0\}$ is a **continuous parameter chain**, since x_t takes on integer values. ▲

Example 3.3. Consider a satellite in a periodic orbit around the earth. Not all forces acting on the satellite are known. We may, in fact, describe the satellite motion via a differential equation with a random forcing term: a *stochastic* differential equation. The altitude of the satellite, h_t, which is part of the solution of that equation, is a (continuous state space, continuous parameter) **stochastic process** $\{h_t, t \in [0, \pi]\}$, where π is the orbital period. The realizations of the process (altitude traces) on the kth and the $(k + 1)$st pass are shown in Fig. 3.2. ▲

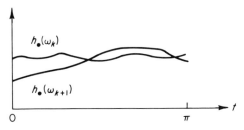

FIG. 3.2. Satellite altitude.

Our treatment of stochastic processes is limited to those concepts needed for the study of stochastic difference and differential equations. More complete formal studies, with numerous applications in various fields, can be found in Parzen [11], Papoulis [10], Cox and Miller [4], and Bartlett [2]. Consult Doob [5] for a complete mathematical treatment.

2. PROBABILITY LAW OF A STOCHASTIC PROCESS

Let $\{x_t, t \in T\}$ be a stochastic process. Its parameter set T may be discrete or continuous. For any finite set $\{t_1, ..., t_n\} \triangleq \{t_i\} \in T$, the joint distribution function of the random variables (vectors) $x_{t_1}, ..., x_{t_n}$ is called a **finite dimensional distribution** of the process. We state without proof that a stochastic process can be characterized (or statistically determined) by specifying the finite dimensional distribution

$$F(x_{t_1}, ..., x_{t_n}) \tag{3.5}$$

for *all* finite sets $\{t_i\} \in T$. By this we mean that, with the distribution functions (3.5) in hand, we can answer all probabilistic questions about the process. Equivalently, the stochastic process can be characterized by specifying the **joint density function**

$$p(x_{t_1}, ..., x_{t_n}) \tag{3.6}$$

or the **joint characteristic function**

$$\varphi_{x_{t_1}, ..., x_{t_n}}(u_1, ..., u_n) \tag{3.7}$$

for *all* finite sets $\{t_i\} \in T$. To specify (3.5), (3.6), or (3.7) for *all* finite sets $\{t_i\} \in T$ is to specify the **probability law** of the stochastic process.

Example 3.4. A sequence of independent rv's $\{w_0, w_1, ...\}$ constitutes a random sequence whose parameter set is $T = \{0, 1, ...\}$. Since the rv's are independent,

$$\varphi_{w_{m_1}, ..., w_{m_n}}(u_1, ..., u_n) = \varphi_{w_{m_1}}(u_1) \cdots \varphi_{w_{m_n}}(u_n)$$

for all $\{m_1, ..., m_n\} \in T$. So, to specify the probability law of this process, it suffices to specify

$$\{\varphi_{w_k}(u_k), k = 0, 1, ...\}.$$

If, moreover, the w_k are identically distributed, it suffices to specify one characteristic function $\varphi_w(u)$. ▲

Example 3.5. A stochastic process $\{x_t, t \geq 0\}$ may be defined by an equation such as

$$x_t = a + bt \tag{3.8}$$

or

$$x_t = a \cos 2\pi t + b \sin 2\pi t, \tag{3.9}$$

where a and b are random variables with specified distribution. The sample functions of the process defined by (3.8) are straight lines, whereas those of the process defined by (3.9) are sine waves. We say that these equations *generate* a stochastic process. Suppose a and b are jointly normally distributed. We can write, from (3.8),

$$\begin{bmatrix} x_{t_1} \\ \vdots \\ x_{t_n} \end{bmatrix} = \begin{bmatrix} 1, t_1 \\ \vdots \\ 1, t_n \end{bmatrix} \begin{bmatrix} a \\ b \end{bmatrix}. \tag{3.10}$$

A similar relationship can be obtained from (3.9), with an appropriately different coefficient matrix. Equation (3.10) represents a linear transformation of the Gaussian random vector $[a, b]^T$, and, by Theorem 2.11, $[x_{t_1}, ..., x_{t_n}]^T$ is a Gaussian random vector with mean and covariance matrix which can be written simply in terms of the mean and covariance matrix of $[a, b]^T$. Therefore, we can write down the probability law of the $\{x_t, t \geqslant 0\}$ process. ▲

Example 3.6. It was seen in Example 3.1 that the stochastic difference equation (3.4) generates a random sequence:

$$x_n = \sum_{i=0}^{n-1} w_i, \qquad n = 1, 2, ... \tag{3.3}$$

Recall that the w_i are independent. Suppose at first that they are not identically distributed, but that the characteristic functions $\varphi_{w_i}(u_i)$, $i = 0, 1, ...,$ are given. We can write

$$\sum_{i=1}^{n} u_i x_i = u_n(x_n - x_{n-1}) + (u_{n-1} + u_n)(x_{n-1} - x_{n-2})$$

$$+ \cdots + (u_1 + \cdots + u_n) x_1$$

$$= u_n w_{n-1} + (u_{n-1} + u_n) w_{n-2}$$

$$+ \cdots + (u_1 + \cdots + u_n) w_0 . \tag{3.11}$$

Therefore,

$$\varphi_{x_1, ..., x_n}(u_1, ..., u_n) = \varphi_{w_{n-1}}(u_n) \varphi_{w_{n-2}}(u_{n-1} + u_n) \cdots \varphi_{w_0}(u_1 + \cdots + u_n), \tag{3.12}$$

and the probability law of the x_n random sequence is specified. If, moreover, the w_i are identically distributed, it suffices to specify the one characteristic function $\varphi_w(u)$ or, equivalently, one density function $p(w)$. ▲

Let $\{x_t, t \in T\}$ be a (discrete or continuous parameter scalar or vector) stochastic process. Although, in general, insufficient for the specification of the probability law of the process, the densities[2]

$$p(x_t), \qquad p(x_t, x_\tau), \qquad t, \tau \in T \tag{3.13}$$

will play a fundamental role in this book.[3] The first density function in

---

[2] As stated in Chapter 2, we assume that all the densities that we write down exist.

[3] They can, of course, be obtained from (3.6) as marginal densities. Similarly, the first density in (3.13) can be obtained from the second.

(3.13) is called the **first-order** density of the process. The second one is called the **second-order** density of the process. Note that, in general, $p(x_t)$ is a function of t, whereas $p(x_t, x_\tau)$ is a function of t and τ. To emphasize this, we may write

$$p(x_t) = p(x, t). \tag{3.14}$$

For each t, $p(\cdot, t)$ is the density $p_{x_t}(\cdot)$ of the random variable (vector) $x_t(\cdot)$. Similarly,

$$p(x_t, x_\tau) = p(x, y, t, \tau). \tag{3.15}$$

For each t and τ, $p(\cdot, \cdot, t, \tau)$ is the joint density $p_{x_t, x_\tau}(\cdot, \cdot)$ of the random variables (vectors) $x_t(\cdot)$ and $x_\tau(\cdot)$. The variables x and y in (3.14) and (3.15) are *dummy* variables. One may think of x as a realization of the rv x_t, and of y as a realization of x_τ. The *conditional* density

$$p(x_t \mid x_\tau) = p(x_t, x_\tau)/p(x_\tau) \tag{3.16}$$

has been defined in (2.100). $p(x_t \mid x_\tau)$ is also a function of t and τ. We sometimes write

$$p(x_t \mid x_\tau) = p(x, t; y, \tau). \tag{3.17}$$

At the risk of boring the reader, we write

$$p_{x_t \mid x_\tau}(x \mid y) = p(x_t \mid x_\tau)$$

to show the relationship with the subscript notation used in Chapter 2. Equation (3.16) can be written as

$$p(x, t; y, \tau) = p(x, y, t, \tau)/p(y, \tau). \tag{3.16}$$

The first- and second-order densities of a stochastic process can answer many important (probabilistic) questions about the process. For two important classes of stochastic processes (Gaussian processes, Section 6; and Markov processes, Section 7), they answer *all* the questions about the process. That is, they specify the probability law of the process.

We now define some parameters (statistics) associated with the first- and second-order densities of a stochastic process. Let $\{x_t, t \in T\}$ be a (discrete or continuous parameter) scalar process. The t function

$$m_x(t) \triangleq \mathscr{E}\{x_t\} \tag{3.18}$$

is called the **mean value function** of the process. The t, τ function

$$\gamma_x(t, \tau) \triangleq \mathscr{E}\{x_t x_\tau\} \tag{3.19}$$

is called the **(auto) correlation function** of the process. The t, τ function

$$c_x(t, \tau) \triangleq \mathscr{E}\{[x_t - m_x(t)][x_\tau - m_x(\tau)]\} \tag{3.20}$$

is the **(auto) covariance function** (*covariance kernel*) of the process. We have that

$$c_x(t, \tau) = \gamma_x(t, \tau) - m_x(t) \, m_x(\tau). \tag{3.21}$$

In case $\{x_t, t \in T\}$ is an n-vector process, we have the **mean value vector** (function)

$$m_x(t) = \mathscr{E}\{x_t\}, \tag{3.18}$$

the **(auto-cross) correlation matrix**

$$\Gamma_x(t, \tau) = \mathscr{E}\{x_t x_\tau^{\mathrm{T}}\}, \tag{3.22}$$

and the **(auto-cross) covariance matrix**

$$C_x(t, \tau) = \mathscr{E}\{[x_t - m_x(t)][x_\tau - m_x(\tau)]^{\mathrm{T}}\} \tag{3.23}$$

and

$$C_x(t, \tau) = \Gamma_x(t, \tau) - m_x(t) \, m_x^{\mathrm{T}}(\tau). \tag{3.24}$$

The matrix t function

$$P_x(t) \triangleq C_x(t, t) \tag{3.25}$$

is, for each t, the *covariance matrix* of the random vector x_t [compare with (2.79)]. Now the diagonal elements of $\Gamma_x(t, \tau)$ and $C_x(t, \tau)$ are, respectively, the (auto) correlation and (auto) covariance functions of $(x_t)_i$, $i = 1,..., n$. The off-diagonal elements are, respectively, the (cross) correlation and (cross) covariance functions of $(x_t)_i$ and $(x_t)_j$, $i \neq j$, $i, j = 1,..., n$. If $\{x_t, t \in T\}$ and $\{y_t, t \in T\}$ are two scalar processes,[4] then their **(cross) correlation** and **(cross) covariance** functions are, respectively,

$$\gamma_{xy}(t, \tau) \triangleq \mathscr{E}\{x_t y_\tau\}, \tag{3.26}$$

$$c_{xy}(t, \tau) \triangleq \mathscr{E}\{[x_t - m_x(t)][y_\tau - m_y(\tau)]\}. \tag{3.27}$$

We shall often omit the subscript x on the (auto) correlation and covariance functions (and matrices) when there is no confusion as to which process we are referring.

---

[4] When speaking of two or more stochastic processes, we always assume that they are defined on the same probability space.

Now the mean value function and correlation function are generally much easier to come by than the probability law of the process. At the same time, they do yield considerable information about the process. Their importance will be emphasized in Sections 4 and 6.

Example 3.7. Consider the process

$$x_t = a + bt \tag{3.8}$$

of Example 3.5. This process has the mean value function

$$m(t) = \mathscr{E}\{a\} + \mathscr{E}\{b\}\, t, \tag{3.28}$$

correlation function

$$\gamma(t, \tau) = \mathscr{E}\{a^2\} + \mathscr{E}\{ab\}(t + \tau) + \mathscr{E}\{b^2\}\, t\tau, \tag{3.29}$$

and covariance function

$$c(t, \tau) = \mathrm{var}\{a\} + \mathrm{cov}\{a, b\}(t + \tau) + \mathrm{var}\{b\}\, t\tau. \tag{3.30}$$

Note that we did not need the joint distribution of a and b, but only their means, variances, and covariance to compute the mean value, correlation, and covariance functions of the x_t process. Suppose a and b are uncorrelated with zero means and unit variances. Then

$$m(t) = 0,$$
$$\gamma(t, \tau) = c(t, \tau) = 1 + t\tau. \quad \blacktriangle \tag{3.31}$$

We now give a brief, formal introduction to the notions of *stationarity* of a stochastic process. Let $\{x_t, t \in T\}$ be a stochastic process with a linear parameter set T.[5] The process is said to be **strictly stationary** (or stationary in the **strict sense**) if it has the same probability law as the process $\{x_{t+\tau}, t \in T\}$ for any $\tau \in T$. This means that

$$p(x_{t_1}, ..., x_{t_n}) = p(x_{t_1+\tau}, ..., x_{t_n+\tau}) \tag{3.32}$$

for *all* finite sets $\{t_i\} \in T$ and for any $\tau \in T$. The process is **strictly stationary of order** k if (3.32) holds for $n \leqslant k$ only. Clearly, if (3.32) holds for $n = k$, then it holds for $n < k$ (marginal densities). A special case of (3.32) $(n = 1)$ is

$$p(x_t) = p(x_{t+\tau})$$

---

[5] T is a linear set if $t, \tau \in T$ implies that $t + \tau \in T$. Examples: $T = \{1, 2, ...\}$, $T = \{t : t \geqslant 0\}$.

for all τ, which means that the first-order density of the process is independent of t:

$$p(x_t) = p(x, t) = p(x). \tag{3.33}$$

As a result, the mean value function of the process, if it exists, is constant:

$$m_x(t) = \text{constant}. \tag{3.34}$$

It is readily seen that the second-order density $[n = 2$ in (3.32)] can depend only on $(t_1 - t_2)$. This, in turn, implies that the correlation function $\gamma(t_1, t_2)$, if it exists, depends only on $(t_1 - t_2)$, or

$$\gamma_x(t + \tau, t) = \mathscr{E}\{x_{t+\tau} x_t\} = \gamma(\tau). \tag{3.35}$$

Equations (3.34) and (3.35) together imply (and are implied by) the covariance function

$$c_x(t + \tau, t) = c(\tau). \tag{3.36}$$

On the basis of the first- and second-order properties of a strictly stationary process, we define a weaker notion of stationarity. The stochastic process $\{x_t, t \in T\}$ is said to be **weakly stationary** (or stationary in the **wide sense**, or **covariance stationary**) if it has finite second moments[6] and (3.34), (3.35) hold.

Finally, we say that the process $\{x_t, t \in T\}$ has **strictly stationary increments** if the process $\{x_{t+s} - x_t, t \in T\}$ is strictly stationary for every $s \in T$.

Example 3.8. The process

$$x_t = a \cos 2\pi t + b \sin 2\pi t \tag{3.9}$$

of Example 3.5, where a and b are uncorrelated, with zero means and unit variances, is weakly stationary:

$$m(t) = 0,$$
$$\gamma(t + \tau, t) = \cos 2\pi\tau. \quad \blacktriangle$$

3. CONVERGENCE OF RANDOM SEQUENCES

Let $\{x_n, n = 1, 2,...\}$ be a scalar random sequence. There are a number of ways in which the sequence might converge (as $n \to \infty$). We

---

[6] This means $\mathscr{E}\{|x_t|^2\} < \infty$ for all $t \in T$. This, in turn, implies that the mean value, correlation, and covariance functions exist. See Example 2.9.

define three modes of convergence here. Other modes of convergence can be found in Loeve [9].

The sequence $\{x_n\}$ is said to converge to x **with probability 1** (wp 1) if

$$\lim_{n \to \infty} x_n(\omega) = x(\omega) \qquad (3.37)$$

for almost all ω (almost all realizations). That is, (3.37) holds except perhaps on an event A such that $\Pr\{A\} = 0$. We then write

$$\lim x_n = x \qquad \text{wp 1.} \qquad (3.38)$$

The sequence $\{x_n\}$ is said to converge to x **in probability** if, for every $\epsilon > 0$,

$$\lim_{n \to \infty} \Pr\{|\,x_n(\omega) - x(\omega)| \geqslant \epsilon\} = 0. \qquad (3.39)$$

We write

$$p \lim x_n = x. \qquad (3.40)$$

The $\{x_n\}$ sequence is said to converge to x **in mean square** if $\mathscr{E}\{|\,x_n\,|^2\} < \infty$ for all n, $\mathscr{E}\{|\,x\,|^2\} < \infty$, and

$$\lim_{n \to \infty} \mathscr{E}\{|\,x - x_n\,|^2\} = 0. \qquad (3.41)$$

We write

$$\text{l.i.m. } x_n = x \qquad (3.42)$$

and call x the **limit in the mean** (or **mean square limit**) of $\{x_n\}$.

Now it can be shown that, in general, wp 1 convergence neither implies nor is implied by mean square convergence. This is shown via counterexamples in Bartlett [2, p. 137]. It can be shown that wp 1 convergence implies convergence in probability, Loeve [9, p. 116]. We show here that *mean square convergence* also *implies convergence in probability*. For every $\epsilon > 0$ and rv y with finite second moment, we have

$$\mathscr{E}\{|\,y\,|^2\} = \int_{-\infty}^{\infty} |\,y\,|^2 p(y)\,dy \geqslant \int_{-\infty}^{-\epsilon} |\,y\,|^2 p(y)\,dy + \int_{\epsilon}^{\infty} |\,y\,|^2 p(y)\,dy$$

$$\geqslant \epsilon^2 \left(\int_{-\infty}^{-\epsilon} p(y)\,dy + \int_{\epsilon}^{\infty} p(y)\,dy \right) = \epsilon^2 \Pr\{|\,y\,| \geqslant \epsilon\}.$$

Substituting $x - x_n$ for y, we get

$$\Pr\{|\,x - x_n\,| \geqslant \epsilon\} \leqslant \mathscr{E}\{|\,x - x_n\,|^2\}/\epsilon^2, \qquad (3.43)$$

which is called the **Chebychev inequality**. Our assertion follows immediately from this inequality.

A necessary and sufficient condition for mean square convergence is the **Cauchy criterion**:

$$\lim_{n,m\to\infty} \mathscr{E}\{|x_n - x_m|^2\} = 0. \tag{3.44}$$

That is, the sequence $\{x_n\}$ has a mean square limit if, and only if, (3.44) holds. Necessity is easy to prove and follows from

$$|x_n - x_m|^2 = |(x_n - x) + (x - x_m)|^2 \leqslant 2|x_n - x|^2 + 2|x - x_m|^2.$$

It can be shown that the Cauchy criterion is also sufficient.[7]

Convergence wp 1 and convergence in probability are more difficult to use than convergence in mean square. On the other hand, the former convergence concepts do not require the existence of second moments, whereas mean square convergence does. In this book, we shall mainly use mean square convergence. We shall indicate, when appropriate, that certain limits also exist wp 1. In that case, it can be shown that both limits must be the same. The properties of mean square (ms) limits are summarized in the following theorem.

Theorem 3.1 (Mean Square Convergence). *$\{x_n\}$, $\{y_n\}$, $\{v_n\}$ are random sequences, z is a random variable, all with finite second moments. $\{c_n\}$ is a sequence of fixed constants, a, b are fixed constants. Let $x =$ l.i.m. x_n, $y =$ l.i.m. y_n, $v =$ l.i.m. v_n, $c = \lim c_n$.[8] Then*

(i) l.i.m. $c_n = \lim c_n$;

(ii) l.i.m. $z = z$;

(iii) l.i.m. $zc_n = zc$;

(iv) l.i.m.$(ax_n + by_n) = ax + by$ (*the* "l.i.m." *operation is linear*);

(v) $\mathscr{E}\{x\} = \lim \mathscr{E}\{x_n\}$ ("$\mathscr{E}$" *and* "l.i.m." *operations commute*);

---

[7] This statement is not completely correct. We assume that all our rv's have density functions, and we define expectation as the Riemann integral (2.27). A Cauchy sequence (sequence satisfying (3.44)) may have as a limit an rv x that does not possess a density, and whose expectation does not exist in the sense of (2.27). If we defined the expectation more generally as a Lebesque integral, then it can be shown (Loeve [9, p. 161]) that the space of rv's with finite second moments, with scalar product $\mathscr{E}\{xy\}$, norm $\mathscr{E}^{1/2}\{|x|^2\}$, and distance $\mathscr{E}^{1/2}\{|x - y|^2\}$ is a *complete* metric space (a Hilbert space). In that case, every Cauchy sequence converges and our claim of sufficiency is correct. The pathological case in which the limit of a Cauchy sequence does not have an expectation, as defined in this book, need not concern an engineer.

[8] This is the usual limit, $|c - c_n| \to 0$ $(n \to \infty)$.

(vi) $\mathscr{E}\{xy\} = \lim \mathscr{E}\{x_n y_n\}$ (*note special case*: $\mathscr{E}\{x^2\} = \lim \mathscr{E}\{x_n{}^2\}$);

(vii) *if* $\mathscr{E}\{x_n y_n\} = \mathscr{E}\{v_n\}$, *then* $\mathscr{E}\{xy\} = \mathscr{E}\{v\}$.

PROOF: (i), (ii), (iii) are trivially true. (iv) follows from

$$\mathscr{E}\{| \, ax_n + by_n - ax - by \, |^2\} = \mathscr{E}\{| \, a(x_n - x) + b(y_n - y)|^2\}$$
$$\leqslant 2a^2\mathscr{E}\{| \, x_n - x \, |^2\} + 2b^2\mathscr{E}\{| \, y_n - y \, |^2\} \to 0 \qquad (n \to \infty).$$

To prove (v), we use Schwarz's inequality (2.61):

$$| \, \mathscr{E}\{x_n - x\}|^2 \leqslant \mathscr{E}\{| \, x_n - x \, |^2\} \to 0 \qquad (n \to \infty).$$

To prove (vi), we write

$$| \, \mathscr{E}\{xy\} - \mathscr{E}\{x_n y_n\}| = | \, \mathscr{E}\{(y - y_n) \, x\} + \mathscr{E}\{(x - x_n) \, y\} - \mathscr{E}\{(x - x_n)(y - y_n)\}|$$
$$\leqslant | \, \mathscr{E}\{(y - y_n) \, x\}| + | \, \mathscr{E}\{(x - x_n) \, y\}| + | \, \mathscr{E}\{(x - x_n)(y - y_n)\}|,$$

and (vi) follows upon applying Schwarz's inequality to the last three terms. (vii) is true because

$$\mathscr{E}\{v\} = \lim \mathscr{E}\{v_n\} = \lim \mathscr{E}\{x_n y_n\} = \mathscr{E}\{xy\},$$

where we have used (v), hypothesis, and (vi), in that order. ∎

Suppose $\{x_n, n = 1, 2,...\}$ is a random m-vector sequence, where

$$x_n{}^T = [(x_n)_1 ,..., (x_n)_m]. \tag{3.45}$$

We take as the *norm* of a (random) m-vector x the usual Euclidean norm

$$| \, x \, | = (x^T x)^{1/2} = \left(\sum_{i=1}^m x_i{}^2 \right)^{1/2}. \tag{3.46}$$

Then the *distance* between the two vectors x and y is

$$| \, x - y \, | = [(x - y)^T (x - y)]^{1/2} = \left[\sum_{i=1}^m (x_i - y_i)^2 \right]^{1/2}. \tag{3.47}$$

We say that the random m-vector sequence $\{x_n\}$ converges in mean square to the random vector x if $\mathscr{E}\{| \, x_n \, |^2\} < \infty$ for all n, $\mathscr{E}\{| \, x \, |^2\} < \infty$, and

$$\lim_{n \to \infty} \mathscr{E}\{| \, x - x_n \, |^2\} = 0. \tag{3.48}$$

It is obvious from the definition of our norm (3.46) that the vector sequence converges if, and only if, the m sequences $\{(x_n)_i, \, n = 1,...\}$ of components of x_n converge. We shall use this fact without further comment.

4. MEAN SQUARE CALCULUS

In the study of stochastic differential equations (Chapter 4), we shall need to consider integrals and derivatives of stochastic processes (random functions). In preparation for this, we present here the elements of mean square calculus. This is an extension of the mean square convergence concepts for random sequences (Section 3), to random functions. A detailed mathematical account of this subject can be found in Loève [9, Chapter X]. We treat here continuity, differentiation, and integration in the mean square sense. These properties and operations can also be defined for the sample functions of the process, that is, with probability one. This more difficult subject is not treated here. We do, however, point out some of the relationships between mean square and wp 1 calculus.

Let $\{x_t(\omega), \, t \in T\}$ be a scalar, continuous parameter, stochastic process with finite second moments. That is, $\mathscr{E}\{|x_t|^2\} < \infty$ for all $t \in T$. In view of Theorems 2.1 and 2.3, the mean value function $m(t) = \mathscr{E}\{x_t\}$ and the correlation function $\gamma(t, \tau) = \mathscr{E}\{x_t x_\tau\}$ exist for all $t, \tau \in T$. Since the covariance function

$$c(t, \tau) = \gamma(t, \tau) - m(t) \, m(\tau), \tag{3.49}$$

it also exists for all $t, \tau \in T$.

The random function x_t is said to be **continuous in mean square** at $t \in T$ if

$$\underset{h \to 0}{\text{l.i.m.}} \; x_{t+h} = x_t \tag{3.50}$$

for $t + h \in T$. Now

$$\mathscr{E}\{|x_{t+h} - x_t|^2\} = \gamma(t + h, t + h) - \gamma(t + h, t) - \gamma(t, t + h) + \gamma(t, t). \tag{3.51}$$

The right-hand side of (3.51) goes to zero with h if $\gamma(t, \tau)$ is continuous at (t, t). Thus, we have a sufficient condition for x_t to be ms continuous at t. This condition is also necessary, since, from Theorem 3.1 (vi), we have

$$\lim_{h, h' \to 0} \mathscr{E}\{x_{t+h} x_{t+h'}\} = \mathscr{E}\{x_t x_t\},$$

which is just

$$\lim_{h,h'\to 0} \gamma(t + h, t + h') = \gamma(t, t). \qquad (3.52)$$

We therefore have

Theorem 3.2. x_t *is mean square continuous at* $t \in T$ *if, and only if,* $\gamma(t, \tau)$ *is continuous at* (t, t).

Corollary 1. *If* $\gamma(t, \tau)$ *is continuous at every diagonal point* (t, t), *then it is continuous at every* (t, τ) *point.*

PROOF: By Theorem 3.2, x_t is ms continuous at every $t \in T$. Thus, l.i.m. $x_{t+h} = x_t (h \to 0)$, and l.i.m. $x_{\tau+h'} = x_\tau (h' \to 0)$ for every $t, \tau \in T$. In view of Theorem 3.1 (vi), then,

$$\lim_{h,h'\to 0} \mathscr{E}\{x_{t+h}x_{\tau+h'}\} = \mathscr{E}\{x_t x_\tau\}, \qquad (3.53)$$

which proves the corollary. ∎

Corollary 2. $\gamma(t, \tau)$ *is continuous at* (t, t) *if, and only if,* $m(t)$ *is continuous at* t *and* $c(t, \tau)$ *is continuous at* (t, t).

PROOF: The *if* part is obvious from (3.49). We now prove the *only if* part. If $\gamma(t, \tau)$ is continuous at (t, t), then, by Theorem 3.2, x_t is ms continuous at t. But then, in view of Theorem 3.1 (v),

$$\lim_{h\to 0} \mathscr{E}\{x_{t+h}\} = \mathscr{E}\{x_t\},$$

which is just

$$\lim_{h\to 0} m(t + h) = m(t). \qquad (3.54)$$

Thus, $m(t)$ is continuous at t. This, together with the continuity of $\gamma(t, t)$ and (3.49), proves the continuity of $c(t, \tau)$ at (t, t). ∎

We point out that, in view of Theorem 3.1 (iv), the sum of mean square continuous random functions is ms continuous. We note that ms continuity on a t interval does not, in general, imply wp 1 continuity on the interval. That is, a random function may be ms continuous on an interval, and yet some, or even every, sample function can be discontinuous at some points in the interval. Finally, we note that by ms continuity of a random vector function we mean the ms continuity of each of the vector elements.

Let $\{x_n(t, \omega), n = 1, 2,...\}$ be a sequence of random functions. That is,

$\{x_n(t, \omega), t \in T\}$ is a random function for each $n \geqslant 1$. Suppose each member of the sequence is mean square continuous at every $t \in T$, and the sequence converges in mean square, uniformly in t. What can we say about the limit random function?

Theorem 3.3. *The uniform mean square limit of mean square continuous random functions is mean square continuous.*

PROOF (*Indirect*):   We have that

$$\mathop{\text{l.i.m.}}_{n \to \infty} x_n(t, \omega) = x(t, \omega) \qquad \text{uniformly in} \quad t \in T.$$

Therefore, by Theorem 3.1 (vi),

$$\lim_{n \to \infty} \mathscr{E}\{x_n(t, \omega)\, x_n(t, \omega)\} = \mathscr{E}\{x(t, \omega)\, x(t, \omega)\}$$

uniformly in t, which is just

$$\lim_{n \to \infty} \gamma_n(t, t) = \gamma(t, t) \tag{3.55}$$

uniformly in t. But since each $x_n(t, \omega)$ is ms continuous at every $t \in T$, then by Theorem 3.2, each $\gamma_n(t, \tau)$ is continuous at every (t, t) point. Therefore, in (3.55) we have the (ordinary) uniform limit of continuous functions, so that $\gamma(t, \tau)$ is continuous at every (t, t) point. But then, by Theorem 3.2, $x(t, \omega)$ is ms continuous at every $t \in T$. ∎

(Theorem 3.3 can readily be proved directly, that is, without going to correlation functions and using the theorem in ordinary calculus which is analogous to Theorem 3.3. We leave this proof for the reader.)

The random function x_t is said to be **mean square differentiable** at $t \in T$ if the following limit, which defines the mean square derivative, exists:

$$\mathop{\text{l.i.m.}}_{h \to 0}(x_{t+h} - x_t)/h = dx_t/dt = \dot{x}_t \tag{3.56}$$

for $t + h \in T$. From (3.56), it is evident that

$$x_{t+h} - x_t = \dot{x}_t h + v,$$

where the remainder v satisfies[9]

$$\mathscr{E}\{v\} = o(h), \qquad \mathscr{E}\{v^2\} = o(h^2).$$

[9] $o(h)$ means that $o(h)/h \to 0$ $(h \to 0)$.

Therefore,

$$\mathscr{E}\{|\, x_{t+h} - x_t\,|^2\} = o(h^2),$$

and we have

Theorem 3.4. *If x_t is mean square differentiable at $t \in T$, then it is mean square continuous at t.*

Theorem 3.5. *x_t is mean square differentiable at $t \in T$, if, and only if, $\partial^2\gamma(t, \tau)/\partial t\, \partial\tau$ exists at (t, t).*

PROOF: In view of the Cauchy criterion (3.44), a sufficient condition for ms differentiability of x_t is the convergence to zero (with h, h') of

$$\mathscr{E}\left\{\left|\frac{x_{t+h} - x_t}{h} - \frac{x_{t+h'} - x_t}{h'}\right|^2\right\}.$$

This does indeed converge to zero if $\partial^2\gamma(t, t)/\partial t\, \partial\tau$ exists, since, in that case,

$$\lim_{h,h'\to 0}\mathscr{E}\left\{\frac{x_{t+h} - x_t}{h} \cdot \frac{x_{t+h'} - x_t}{h'}\right\}$$

$$= \lim_{h,h'\to 0}(hh')^{-1}\{[\gamma(t + h, t + h') - \gamma(t, t + h')]$$

$$- [\gamma(t + h, t) - \gamma(t, t)]\} = \partial^2\gamma(t, t)/\partial t\, \partial\tau. \tag{3.57}$$

This proves the *if* part. To prove the *only if* part, it is sufficient to notice that ms differentiability of x_t implies, via Theorem 3.1 (vi), that

$$\lim_{h,h'\to 0}\mathscr{E}\left\{\frac{x_{t+h} - x_t}{h} \cdot \frac{x_{t+h'} - x_t}{h'}\right\} = \mathscr{E}\{\dot{x}_t\dot{x}_t\}. \tag{3.58}$$

But this, in view of (3.57), implies the existence of the required derivative of γ. ∎

Corollary 1. *If $\partial^2\gamma(t, \tau)/\partial t\, \partial\tau$ exists at every diagonal point (t, t), then it exists at every (t, τ) point.*

PROOF: By Theorem 3.5, x_t is ms differentiable at every $t \in T$. Therefore, in view of Theorem 3.1 (vi),

$$\lim_{h,h'\to 0}\mathscr{E}\left\{\frac{x_{t+h} - x_t}{h} \cdot \frac{x_{\tau+h'} - x_\tau}{h'}\right\} = \mathscr{E}\{\dot{x}_t\dot{x}_\tau\}. \tag{3.59}$$

But an easy computation, similar to that in (3.57), shows that the foregoing limit is

$$\partial^2\gamma(t, \tau)/\partial t\, \partial\tau. \quad \blacksquare \tag{3.60}$$

Corollary 2. $\partial^2\gamma(t, \tau)/\partial t\, \partial\tau$ exists at (t, t) if, and only if, $\dot{m}(t)$ exists at t and $\partial^2 c(t, \tau)/\partial t\, \partial\tau$ exists at (t, t).

PROOF: The *if* part is obvious from (3.49). To prove the *only if* part, we observe that, by Theorem 3.5, x_t is ms differentiable at t. Then from Theorem 3.1 (v) we have

$$\lim_{h\to 0} \mathscr{E}\left\{\frac{x_{t+h} - x_t}{h}\right\} = \mathscr{E}\{\dot{x}_t\}. \tag{3.61}$$

But this limit is just

$$\dot{m}(t). \tag{3.62}$$

Therefore, $\dot{m}(t)$ exists and since, by hypothesis, $\partial^2\gamma(t, \tau)/\partial t\, \partial\tau$ exists at (t, t), then in view of (3.49)

$$\partial c(t, \tau)/\partial t\, \partial\tau$$

also exists at (t, t). ∎

Now suppose x_t is mean square differentiable at every $t \in T$. Then, from Theorem 3.1 (vi),

$$\lim_{h\to 0} \mathscr{E}\left\{\frac{x_{t+h} - x_t}{h} \cdot x_\tau\right\} = \mathscr{E}\{\dot{x}_t x_\tau\}. \tag{3.63}$$

But this limit is nothing more than

$$\partial\gamma(t, \tau)/\partial t. \tag{3.64}$$

Collecting these results together with (3.59)–(3.62), we have

Theorem 3.6. *Let x_t be mean square differentiable at every $t \in T$. Then*

$$m_{\dot{x}}(t) = \mathscr{E}\{\dot{x}_t\} = \frac{d}{dt}\mathscr{E}\{x_t\} = \dot{m}_x(t), \tag{3.65}$$

$$\gamma_{\dot{x}x}(t, \tau) = \mathscr{E}\{\dot{x}_t x_\tau\} = \frac{d}{dt}\mathscr{E}\{x_t x_\tau\} = \frac{\partial\gamma_{xx}(t, \tau)}{\partial t}, \tag{3.66}$$

$$\gamma_{\dot{x}\dot{x}}(t, \tau) = \mathscr{E}\{\dot{x}_t \dot{x}_\tau\} = \frac{d}{dt}\frac{d}{d\tau}\mathscr{E}\{x_t x_\tau\} = \frac{\partial^2\gamma_{xx}(t, \tau)}{\partial t\, \partial\tau}, \tag{3.67}$$

$$c_{\dot{x}x}(t, \tau) = \mathscr{E}\{\dot{x}_t x_\tau\} - \mathscr{E}\{\dot{x}_t\}\,\mathscr{E}\{x_\tau\} = \frac{\partial c_{xx}(t, \tau)}{\partial t}, \tag{3.68}$$

$$c_{\dot{x}\dot{x}}(t, \tau) = \mathscr{E}\{\dot{x}_t \dot{x}_\tau\} - \mathscr{E}\{\dot{x}_t\}\,\mathscr{E}\{\dot{x}_\tau\} = \frac{\partial^2 c_{xx}(t, \tau)}{\partial t\, \partial\tau}. \tag{3.69}$$

That is, the operations of mean square differentiation and expectation commute.

In view of Theorem 3.1 (iv), the sum of ms differentiable functions is ms differentiable, and the ms derivative of the sum equals the sum of the ms derivatives. It should be clear that second- and higher-order ms derivatives can be defined and that these exist if, and only if, appropriate derivatives of the correlation function exist. We add without proof that *ms differentiability implies* wp 1 *continuity. Existence of a second ms derivative implies* wp 1 *existence of the ordinary first derivative*, and so on. Finally, when we say that a random vector function is ms differentiable, we mean that its elements are.

Example 3.9. Let $x_t = at$, where a is a random variable with $\mathscr{E}\{a^4\} < \infty$. Proceeding formally, we have $\dot{x}_t = a$. To verify this, we note that

$$\mathscr{E}\left\{\left|\frac{a(t+h) - at}{h} - a\right|^2\right\} \equiv 0.$$

Now let $y_t = x_t^2$. Formally,

$$\frac{dy_t}{dt} = \frac{dy_t}{dx_t}\frac{dx_t}{dt},$$

which gives $\dot{y}_t = 2x_t a = 2a^2 t$. To verify this,

$$\mathscr{E}\left\{\left|\frac{a^2(t+h)^2 - a^2t^2}{h} - 2a^2t\right|^2\right\} = \mathscr{E}\{a^4\}\,h^2 \to 0 \qquad (h \to 0). \quad \blacktriangle$$

Example 3.10. Let $x_t = \sin at$, where a is the same as in Example 3.9. We show that $\dot{x}_t = a\cos at$:

$$\mathscr{E}\left\{\left|\frac{\sin a(t+h) - \sin at}{h} - a\cos at\right|^2\right\}$$

$$= \mathscr{E}\left\{\left|\frac{\sin at(\cos ah - 1) + \cos at(\sin ah - ah)}{h}\right|^2\right\}$$

$$\leqslant 2\mathscr{E}\left\{\frac{\sin^2 at(\cos ah - 1)^2}{h^2}\right\} + 2\mathscr{E}\left\{\frac{\cos^2 at(\sin ah - ah)^2}{h^2}\right\}.$$

But, since $\cos\alpha \leqslant 1 + \alpha^2$ and $\sin\alpha \leqslant \alpha + \alpha^2$, the foregoing is

$$\leqslant 4h^2\mathscr{E}\{a^4\} \to 0 \qquad (h \to 0). \quad \blacktriangle$$

Example 3.11. Consider the random function e^{x_t}, where x_t is mean square continuous and zero-mean Gaussian for every $t \in T$. We show that e^{x_t} is ms continuous. We have the elementary inequality

$$| e^{x_{t+h}} - e^{x_t} | \leqslant \tfrac{1}{2} | x_{t+h} - x_t | [e^{x_{t+h}} + e^{x_t}]. \tag{3.70}$$

Therefore,

$$\mathscr{E}\{| e^{x_{t+h}} - e^{x_t} |^2\} \leqslant \tfrac{1}{4}\mathscr{E}\{| x_{t+h} - x_t |^2 | e^{x_{t+h}} + e^{x_t} |^2\}$$

$$\leqslant \tfrac{1}{4}[\mathscr{E}\{| x_{t+h} - x_t |^4\}]^{1/2} [\mathscr{E}\{| e^{x_{t+h}} + e^{x_t} |^4\}]^{1/2}.$$

But

$$\mathscr{E}\{| x_{t+h} - x_t |^4\} = 3\mathscr{E}\{| x_{t+h} - x_t |^2\}^2$$

by Example 2.6. Also, an easy computation shows that

$$\mathscr{E}\{\exp(vx_t)\} = \exp(v^2\mathscr{E}\{x_t^2\}/2)$$

for every finite real v. (This is just the *moment generating function.* See Parzen [11, p. 11].) Therefore,

$$[\mathscr{E}\{| e^{x_{t+h}} + e^{x_t} |^4\}]^{1/2} \leqslant M < \infty,$$

and so

$$\mathscr{E}\{| e^{x_{t+h}} - e^{x_t} |^2\} \leqslant \frac{3^{1/2}M}{4} \mathscr{E}\{| x_{t+h} - x_t |^2\} \to 0 \qquad (h \to 0). \quad \blacktriangle$$

Let $a, b \in T$, $a \leqslant b$, and partition the interval $[a, b]$:

$$a = t_0 < t_1 < \cdots < t_n = b.$$

Set $\rho = \max_i(t_{i+1} - t_i)$, $t_i \leqslant t_i' < t_{i+1}$. We say that the random function x_t is **mean square Riemann integrable** over $[a, b]$ if the following limit, which then defines the integral, exists:

$$\operatorname*{l.i.m.}_{\rho \to 0} \sum_{i=0}^{n-1} x_{t_i'}(t_{i+1} - t_i) = \int_a^b x_t \, dt. \tag{3.71}$$

Theorem 3.7. x_t *is mean square Riemann integrable over* $[a, b]$ *if, and only if,* $\gamma(t, \tau)$ *is Riemann integrable over* $[a, b] \times [a, b]$.

PROOF: Let

$$a = \bar{t}_0 < \bar{t}_1 < \cdots < \bar{t}_m = b$$

be another partition of the interval with

$$\bar{\rho} = \max_j(_j\bar{t}_{+1} - \bar{t}_j), \bar{t}_j \leqslant \bar{t}_j' < \bar{t}_{j+1}.$$

According to the Cauchy criterion (3.44), a sufficient condition for the existence of the limit in (3.71) is the convergence to zero (with $\rho, \bar{\rho}$) of

$$\mathcal{E}\left\{\left|\sum_{i=0}^{n-1} x_{t_i'}(t_{i+1} - t_i) - \sum_{j=0}^{m-1} x_{\bar{t}_j'}(\bar{t}_{j+1} - \bar{t}_j)\right|^2\right\}. \qquad (3.72)$$

Now

$$\lim_{\rho,\bar{\rho}\to 0} \mathcal{E}\left\{\sum_{i=0}^{n-1} x_{t_i'}(t_{i+1} - t_i) \cdot \sum_{j=0}^{m-1} x_{\bar{t}_j'}(\bar{t}_{j+1} - \bar{t}_j)\right\}$$

$$= \lim_{\rho,\bar{\rho}\to 0} \sum_{i=0}^{n-1} \sum_{j=0}^{m-1} \mathcal{E}\{x_{t_i'} x_{\bar{t}_j'}\}(t_{i+1} - t_i)(\bar{t}_{j+1} - \bar{t}_j) \qquad (3.73)$$

is just the Riemann integral of $\gamma(t, \tau)$, if it exists. Thus, if $\gamma(t, \tau)$ is Riemann integrable over $[a, b] \times [a, b]$, the limit in (3.73) exists, and (3.72) converges to zero. This proves the *if* part. To prove the *only if* part, we note that, if x_t is ms Riemann integrable, then the limit in (3.73) exists [Theorem 3.1 (vi)], which means that $\gamma(t, \tau)$ is Riemann integrable. ∎

We note that, in fact [Theorem 3.1 (vi)],

$$\int_a^b \int_a^b \gamma(t, \tau) \, dt \, d\tau = \mathcal{E}\left\{\left|\int_a^b x_t \, dt\right|^2\right\},$$

and even more generally

$$\int_a^b \int_c^d \gamma(t, \tau) \, dt \, d\tau = \mathcal{E}\left\{\int_a^b x_t \, dt \cdot \int_c^d x_\tau \, d\tau\right\} \qquad (3.74)$$

if x_t is ms Riemann integrable over $[a, b]$ and $[c, d]$. Also, from Theorem 3.1 (v),

$$\int_a^b m(t) \, dt = \mathcal{E}\left\{\int_a^b x_t \, dt\right\}. \qquad (3.75)$$

Corollary 1. $\gamma(t, \tau)$ *is Riemann integrable over* $[a, b] \times [a, b]$ *if, and only if,* $m(t)$ *is R-integrable over* $[a, b]$ *and* $c(t, \tau)$ *is R-integrable over* $[a, b] \times [a, b]$.

PROOF: The *if* part is obvious from (3.49). The *only if* part follows from (3.75) via Theorem 3.7, and from (3.49). ∎

From Theorem 3.2, Corollary 1 of that theorem, and from Theorem 3.7, we have

Corollary 2. x_t *is ms Riemann integrable over* $[a, b]$ *if* x_t *is ms continuous at every* $t \in [a, b]$.

We now collect computational results (3.74) and (3.75), and their immediate consequences, in

Theorem 3.8. *Let x_t and y_t be mean square Riemann integrable over* T, $[a, b] \in T$, $[c, d] \in T$. *Then*

$$\mathscr{E}\left\{\int_a^b x_t \, dt\right\} = \int_a^b m(t) \, dt, \tag{3.75}$$

$$\mathscr{E}\left\{\int_a^b x_t \, dt \cdot \int_c^d x_\tau \, d\tau\right\} = \int_a^b \int_c^d \gamma(t, \tau) \, dt \, d\tau, \tag{3.74}$$

$$\text{cov}\left\{\int_a^b x_t \, dt, \int_c^d x_\tau \, d\tau\right\} = \int_a^b \int_c^d c(t, \tau) \, dt \, d\tau. \tag{3.76}$$

That is, the operations of integration and expectation may be interchanged. Also,

$$\left(\int_a^b \int_c^d \gamma_{xy}(t, \tau) \, dt \, d\tau\right)^2 \leqslant \int_a^b \int_a^b \gamma_{xx}(t, \tau) \, dt \, d\tau$$

$$\cdot \int_c^d \int_c^d \gamma_{yy}(t, \tau) \, dt \, d\tau. \tag{3.77}$$

[To prove (3.77), use (3.74) and Schwarz's inequality.]

We may note that, in view of Theorem 3.1 (iv), the ms Riemann integral is linear and additive in the domain of integration. We state without proof that, *if x_t is ms continuous, then it is also integrable* wp 1. That is, almost all sample functions are integrable. Also, the wp 1 integral equals the ms integral. Finally, by the integral of a random vector function, we mean the vector whose elements are the integrals of the elements of the random vector function.

Suppose x_t is mean square Riemann integrable over $[a, t]$ for every $t \in [a, b]$. Then

$$y_t = \int_a^t x_\tau \, d\tau \tag{3.78}$$

is a random function (of t) defined on $[a, b]$. Clearly [see (3.74)], $\mathscr{E}\{y_t^2\} < \infty$ for every $t \in [a, b]$. We now study some of the properties of this random function.

Theorem 3.9. *Let x_t be ms Riemann integrable over $[a, t]$ for every* $t \in [a, b]$. *Then y_t (3.78) is ms continuous on $[a, b]$. Furthermore, if x_t is ms continuous on $[a, b]$, then y_t is ms differentiable on (a, b) and $\dot{y}_t = x_t$.*

PROOF (*continuity*):

$$\mathscr{E}\left\{\left|\int_a^{t+h} x_\tau\,d\tau - \int_a^t x_\tau\,d\tau\right|^2\right\} = \mathscr{E}\left\{\left|\int_t^{t+h} x_\tau\,d\tau\right|^2\right\}$$

$$= \int_t^{t+h}\int_t^{t+h}\gamma(s,\tau)\,ds\,d\tau \to 0 \qquad (h \to 0),$$

since the last Riemann integral is a continuous function of its upper limit.

PROOF (*differentiability*):

$$\mathscr{E}\left\{\left|h^{-1}\int_t^{t+h} x_\tau\,d\tau - x_t\right|^2\right\} = (h^2)^{-1}\mathscr{E}\left\{\left|\int_t^{t+h}(x_\tau - x_t)\,d\tau\right|^2\right\}$$

$$= (h^2)^{-1}\int_t^{t+h}\int_t^{t+h}\mathscr{E}\{(x_s - x_t)(x_\tau - x_t)\}\,ds\,d\tau$$

$$\leqslant (h^2)^{-1}\left(\int_t^{t+h}\mathscr{E}^{1/2}\{(x_\tau - x_t)^2\}\,d\tau\right)^2.$$

Since x_t is ms continuous on $[a, b]$, by Theorem 3.2 $\mathscr{E}^{1/2}\{(x_\tau - x_t)^2\}$ is continuous on $[a, b]$. Then by the mean value theorem for Riemann integrals,

$$\int_t^{t+h}\mathscr{E}^{1/2}\{(x_\tau - x_t)^2\}\,d\tau = h\mathscr{E}^{1/2}\{(x_{\tau_0} - x_t)^2\}, \qquad \tau_0 \in [t, t+h].$$

Therefore,

$$\mathscr{E}\left\{\left|h^{-1}\int_t^{t+h} x_\tau\,d\tau - x_t\right|^2\right\} \leqslant \mathscr{E}\{(x_{\tau_0} - x_t)^2\} \to 0 \qquad (h \to 0). \quad \blacksquare$$

Theorem 3.10 (Fundamental Theorem of ms Calculus). *Let $\dot{x}_t$ be ms Riemann integrable on $[a, b]$. Then*

$$x_t - x_a = \int_a^t \dot{x}_\tau\,d\tau$$

with probability 1.

PROOF: We show that

$$\mathscr{E}\left\{\left|x_t - x_a - \int_a^t \dot{x}_\tau\,d\tau\right|^2\right\} = 0,$$

and then the conclusion follows via Example 2.5. Now

$$\mathscr{E}\left\{\left|x_t - x_a - \int_a^t \dot{x}_\tau \, d\tau\right|^2\right\} = \mathscr{E}\{(x_t - x_a)^2\}$$

$$- 2\int_a^t \mathscr{E}\{(x_t - x_a)\,\dot{x}_\tau\}\, d\tau$$

$$+ \int_a^t \int_a^t \mathscr{E}\{\dot{x}_s \dot{x}_\tau\}\, ds\, d\tau.$$

By Theorem 3.6, this equals

$$\gamma_{xx}(t, t) - 2\gamma_{xx}(t, a) + \gamma_{xx}(a, a) - 2\int_a^t \left[\frac{\partial \gamma_{xx}(t, \tau)}{\partial \tau}\right.$$

$$\left. - \frac{\partial \gamma_{xx}(a, \tau)}{\partial \tau}\right] d\tau + \int_a^t \int_a^t \frac{\partial^2 \gamma_{xx}(s, \tau)}{\partial s \, \partial \tau} \, ds \, d\tau.$$

Carrying out the foregoing integrations shows that this is zero. ∎

Example 3.12.   Let $z_t = 2a^2 t$, where a is an rv with $\mathscr{E}\{a^4\} < \infty$. Formally,

$$\int_0^t z_\tau \, d\tau = a^2 t^2.$$

We verify this:

$$\int_0^t z_\tau \, d\tau = \text{l.i.m.} \ 2a^2 \sum_{i=0}^{n-1} \tau_i{}^2(\tau_{i+1} - \tau_i)$$

$$= 2a^2 \lim \sum_{i=0}^{n-1} \tau_i{}^2(\tau_{i+1} - \tau_i),$$

the last equality following from Theorem 3.1 (iii). But the last limit is the limit of an ordinary Riemann sum and is just $t^2/2$. Thus the formal result is verified. Compare this with Example 3.9. ▲

5. INDEPENDENCE, CONDITIONING,
THE BROWNIAN MOTION PROCESS

Two stochastic processes $\{x_t, t \in T\}$ and $\{y_t, t \in T\}$ are said to be **independent** if, for all finite parameter sets $\{t_i\} \in T$, the random vectors

$$[x_{t_1}, ..., x_{t_n}]^T, \qquad [y_{t_1}, ..., y_{t_n}]^T$$

are independent.

We may consider the conditional probability density

$$p(x_{t_1},...,x_{t_n} \mid y_{t_1},...,y_{t_n}) \qquad (3.79)$$

or

$$p(x_{t_n} \mid y_{t_1},...,y_{t_n}) \qquad (3.80)$$

and associated expectations. These have already been defined in Section 2.5, and their properties are given in Theorems 2.8 and 2.9. If $\{y_t, t \in T\}$ is a continuous parameter process, we can consider

$$p(x_t \mid y_\tau, \tau \in T), \qquad (3.81)$$

the probability density of x_t, given the realization $y_\tau, \tau \in T$ of the $\{y_t, t \in T\}$ process. The rigorous definition of the conditional density in (3.81) is beyond the scope of this book. Let $T = [a, b]$. Partition this interval

$$a = t_1 < t_2 < \cdots < t_n = b$$

and let $\rho = \max_i(t_{i+1} - t_i)$. Then we shall consider the conditional density in (3.81) as the formal limit

$$p(x_t \mid y_\tau, \tau \in T) = \underset{\substack{\rho \to 0 \\ n \to \infty}}{\text{l.i.m.}} \, p(x_t \mid y_{t_1},...,y_{t_n}), \qquad (3.82)$$

assuming the limit exists. Clearly, if the $\{x_t\}$ and $\{y_t\}$ processes are independent, then the densities in (3.79), (3.80), and (3.81) are equal to

$$p(x_{t_1},...,x_{t_n}), \qquad p(x_{t_n}), \qquad p(x_t),$$

respectively.

A continuous parameter process $\{x_t, t \in T\}$ has **independent increments** if, for all finite sets $\{t_i : t_i < t_{i+1}\} \in T$, the random variables (vectors)

$$x_{t_2} - x_{t_1}, x_{t_3} - x_{t_2},..., x_{t_n} - x_{t_{n-1}}$$

are independent. It follows directly from Example 3.6 that the joint characteristic function

$$\varphi_{x_{t_1},...,x_{t_n}}(u_1,...,u_n) = \varphi_{x_{t_1}}(u_1 + \cdots + u_n) \prod_{i=2}^{n} \varphi_{x_{t_i}-x_{t_{i-1}}}(u_i + \cdots + u_n). \qquad (3.83)$$

Therefore, to specify the probability law of a process with independent increments, it suffices to specify the distribution of x_t and $x_t - x_\tau$

for all $t > \tau \in T$. The $\{x_t\}$ process is said to have **stationary independent increments** (see Section 2) if, in addition,

$$x_{t+h} - x_{\tau+h}$$

has the same distribution as $x_t - x_\tau$ for every $t > \tau \in T$ and every $h > 0$.

A stochastic process that is of great importance in theory and applications, and that will play a prominent role in this book, is the **Brownian motion** (or **Wiener**, or **Wiener–Lévy**) process. A continuous parameter process $\{x_t, t \geqslant 0\}$ is a Brownian motion process if

  (i) $\{x_t, t \geqslant 0\}$ has stationary independent increments;
  (ii) for every $t \geqslant 0$, x_t is normally distributed;
  (iii) for every $t \geqslant 0$, $\mathscr{E}\{x_t\} = 0$;
  (iv) $\Pr\{x_0 = 0\} = 1$.

Examples of Brownian motion processes include particles undergoing Brownian motion and thermal noise in electric circuits.

It follows from Theorem 2.11 that $x_t - x_\tau$ is also normally distributed for every $t, \tau \geqslant 0$. In view of (3.83), to specify the probability law of the Brownian motion process, it remains to specify the distribution of $x_t - x_\tau$ for all $t > \tau \geqslant 0$. Since $x_t - x_\tau$ is Gaussian, its distribution is determined by its mean and variance. Clearly,

$$\mathscr{E}\{x_t - x_\tau\} = 0. \tag{3.84}$$

It can be shown (Parzen [11, p. 28]), from the fact that the Brownian motion process has stationary independent increments, that

$$\operatorname{var}\{(x_t - x_\tau)\} = \sigma^2(t - \tau), \qquad t > \tau, \tag{3.85}$$

where σ^2 (called the *variance parameter*) is an empirical positive constant.

Let us compute the correlation function of the Brownian motion process. (Its mean value function is identically zero.) Let $t > \tau$:

$$\gamma(t, \tau) = \mathscr{E}\{x_t x_\tau\} = \mathscr{E}\{[(x_t - x_\tau) + x_\tau] x_\tau\}$$
$$= \mathscr{E}\{(x_t - x_\tau) x_\tau\} + \mathscr{E}\{x_\tau^2\} = 0 + \sigma^2\tau.$$

[In view of (iv), $x_\tau = x_\tau - x_0$.] Therefore,

$$\gamma(t, \tau) = \sigma^2 \min(t, \tau) \qquad \text{for all} \quad t, \tau > 0.^{10} \tag{3.86}$$

---

[10] $\min(t, \tau)$ equals t if $t \leqslant \tau$, and equals τ if $\tau \leqslant t$.

Note that $\min(t, \tau)$ is continuous at every (t, τ), but $\partial^2 \min(t, \tau)/\partial t \, \partial \tau$ exists at no (t, t). Therefore, in view of Theorem 3.2, the Brownian motion process is mean square continuous on $[0, \infty)$. In view of Theorem 3.7, it is ms Riemann integrable on every t interval. However, according to Theorem 3.5, it is ms differentiable nowhere. It can also be shown (Doob [5, p. 393]) that almost all sample functions of the Brownian motion process are continuous (continuous wp 1). Otherwise, the sample functions of the Brownian motion process are very irregular. They are also differentiable nowhere wp 1. Furthermore, they are of unbounded variation wp 1.

Example 3.13 (*Brownian Motion as Limit of a Random Walk*). Consider the random walk of Example 3.1 with discrete random steps as defined in (3.1) and (3.2). We shall derive a continuous state space, continuous parameter process by considering infinitesimal steps Δx in infinitesimal time increments Δt, and passing to the $\Delta x = \Delta t = 0$ limit. We first compute the characteristic function of a random step w_i. By expanding e^{iuw_i} and averaging, it is easy to compute

$$\varphi_{w_i}(u) = p e^{iu\Delta x} + q e^{-iu\Delta x}.$$

In view of (3.3), and using the fact that the w_i are independent,

$$\varphi_{x_n}(u) = [p e^{iu\Delta x} + q e^{-iu\Delta x}]^n.$$

Now in time t there will be $n = t/\Delta t$ steps; therefore,

$$\varphi_{x_t}(u) = [p e^{iu\Delta x} + q e^{-iu\Delta x}]^{t/\Delta t}. \tag{3.87}$$

Let $p = q = \frac{1}{2}$. By differentiating the characteristic function [see (2.37)], we compute the mean and variance of x_t

$$\mathscr{E}\{x_t\} = 0, \qquad \mathrm{var}\{x_t\} = t(\Delta x)^2/\Delta t.$$

Now we wish to pass to the $\Delta x = \Delta t = 0$ limit. In order to get a sensible result, we must require that $(\Delta x)^2/\Delta t$ have a limit, say σ^2. In that way the limiting process will have finite variance. With this restriction, then, (3.87) becomes

$$\varphi_{x_t}(u) = \{\tfrac{1}{2}[\exp(iu\sigma \, \Delta t^{1/2}) + \exp(-iu\sigma \, \Delta t^{1/2})]\}^{t/\Delta t}. \tag{3.88}$$

We note that for Δt small, $\Delta x = O(\Delta t^{1/2})$, so that the displacement Δx

must be orders of magnitude larger than Δt, the time interval in which it occurs. Taking the logarithm in (3.88), we find that

$$\ln \varphi_{x_t}(u) = -\tfrac{1}{2}u^2\sigma^2 t + O(\Delta t),$$

and in the limit

$$\varphi_{x_t}(u) = \exp(-\tfrac{1}{2}u^2\sigma^2 t). \tag{3.89}$$

Comparing this with (2.44) of Example 2.6, we see that in the limit x_t is Gaussian with zero mean and variance $\sigma^2 t$. Since the random walk has stationary independent increments (w_i), our limiting process x_t does also. Our limiting process is, therefore, a Brownian motion process. ▲

6. GAUSSIAN PROCESSES

A stochastic process $\{x_t,\, t \in T\}$ is a **Gaussian** (or **normal**) process if the probability law [(3.5), (3.6), or (3.7)] of the process is normal. Since the Gaussian density function is completely characterized by the first two moments, it is clear that one can obtain the probability law of a normal process from its mean value and covariance functions. These, in turn, can be obtained from the first- and second-order densities of the process. For the same reason, a weakly stationary Gaussian process is strictly stationary.

Example 3.14. Consider the process

$$x_t = a + bt$$

of Example 3.5, where the rv's a and b are Gaussian, uncorrelated, with zero means and unit variances. We saw in Example 3.5 that the probability law of the x_t process can be obtained via Theorem 2.11; the x_t process is Gaussian. From (3.10), we have

$$\mathscr{E}[x_{t_1},...,x_{t_n}]^{\mathrm{T}} = 0,$$

$$P \equiv \mathscr{E}\begin{bmatrix} x_{t_1} \\ \vdots \\ x_{t_n} \end{bmatrix} [x_{t_1},...,x_{t_n}]$$

$$= \begin{bmatrix} 1 + t_1^2, & 1 + t_1 t_2, & ..., & 1 + t_1 t_n \\ 1 + t_2 t_1, & 1 + t_2^2, & ..., & 1 + t_2 t_n \\ \vdots & \vdots & \vdots & \vdots \\ 1 + t_n t_1, & 1 + t_n t_2, & ..., & 1 + t_n^2 \end{bmatrix}.$$

Now the mean value and covariance functions of this process have been computed in Example 3.7 (3.31) and are

$$m(t) = 0, \qquad c(t, \tau) = \gamma(t, \tau) = 1 + t\tau.$$

It is easy to see that the covariance matrix P can be generated from the covariance function $c(t, \tau)$. The probability law of the x_t process can now be written in terms of the characteristic function

$$\varphi_{x_{t_1},\ldots,x_{t_n}^{(u)}} = \exp(-\tfrac{1}{2}u^\mathrm{T}Pu),$$

where $u^\mathrm{T} = [u_1, \ldots, u_n]$. In a similar way, it follows by Theorem 2.11 that the process in (3.9) of Example 3.5 is Gaussian if a and b are jointly Gaussian. The correlation function of that process, in case a and b are uncorrelated, is given in Example 3.8. As was noted in Example 3.8, that process is weakly stationary. Let the reader convince himself that it is also strictly stationary. ▲

Example 3.15. The Brownian motion process is a normal process. This can be seen from the following reasoning. Given n times $t_1 < t_2 < \cdots < t_n$ one can write

$$x_{t_k} = (x_{t_k} - x_{t_{k-1}}) + (x_{t_{k-1}} - x_{t_{k-2}}) + \cdots + x_{t_1}$$

for any $1 \leqslant k \leqslant n$. Now the Brownian motion increments on the right-hand side are independent and Gaussian, and are therefore jointly Gaussian. As a result, the x_{t_k}'s are linear combinations of jointly normal rv's, and by Theorem 2.11, are themselves jointly normal. ▲

We have already seen (here and in Chapter 2) that Gaussian random variables and, as a result, Gaussian processes have many nice mathematical properties. One of these outstanding properties is embodied in Theorem 2.11, which essentially says that when we perform linear algebraic operations on Gaussian processes we get back a Gaussian process. This has already been demonstrated in Example 3.14. A very important application of this theorem is to linear random difference equations. The difference equation

$$x_{n+1} = x_n + w_n\,; \qquad n = 0, 1,\ldots; \qquad x_0 = 0, \tag{3.90}$$

which we have already encountered in Example 3.1, generates a Gaussian process (random sequence) if the w_i are independent Gaussian random variables. This is easy to see, since the solution of (3.90) is

$$x_n = \sum_{i=0}^{n-1} w_i \,,$$

and an argument identical to that in Example 3.15 shows that the $\{x_n, n = 0, 1,...\}$ process is Gaussian. We shall have much more to say about random difference equations in Section 9 of this chapter and throughout this book.

Another outstanding property of Gaussian processes is contained in the following theorem.

Theorem 3.11. *Let* $\{x_n, n = 1, 2,...\}$ *be a Gaussian random sequence, and let* l.i.m. $x_n = x$. *Then* x *is Gaussian.*

PROOF: To prove that x is Gaussian, we show that its characteristic function is given by

$$\varphi_x(u) = \exp(iu\mathscr{E}\{x\} - \tfrac{1}{2}u^2 \operatorname{var}\{x\}).$$

Since x_n is Gaussian for each n (Theorem 2.10), we have

$$\varphi_{x_n}(u) = \exp(iu\mathscr{E}\{x_n\} - \tfrac{1}{2}u^2 \operatorname{var}\{x_n\}).$$

Thus we have to show that (a) $\mathscr{E}\{x_n\} \to \mathscr{E}\{x\}$, (b) $\operatorname{var}\{x_n\} \to \operatorname{var}\{x\}$, and (c) $\varphi_{x_n} \to \varphi_x$ $(n \to \infty)$. But (a) is just the statement of Theorem 3.1 (v); (b) follows from (a) and Theorem 3.1 (vi). As for (c),

$$| \varphi_{x_n} - \varphi_x | = | \mathscr{E}\{\exp(iux_n) - \exp(iux)\}|$$
$$\leqslant \mathscr{E}\{| \exp(iux_n) - \exp(iux)|\}$$
$$= \sqrt{2}\, \mathscr{E}\{| 1 - \cos u(x_n - x)|^{1/2}\} \leqslant | u |\, \mathscr{E}\{| x_n - x |\} \to 0. \quad \blacksquare$$

The consequence of this theorem is that ms derivatives and integrals (Section 4) of Gaussian processes are themselves Gaussian processes. This will have important application in the study of linear stochastic (random) differential equations in Chapter 4, and in our study of filtering theory.

The engineering importance of Gaussian processes stems from the fact that many physical processes are approximately Gaussian. We have already studied the Brownian motion process which, among other things, provides a good model for particles undergoing Brownian motion and for stellar motion in the universe. It often turns out that when a large number of small independent random effects are superimposed, then, regardless of their individual distribution, the distribution of the sum of these effects is approximately Gaussian. This is essentially the content of the central limit theorem (see, for example, Papoulis [10, p. 266]). Let x_i be independent random variables. Consider the sum

$$x = x_1 + \cdots + x_n .$$

Then under certain general conditions, and if x is properly scaled, the distribution of x approaches the Gaussian distribution as $n \to \infty$. We have already performed such a central limiting operation when we obtained the Brownian motion process as the limit of a random walk in Example 3.13. It is actually amazing how well the Gaussian density is approximated for very modest n in some cases. Papoulis [10, p. 268] has pictured an example in which the x_i are uniformly and identically distributed. The approximation is surprisingly good for $n = 3$.

Because of the content of the central limit theorem just discussed, a Gaussian process is often a good model for the "noise" in physical devices such as radio transmitters and receivers, radars, and control systems. This will be formalized in Section 8.

7. MARKOV PROCESSES

The reader is familiar with the prominent role played by ordinary differential equations in the analysis of deterministic dynamical systems. The importance of ordinary differential equations stems in large part from certain nice mathematical properties that they enjoy. The differential equation

$$dx(t)/dt = f(x(t))$$

says that the rate of change of x at time t depends only on x at t (*now*), and *not* on $x(\tau)$, $\tau < t$. As a result of this, and as is well-known, for $t_1 < t_2$,

$$x(t_2) = g(t_2 ; x(t_1), t_1).$$

That is, the solution at t_2 is a function of $x(t_1)$ and does not depend on $x(\tau)$, $\tau < t_1$.

This property of ordinary differential equations has its stochastic analog in a class of processes called Markov processes. A discrete or continuous parameter stochastic process $\{x_t , t \in T\}$ is called a **Markov process** if, for any finite parameter set $\{t_i : t_i < t_{i+1}\} \in T$, and for every real λ,

$$\Pr\{x_{t_n}(\omega) \leqslant \lambda \mid x_{t_1} ,..., x_{t_{n-1}}\} = \Pr\{x_{t_n}(\omega) \leqslant \lambda \mid x_{t_{n-1}}\}. \qquad (3.91)$$

The Markov property (3.91) says that the probability law of the process in the future, once it is in a given state, does not depend on how the process arrived at the given state. This property is sometimes referred to as the generalized causality principle: the future can be predicted from a knowledge of the present. As we shall subsequently see, the Markov property is a basic assumption that is made in the study of stochastic dynamical systems.

For a continuous parameter process, Eq. (3.91) implies that for $t_1 < t_2$ and all λ,

$$\Pr\{x_{t_2}(\omega) \leqslant \lambda \mid x_\tau, \tau \leqslant t_1\} = \Pr\{x_{t_2}(\omega) \leqslant \lambda \mid x_{t_1}\}. \tag{3.92}$$

The Markov property (3.91) can be written in terms of density functions

$$p(x_{t_n} \mid x_{t_1}, ..., x_{t_{n-1}}) = p(x_{t_n} \mid x_{t_{n-1}}), \tag{3.93}$$

where $t_1 < t_2 < \cdots < t_n$. Equation (3.92) becomes

$$p(x_{t_2} \mid x_\tau, \tau \leqslant t_1) = p(x_{t_2} \mid x_{t_1}). \tag{3.94}$$

Now consider the joint density function $p(x_{t_n}, x_{t_{n-1}}, ..., x_{t_1})$. By definition of the conditional density function (2.100), this can be written as

$$p(x_{t_n}, ..., x_{t_1}) = p(x_{t_n} \mid x_{t_{n-1}}, ..., x_{t_1}) \, p(x_{t_{n-1}}, ..., x_{t_1}).$$

Now, using the Markov property (3.93),

$$p(x_{t_n}, ..., x_{t_1}) = p(x_{t_n} \mid x_{t_{n-1}}) \, p(x_{t_{n-1}}, ..., x_{t_1}).$$

Continuing this process, we finally obtain

$$p(x_{t_n}, ..., x_{t_1}) = p(x_{t_n} \mid x_{t_{n-1}}) \cdot p(x_{t_{n-1}} \mid x_{t_{n-2}}) \cdots p(x_{t_2} \mid x_{t_1}) \cdot p(x_{t_1}). \tag{3.95}$$

Therefore, the probability law of a Markov process can be specified by specifying $p(x_t)$ and $p(x_t \mid x_\tau)$ for all $t > \tau \in T$. The conditional densities $p(x_t \mid x_\tau)$ are called the **transition probability densities** of the Markov process.

Example 3.16. The random sequence $\{x_n, n = 1, 2,...\}$, where the x_i are mutually independent, is a Markov sequence since

$$p(x_n \mid x_{n-1}, ..., x_1) = p(x_n \mid x_{n-1}) = p(x_n). \quad \blacktriangle$$

Example 3.17. The Brownian motion process is a Markov process, since

$$p(x_{t_n} \mid x_{t_{n-1}}, ..., x_{t_1}) = p(x_{t_n} - x_{t_{n-1}} + x_{t_{n-1}} - x_0 \mid x_{t_{n-1}} - x_0, ..., x_{t_1} - x_0)$$
$$= p(x_{t_n} - x_{t_{n-1}} + x_{t_{n-1}} - x_0 \mid x_{t_{n-1}} - x_0)$$
$$= p(x_{t_n} \mid x_{t_{n-1}}).$$

Since the Brownian motion process is also Gaussian (Example 3.15), it is a *Gauss–Markov* process. Clearly, the process of Brownian motion increments is a Markov process (by Example 3.16), since the increments are independent. ▲

Example 3.18. The process of Example 3.14 is *not* a Markov process. We can write

$$x_{t_n} = x_{t_{n-1}} + b(t_n - t_{n-1})$$

and

$$x_{t_n} = x_{t_{n-1}} + (x_{t_{n-1}} - x_{t_{n-2}})(t_n - t_{n-1})/(t_{n-1} - t_{n-2}).$$

Therefore, we see that $p(x_{t_n} \mid x_{t_{n-1}})$ depends on the density of b, whereas $p(x_{t_n} \mid x_{t_{n-1}}, x_{t_{n-2}})$ is a Dirac delta function. We can, however, construct a Markov process by increasing the dimension of the process. Let the reader show that the vector process

$$x_t = a + bt, \qquad y_t = a + b(t - 1)$$

is Markov. ▲

Example 3.19. Consider the difference equation

$$x_{n+1} = x_n + w_n ; \qquad n = 0, 1, ..., \qquad (3.90)$$

where the w_i are independent Gaussian random variables, and the initial condition x_0 is Gaussian, independent of w_i. The sequence generated by (3.90) is clearly Markov since, *given* x_n, x_{n+1} depends only on w_n, which is independent of $x_{n-1}, ..., x_0$. The sequence $\{x_n, n = 0, 1, ...\}$ is therefore Gauss–Markov. We compute its transition probability densities via Theorem 2.7. Given x_n, x_{n+1} is a linear function of w_n and, from Theorem 2.7,

$$p(x_{n+1} \mid x_n) = p_{w_n}(x_{n+1} - x_n). \qquad (3.96)$$

The density on the right-hand side is the density of w_n evaluated at $x_{n+1} - x_n$. ▲

In Example 3.19, we wrote down the transition probability densities between adjacent time steps for the sequence generated by difference equation (3.90). It turns out that all other transition probability densities can be obtained from these via an integral difference equation. We develop that equation now. We have

$$p(x_n \mid x_{n-2}) = \int p(x_n, x_{n-1} \mid x_{n-2}) \, dx_{n-1} \qquad (3.97)$$

as a marginal density [see (2.53)]. Now the integrand can be written

$$p(x_n, x_{n-1} \mid x_{n-2}) = p(x_n \mid x_{n-1}, x_{n-2}) \, p(x_{n-1} \mid x_{n-2})$$
$$= p(x_n \mid x_{n-1}) \, p(x_{n-1} \mid x_{n-2}),$$

where the last equality follows from the Markov property. Therefore,

$$p(x_n \mid x_{n-2}) = \int p(x_n \mid x_{n-1}) \, p(x_{n-1} \mid x_{n-2}) \, dx_{n-1} . \qquad (3.98)$$

Equation (3.98) is called the **Chapman–Kolmogorov** equation. Note that the x_n sequence may be a vector sequence. In that case, the integral in (3.98) is over E^n. Note also that (3.98) holds for a continuous parameter process as well. In that case $n = t_n$, and so on.

It is clear that we can replace x_{n-2} in (3.98) by x_m, $m \leqslant n - 2$, and

$$p(x_n \mid x_m) = \int p(x_n \mid x_{n-1}) \, p(x_{n-1} \mid x_m) \, dx_{n-1} . \qquad (3.99)$$

The transition probability density $p(x_n \mid x_{n-1})$ is the kernel of (3.99). The initial condition for (3.99) is $p(x_m \mid x_m) = \delta(x_m)$, a Dirac delta function.

Now taking the expectation in (3.99) (see Theorem 2.8), we have

$$p(x_n) = \int p(x_n \mid x_{n-1}) \, p(x_{n-1}) \, dx_{n-1} . \qquad (3.100)$$

Thus the probability density of x_n, given only the density of x_0 (the initial condition), also satisfies the Chapman–Kolmogorov equation. The density $p(x_n)$ is of great interest in the study of discrete stochastic dynamical systems, such as the one of Example 3.19. We shall make use of (3.100) in the study of stochastic difference equations in Section 9.

All the stochastic processes we shall encounter in the remainder of this book are Markov. As we have already seen, Markov processes arise in the study of stochastic difference equations (discrete dynamical systems).

They also arise in the study of stochastic differential equations (continuous dynamical systems). There is a large body of theory associated with Markov processes and voluminous applications to various physical problems. These topics are not germane to the subject of this book, and we do not treat them here. Formal treatments of Markov processes with many applications may be found in Bharucha–Reid [3], Parzen [11], Papoulis [10], and Cox and Miller [4]. Mathematical treatments may be found in Doob [5], Itô [7], and Dynkin [6].

8. WHITE NOISE

In modeling any physical process or device, be it as simple as a pendulum or as sophisticated as a spacecraft guidance and control system, the engineer proceeds by delineating the variables describing the process and then connecting them via certain causal relationships or physical laws. He then simulates his model to determine how well it can predict the physical process. He often finds that his predictions are inexact. At this point he may, if he is willing, remodel the process in order to obtain better predictability. He may or may not succeed. If he repeats this process many times, he will reach a point where further modeling does not improve the prediction. This may be because there are actually unpredictable fluctuations in the process for which no causal relationships exist, or because he has reached the level of random errors in the instruments he is using to measure the physical process, or both. In any case, he has reached the "noise level." At this point, if our engineer is ambitious, he might attempt to determine the statistics of the noise (its probability law) by repeated experimentation with the process or device and statistical testing.

A useful concept, and in many cases a good model for the "noise" we discussed previously, is white noise. A **white random sequence** $\{x_n, n = 1, 2, ...\}$ is a Markov sequence for which

$$p(x_k \mid x_l) = p(x_k) \qquad (k > l). \tag{3.101}$$

That is, all the x_k's are mutually independent. As a result, knowing the realization of x_l in no way helps in predicting what x_k will be. A white sequence is completely random or totally unpredictable. If the x_k's are all normally distributed, the $\{x_n\}$ sequence is called a **white Gaussian random sequence**. We have already noted (Section 6) that noise due to the superposition of a large number of small, independent, random effects is often Gaussian, and that this is the statement of the central limit theorem. The importance and usefulness of white Gaussian sequences stem from this fact. The motivation for calling the sequence white will become apparent when we discuss the analogous continuous parameter process.

Let $\{x_n, n = 1, 2,...\}$ be a white Gaussian random vector sequence. Because the sequence is Gaussian, its probability law is specified by the mean value vector

$$\mathscr{E}\{x_n\}, \quad \text{all} \quad n \geqslant 1, \tag{3.102}$$

and the covariance matrix

$$\mathscr{E}\{(x_n - \mathscr{E}\{x_n\})(x_m - \mathscr{E}\{x_m\})^{\mathrm{T}}\}, \quad \text{all} \quad n, m \geqslant 1. \tag{3.103}$$

Because the sequence is white,

$$\mathscr{E}\{(x_n - \mathscr{E}\{x_n\})(x_m - \mathscr{E}\{x_m\})^{\mathrm{T}}\} = Q_n \delta_{nm}, \tag{3.104}$$

where

$$\delta_{nm} = \begin{cases} 1, & n = m \\ 0, & n \neq m \end{cases} \tag{3.105}$$

is the Kronecker delta and Q_n is a positive semidefinite matrix. Therefore, the probability law of the sequence is specified by the mean value vector and the covariance matrix Q_n, all $n \geqslant 1$.

We wish to extend the notion of white noise to the continuous parameter case. Our motivation is as follows. Since the white Gaussian sequence serves as a good model for the noise in discrete physical processes or dynamical systems, its continuous analog may serve as a good model for the noise in continuous dynamical systems, that is, those which evolve in continuous time. We might formally define a white process $\{x_t, t \in T\}$ as a Markov process for which

$$p(x_t \mid x_\tau) = p(x_t), \quad t > \tau \in T. \tag{3.106}$$

That is, the x_t's are mutually independent for all $t \in T$. If the x_t's are normally distributed for each $t \in T$, then the process is a white Gaussian process.

This formal definition, though expressing the properties we wish the white Gaussian process to possess, is not useful in characterizing the process. We attempt to characterize the process indirectly. Consider a scalar zero-mean stationary Gaussian process with correlation function

$$\gamma^\rho(t + \tau, t) = \sigma^2(\rho/2)\, e^{-\rho|\tau|}. \tag{3.107}$$

Now, for ρ large, this process approximates the properties we wish for our white Gaussian process. We note that if ρ is an integer, then the sequence $\{\gamma^{\rho_i}/\sigma^2, \rho_1 < \rho_2 < \cdots\}$ defines the Dirac delta function. Therefore,

$$\gamma^\infty = \sigma^2 \delta(\tau). \tag{3.108}$$

We therefore *formally* define a **white Gaussian process** $\{x_t, t \in T\}$ as a Gaussian process with

$$\mathscr{E}\{(x_t - \mathscr{E}\{x_t\})(x_\tau - \mathscr{E}\{x_\tau\})^{\mathrm{T}}\} = Q(t)\,\delta(t - \tau), \tag{3.109}$$

where $Q(t)$ is a positive semidefinite covariance matrix, and $\delta(t - \tau)$ is the Dirac delta function. For obvious reasons this process is often called a *delta-correlated* process. Since the Dirac delta function is not an ordinary function, we see that, although useful, the white Gaussian process is a mathematical fiction.

To gain more insight into the properties of white Gaussian noise, consider the *power spectral density function*,[11] which is defined as the Fourier transform of the correlation function

$$f(\omega) = \int_{-\infty}^{\infty} e^{-i\tau\omega}\gamma(t + \tau, t)\,d\tau.^{12} \tag{3.110}$$

For the process with correlation function (3.107), the power spectral density function is computed to be

$$f^\rho(\omega) = \frac{\sigma^2}{1 + (\omega/\rho)^2}. \tag{3.111}$$

With $\rho = \infty$, $f^\rho(\omega) = \sigma^2$, a positive constant. This is the origin of the adjective *white* in white Gaussian process. It is used by analogy with *white light*, which contains all frequency components. But a constant power spectral density requires infinite power and, therefore, white Gaussian noise is not physically realizable. If the power spectral density is essentially flat up to very high frequencies, the process is called a *wide band* process. Its correlation function is almost a delta function. The white Gaussian process is then a useful approximation.

Example 3.20 (White Gaussian process as limit of a white Gaussian sequence).[13]   Let $\{x_t^{(\Delta)}, t \geq 0\}$ be the Gaussian vector process defined by

$$x_t^{(\Delta)} = x_n, \qquad n\,\Delta \leq t < (n + 1)\,\Delta, \tag{3.112}$$

---

[11] The spectral density and related concepts useful in the frequency domain analysis of stationary stochastic processes are treated in Papoulis [10, Chapter 10] and Parzen [11, Chapter 3].

[12] Here ω is the frequency and is not to be confused with the probability parameter.

[13] This example is due to Kalman *et al.* [8].

where $\{x_n, n = 0, 1,...\}$ is a zero-mean, white Gaussian random vector sequence with

$$\mathscr{E}\{x_k x_l^{\mathrm{T}}\} = Q\delta_{kl},$$

where Q is a constant matrix, and $1/\varDelta$ is an integer. A sample function of this process is shown in Fig. 3.3. Consider the random vector defined

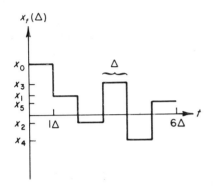

FIG. 3.3. White noise sequence.

for each sample function of the $x_t^{(\varDelta)}$ process (wp 1) by the ordinary (Riemann) integral

$$y = \int_0^1 x_t^{(\varDelta)}\, dt.$$

Clearly, $\mathscr{E}\{y\} = 0$. Now compute

$$\mathrm{var}\{|\, y\, |\} = \mathscr{E}\left\{\left[\int_0^1 x_t^{(\varDelta)}\, dt\right]^{\mathrm{T}}\left[\int_0^1 x_t^{(\varDelta)}\, dt\right]\right\}$$

$$= \mathrm{tr}\int_0^1\int_0^1 \mathscr{E}\{x_t^{(\varDelta)} x_\tau^{(\varDelta)\mathrm{T}}\}\, dt\, d\tau$$

$$= \mathrm{tr}\sum_{i=1}^{1/\varDelta} Q\, \varDelta^2 = \mathrm{tr}\, Q\, \varDelta.^{14} \tag{3.113}$$

This expression goes to zero with $\varDelta$ if Q is kept constant. This says that the effect of white noise in the differential equation

$$dy/dt = x_t^{(\varDelta)}$$

reduces to zero with $\varDelta$. Now this is physically absurd. Hence, to keep $\mathrm{var}\{|\, y\, |\}$ constant as $\varDelta \to 0$, we must replace Q by $Q/\varDelta$. This means

[14] tr A is the trace of the matrix A.

that the amplitude of the random steps in the sample functions of the $x_t^{(\Delta)}$ process increases as $\Delta^{-1/2}$, whereas the area under a random step tends to zero as $\Delta^{1/2}$. This example gives a heuristic justification for (3.109). Q/Δ goes over to $Q \cdot$ (Dirac delta). White Gaussian noise sample functions may be formally regarded as delta functions of vanishingly small area. ▲

Example 3.21 (White Gaussian noise as derivative of Brownian motion). Although we know (Section 5) that the Brownian motion process is not differentiable in any sense, let us proceed formally. Let $\{x_t, t \geq 0\}$ be a Brownian motion process. We know that its covariance function is (3.86)

$$c_x(t, \tau) = \sigma^2 \min(t, \tau).$$

From Theorem 3.6, the covariance function of the process $\{dx_t/dt, t \geq 0\}$ is

$$c_{\dot{x}}(t, \tau) = \frac{\partial^2 c_x(t, \tau)}{\partial t \, \partial \tau}.$$

Therefore,

$$c_{\dot{x}}(t, \tau) = \sigma^2 \frac{\partial^2 \min(t, \tau)}{\partial t \, \partial \tau}.$$

Now

$$\min(t, \tau) = \begin{cases} \tau, & \tau < t \\ t, & \tau > t \end{cases}$$

so that

$$\frac{\partial}{\partial t} \min(t, \tau) = \begin{cases} 0, & \tau < t \\ 1, & \tau > t \end{cases}$$

which is the Heaviside unit step function (in τ), and its derivative with respect to τ is the Dirac delta $\delta(t - \tau)$. As a result,

$$c_{\dot{x}}(t, \tau) = \sigma^2 \delta(t - \tau),$$

and $\dot{x}_t$ is a delta-correlated process. Thus, formally, white Gaussian noise is the derivative of Brownian motion. ▲

9. STOCHASTIC DIFFERENCE EQUATIONS

Many physical processes and devices subject to random disturbances, whose state can be represented as a finite-dimensional vector, can be modeled via a vector difference equation. Let x_k be the n-dimensional state vector at time t_k and w_k the m-dimensional vector of random disturbances ($m \leqslant n$). We can write such a difference equation as

$$x_{k+1} = f(x_k, w_{k+1}, t_k), \qquad k = 0, 1,..., \tag{3.114}$$

where f is a real, in general nonlinear, n-vector function, which, we suppose, is a continuously differentiable function of its arguments. Equation (3.114) is a nonlinear *stochastic vector difference equation* and represents (is sometimes called) a *discrete stochastic dynamical system*. The disturbance w_k is often called the *random noise input* to the system. The sequence $\{w_k, k = 1, 2,...\}$ is a random vector sequence, and, as a result, so is $\{x_k, k = 0, 1,...\}$. The initial condition for (3.114), that is, x_0, may be a given constant vector or a random vector with a specified distribution. The probability law of $\{w_k, k = 1, 2,...\}$ must also be specified. We have already encountered a special case of (3.114) in Example 3.19.

If the random input w_k is absent from (3.114), we have an ordinary difference equation, and we speak of x_k as its solution. In the presence of the random input, it is the density of x_k, $p(x_k)$, which is usually of interest. From it we can compute $\mathscr{E}\{x_k\}$, var$\{x_k\}$, and so on, which give us much information about the statistical behavior of x_k. Probabilities such as $\Pr\{|x_k| \leqslant M\}$ are also often of interest and can be computed from $p(x_k)$.

If the probability law of $\{w_k\}$ is arbitrary, little can be said about the dynamical system (3.114). Let $\{w_k\}$ be a white Gaussian sequence, independent of the initial condition x_0. Then, given x_k, x_{k+1} depends only on w_{k+1}, which is independent of $x_{k-1},..., x_0$. Therefore, the solution of (3.114) is a *Markov* sequence.[15]

Suppose that $n = m$ and (3.114) can be solved for w_{k+1}. That is, for fixed x_k and t_k, $f(x_k, \cdot, t_k)$ has an inverse, say f^{-1}, which is continuously differentiable. Then, according to Theorem 2.7, we have the transition probability density

$$p(x_{k+1} \mid x_k) = p_{w_{k+1}}(f^{-1}(x_k, x_{k+1}, t_k)) \left\| \frac{\partial f^{-1}}{\partial x_{k+1}} \right\|. \tag{3.115}$$

With these transition probability densities in hand, $p(x_k)$, $k = 1, 2,...$ can be computed from (3.100), namely,

$$p(x_{k+1}) = \int p(x_{k+1} \mid x_k) \, p(x_k) \, dx_k. \tag{3.116}$$

$p(x_0)$ is given. If $x_0 = c$, a fixed constant, then $p(x_0) = \delta(x_0 - c)$.

---

[15] Actually, $\{x_k, k = 0, 1,...\}$ is Markov if $\{w_k, k = 1, 2,...\}$ is just white, not necessarily Gaussian.

Now if $w_{k+1} \sim N(0, Q_{k+1})$ and $Q_{k+1} > 0$, then

$$p_{w_{k+1}}(f^{-1}(x_k, x_{k+1}, t_k)) = [(2\pi)^{n/2} \, |Q_{k+1}|^{1/2}]^{-1}$$

$$\cdot \exp\{-\tfrac{1}{2}(f^{-1}(x_k, x_{k+1}, t_k))^{\mathrm{T}} Q_{k+1}^{-1}(f^{-1}(x_k, x_{k+1}, t_k))\}. \quad (3.117)$$

If Q_{k+1} is only semidefinite, say of rank r, then according to Anderson [1, Theorem 6, p. 344] there exists a nonsingular $n \times n$ matrix A_{k+1} such that

$$A_{k+1} Q_{k+1} A_{k+1}^{\mathrm{T}} = \begin{bmatrix} I_{r \times r}, & 0 \\ 0, & 0 \end{bmatrix}. \quad (3.118)$$

Let $v_{k+1} = A_{k+1} w_{k+1}$. If we partition

$$v_{k+1}^{\mathrm{T}} = [\overset{(1 \times r)}{v_{k+1}^{(1)\mathrm{T}}}, \quad v_{k+1}^{(2)\mathrm{T}}],$$
$$A_{k+1}^{-1} = [\overset{(n \times r)}{B_{k+1}}, \quad C_{k+1}], \quad (3.119)$$

then, in view of (3.118), $v_{k+1}^{(1)} \sim N(0, I_{r \times r})$ and

$$\mathscr{E}\{v_{k+1}^{(2)} v_{k+1}^{(2)\mathrm{T}}\} = 0. \quad (3.120)$$

But then, according to Example 2.5, $v_{k+1}^{(2)} = 0$ wp 1, and therefore

$$w_{k+1} = B_{k+1} v_{k+1}^{(1)}. \quad (3.121)$$

If we substitute (3.121) in (3.114), then we have a problem with noise of lower dimension than the state, where the noise covariance matrix is positive definite. This is a special case of the following analysis.

Suppose $m < n$ (which is an important case), the matrix $\partial f/\partial w_{k+1}$ is of rank m, and $w_{k+1} \sim N(0, Q_{k+1})$, $Q_{k+1} > 0$.[16] Then, with the aid of an implicit function theorem (and perhaps a change of variables), Eq. (3.114) can be put in the form

$$x_{k+1}^{(1)} = f^{(1)}(x_k, w_{k+1}, t_k),$$
$$x_{k+1}^{(2)} = f^{(2)}(x_k, x_{k+1}^{(1)}, t_k), \quad (3.122)$$

where x_k is partitioned,

$$x_k^{\mathrm{T}} = [\overset{(1 \times m)}{x_k^{(1)\mathrm{T}}}, \quad x_k^{(2)\mathrm{T}}],$$

[16] If ever rank $\partial f/\partial w_{k+1} <$ rank Q_{k+1}, then we can make them equal by augmenting the state x_k. Add dummy state variables $(z_{k+1})_i = (w_{k+1})_i$.

and $\partial f^{(1)}/\partial w_{k+1}$ is nonsingular. Now we have

$$p(x_{k+1} \mid x_k) = p(x_{k+1}^{(1)}, x_{k+1}^{(2)} \mid x_k)$$

$$= p(x_{k+1}^{(2)} \mid x_{k+1}^{(1)}, x_k) \, p(x_{k+1}^{(1)} \mid x_k). \tag{3.123}$$

Since $f^{(1)}$ has an inverse, we can apply Theorem 2.7 to the first of equations (3.122) and write $p(x_{k+1}^{(1)} \mid x_k)$ as in (3.115) and (3.117). Now

$$p(x_{k+1}^{(2)} \mid x_{k+1}^{(1)}, x_k) = \delta(x_{k+1}^{(2)} - f^{(2)}), \tag{3.124}$$

a Dirac delta function. The specification of $p(x_{k+1} \mid x_k)$ in (3.123) is complete.

Example 3.22. Consider the linear system

$$x_{k+1} = \Phi_{k+1,k} x_k + \bar{\Gamma}_{k+1} \bar{w}_{k+1}, \tag{3.125}$$

where x_k is the n-vector state, $\Phi_{k+1,k}$ is the $n \times n$ (nonsingular) *state transition matrix*, $\bar{\Gamma}_{k+1}$ is $n \times m$, and $\bar{w}_{k+1}$ is $m \times 1$. Suppose $\bar{\Gamma}_{k+1}\bar{w}_{k+1}$ has a singular covariance matrix

$$\bar{\Gamma}_{k+1} \bar{Q}_{k+1} \bar{\Gamma}_{k+1}^{\mathrm{T}}.$$

Then, as we have already noted below (3.117),

$$\bar{\Gamma}_{k+1} \bar{w}_{k+1} = \Gamma_{k+1} w_{k+1},$$

where w_{k+1} is of dimension $r \leqslant m$ and has a positive definite covariance matrix. Γ_{k+1} is $n \times r$ and of rank r. System (3.125) therefore becomes

$$x_{k+1} = \Phi_{k+1,k} x_k + \Gamma_{k+1} w_{k+1}. \tag{3.126}$$

Now partition

$$\Gamma_{k+1} = \begin{bmatrix} \Gamma_{k+1}^{(1)} \\ \Gamma_{k+1}^{(2)} \end{bmatrix},$$

where $\Gamma_{k+1}^{(1)}$ is $r \times r$, and assume, without loss of generality, that $\Gamma^{(1)}$ is nonsingular. Let $y_k = T_k x_k$, where T_k is the nonsingular transformation

$$T_k = \begin{bmatrix} I^{(r \times r)}, & 0 \\ -\Gamma_k^{(2)} \Gamma_k^{(1)-1}, & I \end{bmatrix}, \qquad T_k^{-1} = \begin{bmatrix} I, & 0 \\ \Gamma_k^{(2)} \Gamma_k^{(1)-1}, & I \end{bmatrix}.$$

Let

$$\Psi_{k+1,k} = T_{k+1} \Phi_{k+1,k} T_k^{-1}.$$

Then system (3.126) transforms into

$$\begin{bmatrix} y_{k+1}^{(1)} \\ y_{k+1}^{(2)} \end{bmatrix} = \Psi_{k+1,k} \begin{bmatrix} y_k^{(1)} \\ y_k^{(2)} \end{bmatrix} + \begin{bmatrix} \Gamma_{k+1}^{(1)} w_{k+1} \\ 0 \end{bmatrix}, \tag{3.127}$$

which is of the form (3.122). ▲

An important special case of (3.114) is the *linear* discrete stochastic dynamical system

$$x_{k+1} = \Phi_{k+1,k} x_k + \Gamma_{k+1} w_{k+1}, \qquad k = 0, 1,\dots . \tag{3.128}$$

x_k is the n-vector state of the system, $\Phi_{k+1,k}$ is the $n \times n$ state transition matrix, Γ_{k+1} is the $n \times m$ noise coefficient matrix, and $\{w_k, k = 1, 2,\dots\}$ is a white, Gaussian, random m-vector sequence, $w_k \sim N(0, Q_k)$. The initial condition x_0 is assumed independent of the $\{w_k\}$ sequence.

With the machinery we have just developed, we can compute

$$p(x_{k+1} \mid x_k)$$

and obtain $p(x_k)$, $k = 1, 2,\dots$ from (3.116). From (3.116), we can compute difference equations for

$$\hat{x}_k = \mathscr{E}\{x_k\} \tag{3.129}$$

and

$$P_k = \mathscr{E}\{(x_k - \hat{x}_k)(x_k - \hat{x}_k)^{\mathrm{T}}\}. \tag{3.130}$$

This is a laborious algebraic exercise, especially if we consider the possibility that $\Gamma_{k+1} w_{k+1}$ may have a singular covariance matrix and have to use the transformation developed in Example 3.22. We leave that exercise for the reader. Instead we notice that, if x_0 is a fixed constant, or if x_0 is Gaussian, $x_0 \sim N(\hat{x}_0, P_0)$, then, by Theorem 2.11, Eq. (3.128) generates a *Gauss–Markov* sequence. Then $x_k \sim N(\hat{x}_k, P_k)$; $\hat{x}_k$ and P_k completely determine $p(x_k)$.

We compute difference equations for $\hat{x}_k$ and P_k directly from (3.128). Taking the expectation in (3.128), we have

$$\hat{x}_{k+1} = \Phi_{k+1,k} \hat{x}_k, \tag{3.131}$$

since w_k is zero-mean. Subtracting (3.131) from (3.128),

$$x_{k+1} - \hat{x}_{k+1} = \Phi_{k+1,k}(x_k - \hat{x}_k) + \Gamma_{k+1} w_{k+1}.$$

Squaring this and taking the expectation,

$$P_{k+1} = \Phi_{k+1,k}P_k\Phi_{k+1,k}^{\mathrm{T}} + \Gamma_{k+1}Q_{k+1}\Gamma_{k+1}^{\mathrm{T}},$$

since x_k and w_{k+1} are independent. To summarize, the mean vector and covariance matrix of x_k evolve according to

$$\hat{x}_{k+1} = \Phi_{k+1,k}\hat{x}_k,$$

$$P_{k+1} = \Phi_{k+1,k}P_k\Phi_{k+1,k}^{\mathrm{T}} + \Gamma_{k+1}Q_{k+1}\Gamma_{k+1}^{\mathrm{T}}.$$

(3.132)

Note that Eqs. (3.132) describe the evolution of $\hat{x}_k$ and P_k even if x_0 is not Gaussian (or constant). In that case, however, the $\{x_k\}$ sequence, though Markov, is not Gaussian.

We observe that the effect of the noise in (3.128) is to add uncertainty to the system. $\Gamma_{k+1}Q_{k+1}\Gamma_{k+1}^{\mathrm{T}}$ is positive semidefinite and tends to make the eigenvalues of P_{k+1} larger. This, in turn, tends to "flatten" the density $p(x_{k+1})$. If the noise is absent from (3.128), we can think of w_{k+1} as a random vector with a density that is a Dirac delta function concentrated at 0. But this is the limiting case of $N(0, Q_{k+1})$ as $Q_{k+1} \to 0$. Therefore, we need only set $Q_{k+1} \equiv 0$ in (3.132). We could have, of course, rederived this result from (3.128) after setting $w_{k+1} \equiv 0$. We also could have applied Theorem 2.7 and obtained $p(x_{k+1})$, and subsequently its parameters [(3.132) with $Q_{k+1} \equiv 0$], from (3.128). Alternatively, we could compute $p(x_{k+1} \mid x_k)$ (it is a delta function) from (3.128) and use (3.116). The reader may pursue the details of these alternate derivations.

Now if, in addition to the absence of noise in (3.128), the initial condition x_0 is certain (say $x_0 = c$), then $p(x_k) = \delta(x_k - c)$, a delta function for all k. In that case, Eqs. (3.132) reduce to

$$\hat{x}_{k+1} = \Phi_{k+1,k}\hat{x}_k, \qquad \hat{x}_0 = c, \qquad P_k \equiv 0,$$

(3.133)

a completely deterministic (nonrandom) system.

Consider the scalar, nonlinear, stochastic difference equation with *additive* noise

$$x_{k+1} = f(x_k) + g(x_k)\,w_{k+1}, \qquad k = 0, 1,\ldots,$$

(3.134)

where $\{w_k\}$ is a white Gaussian sequence, $w_k \sim N(0, q_k)$, $p(x_0)$ is given, and x_0 is assumed independent of $\{w_k\}$. $\{x_k\}$ is clearly a Markov sequence. Assuming $g(x_k) \neq 0$, we can compute $p(x_k)$ using the theory that we have developed previously. Instead, however, we attempt to determine the equations of evolution of $\hat{x}_k$ (3.129) and P_k (3.130) directly from

(3.134). We note that these two parameters *do not* determine $p(x_k)$, since x_k is not, in general, Gaussian. They do, however, give information about the mean path of system (3.134) and the dispersion about that path.

Taking the expectation in (3.134) and assuming that it, and subsequent expectations, exist,

$$\hat{x}_{k+1} = \mathscr{E}\{f(x_k)\}. \tag{3.135}$$

Subtracting (3.135) from (3.134), squaring, and averaging,

$$P_{k+1} = \mathscr{E}\{f^2(x_k)\} - \mathscr{E}^2\{f(x_k)\} + \mathscr{E}\{g^2(x_k)\}\, q_{k+1}. \tag{3.136}$$

Now the expectations on the right-hand sides of (3.135) and (3.136) require the knowledge of $p(x_k)$ for their evaluation. We can evaluate these expectations approximately by making certain assumptions about $p(x_k)$. Suppose $p(x_k)$ is assumed symmetric and "close to the mean." Then its third central moment is zero, and fourth and higher-order central moments might be neglected. If f and g are sufficiently smooth, we can expand f, f^2, and g^2 in Taylor series around $\hat{x}_k$ and then average. For example,

$$f(x_k) = f(\hat{x}_k) + f_x(\hat{x}_k)(x_k - \hat{x}_k) + \tfrac{1}{2} f_{xx}(\hat{x}_k)(x_k - \hat{x}_k)^2 + \cdots$$

and

$$\mathscr{E}\{f(x_k)\} \cong f(\hat{x}_k) + \tfrac{1}{2} f_{xx}(\hat{x}_k)\, P_k,$$

since $\mathscr{E}\{x_k - \hat{x}_k\} = 0$, and in view of our assumption about $p(x_k)$.[17] We thus obtain approximate difference equations for $\hat{x}_k$ and P_k:

$$\hat{x}_{k+1} = f(\hat{x}_k) + (P_k/2) f_{xx}(\hat{x}_k),$$

$$P_{k+1} = P_k f_x^2(\hat{x}_k) - \tfrac{1}{4} P_k^2 f_{xx}^2(\hat{x}_k) \tag{3.137}$$

$$+ q_{k+1}[g^2(\hat{x}_k) + P_k g_x^2(\hat{x}_k) + P_k g(\hat{x}_k)\, g_{xx}(\hat{x}_k)].$$

Note that these difference equations are nonlinear and *coupled*, unlike (3.132) for the linear system. They do reduce to (3.132) if f is linear and g is independent of x.

We point out that other approximating assumptions on $p(x_k)$ are possible. We could suppose that $p(x_k)$ is close to the mean and Gaussian

---

[17] Strictly speaking, we should use the Taylor series with a remainder and assume, for example, that f_{xxx} is uniformly bounded, or bounded in mean square. Then we would need only that the sixth central moment of $p(x_k)$ be small, and we would not need to say anything about very high-order moments. Recall from Example 2.6 that, even if $p(x_k)$ is close to the mean (variance small), the moments can grow without bound as the order of the moment gets large.

and retain fourth central moments, writing these in terms of the second central moments as in Example 2.6. We note that the analysis can be readily extended to the vector case. The notation, in that case, becomes rather cumbersome.

REFERENCES

1. T. W. Anderson, "Introduction to Multivariate Statistical Analysis." Wiley, New York, 1958.
2. M. S. Bartlett, "An Introduction to Stochastic Processes." Cambridge Univ. Press, London and New York, 1956.
3. A. T. Bharucha-Reid, "Elements of the Theory of Markov Processes and their Applications." McGraw-Hill, New York, 1960.
4. D. R. Cox and H. D. Miller, "The Theory of Stochastic Processes." Wiley, New York, 1965.
5. J. L. Doob, "Stochastic Processes." Wiley, New York, 1953.
6. E. B. Dynkin, "Markov Processes." Vols. 1 and 2. Academic Press, New York, 1965.
7. K. Itô, Lectures on Stochastic Processes (mimeographed notes), Tata Institute of Fundamental Research, Bombay, India, 1961.
8. R. E. Kalman, T. S. Englar, and R. S. Bucy, Fundamental Study of Adaptive Control Systems, ASD-TR-61-27, Vol. 1, 1962.
9. M. Loeve, "Probability Theory." Van Nostrand, Princeton, New Jersey, 1963.
10. A. Papoulis, "Probability, Random Variables, and Stochastic Processes." McGraw-Hill, New York, 1965.
11. E. Parzen, "Stochastic Processes." Holden-Day, San Francisco, California, 1962.

4

Stochastic Differential Equations

1. INTRODUCTION

Continuous dynamical systems with a finite-dimensional state, which are subject to random disturbances, can often be represented by a nonlinear ordinary differential equation. Let x_t and w_t be, respectively, the n-dimensional vector state and the m-vector random disturbance at time t. Then a general differential equation of the type described can be written as

$$dx_t/dt = f(x_t, w_t, t), \qquad t \geqslant t_0, \tag{4.1}$$

where f is a nonlinear, real n-vector function. Equation (4.1) is the continuous-time analog of the stochastic difference equation that we studied in Section 3.9. It is called a *stochastic differential equation*, or a *continuous stochastic dynamical system*. The random disturbing function w_t is called a *random forcing* (*driving, input*) *function*. The initial condition for (4.1) can be a fixed constant or a random variable x_{t_0} with a specified distribution. The probability law of the $\{w_t, t \geqslant t_0\}$ process is assumed specified.

Now if the function f and the $\{w_t\}$ process are suitably restricted so that the integral

$$\int_{t_0}^{t} f(x_\tau, w_\tau, \tau) \, d\tau$$

93

is well defined, say in the mean square sense (Section 3.4), and the derivative dx_t/dt is also understood in the mean square sense, then by Theorem 3.10

$$x_t - x_{t_0} = \int_{t_0}^t f(x_\tau, w_\tau, \tau)\, d\tau. \tag{4.2}$$

By Theorem 3.9, (4.2) implies (4.1). Thus (4.1) and (4.2) are equivalent. We might now inquire as to the existence, uniqueness, and other properties of the stochastic process generated by (4.2). As in the case of stochastic difference equations, we are interested in the density function $p(x_t)$, $t \geq t_0$.

An important special case of (4.1) is the stochastic differential equation with an *additive* white Gaussian forcing function

$$dx_t/dt = f(x_t, t) + G(x_t, t)w_t, \qquad t \geq t_0, \tag{4.3}$$

where G is an $n \times m$ matrix function, and with initial condition x_{t_0} that is independent of the white Gaussian process $\{w_t, t \geq t_0\}$. Equation (4.3) is sometimes called a *Langevin* equation. The importance of the white noise model for random disturbances in dynamical systems has already been noted (Section 3.8).

Now the $\{w_t\}$ process is delta-correlated, and therefore, in view of Theorem 3.7, w_t is not mean square Riemann integrable. Its sample functions are delta functions (Example 3.20), and as a result, w_t is not integrable wp 1. Consequently, (4.3) has no mathematical meaning. Recall from Example 3.21 that white Gaussian noise is the *formal* derivative of Brownian motion. Let $\{\beta_t, t \geq t_0\}$ be a vector process of independent Brownian motions. Then, formally,

$$w_t \sim d\beta_t/dt,$$

and (4.3) can be considered to be *formally* equivalent to

$$dx_t = f(x_t, t)\, dt + G(x_t, t)\, d\beta_t, \qquad t \geq t_0. \tag{4.4}$$

Now (4.4) is meaningful only insofar as its integral is defined:

$$x_t - x_{t_0} = \int_{t_0}^t f(x_\tau, \tau)\, d\tau + \int_{t_0}^t G(x_\tau, \tau)\, d\beta_\tau. \tag{4.5}$$

The first integral can be defined as an ms Riemann integral (Section 3.4), or as an ordinary integral for the sample functions. The second integral has not yet been defined. Because of the erratic properties of the

Brownian motion process (unbounded variation; see Section 3.5), it cannot be defined for the sample functions. It has been defined in a mean square sense by Itô [9]. Another definition, also in the mean square sense, has recently been proposed by Stratonovich [17]. We shall be largely concerned here with the second integral in (4.5) as defined in the Itô sense. That integral is then called an *Itô stochastic integral*, and (4.4) is an *Itô stochastic differential equation*.

For the moment, consider the differentials in (4.4) as small increments and write (4.4) as

$$x_{t+\delta t} - x_t = f(x_t, t)\,\delta t + G(x_t, t)(\beta_{t+\delta t} - \beta_t).$$

Then, given x_t, $x_{t+\delta t}$ depends only on the Brownian motion increment $\beta_{t+\delta t} - \beta_t$. Now Brownian motion increments are independent, and by assumption, $\{d\beta_t, t \geqslant t_0\}$ is independent of x_{t_0}. Therefore, with x_t given, $x_{t+\delta t}$ is independent of $\{x_\tau, \tau \leqslant t\}$. Thus the $\{x_t, t \geqslant t_0\}$ process generated by (4.4) (if it exists) is a *Markov* process.

This chapter is devoted to a study of the stochastic integral, the stochastic differential equation (4.4), and the properties of its solution process. That process is Markov, and we shall derive the laws of evolution for its transition probability densities and for the density $p(x_t)$. We shall also be concerned briefly with the relationship between the solutions of (4.4) and physically realizable processes, that is, with the problem of modeling.

2. MODELING THE PROCESS e^{β_t}

The solution x_t of the stochastic differential equation (4.4) will clearly be a function of the Brownian motion process $\{\beta_t\}$. It is instructive to consider such a process x_t and ask what differential equation that process satisfies.

Consider the scalar process defined by

$$x_t = e^{\beta_t}, \qquad t \geqslant 0, \qquad \beta_0 = 0 \qquad \text{wp 1}, \tag{4.6}$$

where $\{\beta_t, t \geqslant 0\}$ is a scalar Brownian motion process with variance parameter σ^2. Now β_t is differentiable nowhere wp 1 and in mean square (Section 3.5). As a result, we cannot differentiate (4.6) to obtain a differential equation for x_t. Consider a power series expansion of (4.6):

$$\delta x_t = e^{\beta_t + \delta\beta_t} - e^{\beta_t} = x_t\left(\delta\beta_t + \frac{\delta\beta_t^2}{2} + \cdots\right), \qquad \delta\beta_t^2 \equiv (\delta\beta_t)^2, \tag{4.7}$$

where $\delta\xi_t = \xi_{t+\delta t} - \xi_t$, $\delta t > 0$. Now if we truncate the expansion at the first-order term, we have

$$\mathscr{E}\{\delta x_t - x_t\,\delta\beta_t\} = \mathscr{E}\left\{x_t\left(\frac{\delta\beta_t^2}{2} + \cdots\right)\right\} = \mathscr{E}\left\{\mathscr{E}\left\{x_t\left(\frac{\delta\beta_t^2}{2} + \cdots\right)\,\Big|\,x_t\right\}\right\}$$

$$= \mathscr{E}\left\{x_t\mathscr{E}\left\{\left(\frac{\delta\beta_t^2}{2} + \cdots\right)\,\Big|\,x_t\right\}\right\} = \mathscr{E}\left\{x_t\mathscr{E}\left\{\left(\frac{\delta\beta_t^2}{2} + \cdots\right)\right\}\right\}$$

$$= \mathscr{E}\{x_t\}\,\mathscr{E}\left\{\left(\frac{\delta\beta_t^2}{2} + \cdots\right)\right\} = O(\delta t),^1 \qquad (4.8)$$

and, similarly,

$$\mathscr{E}\{|\,\delta x_t - x_t\,\delta\beta_t\,|^2\} = O(\delta t^2), \qquad (4.9)$$

since

$$\mathscr{E}\{[\delta\beta_t]^n\} = \begin{cases} 0, & \text{all odd } n \geqslant 1 \\ 1 \cdot 3 \cdot 5 \cdots (n-1)(\sigma\,\delta t^{1/2})^n, & \text{all even } n \geqslant 2, \end{cases} \qquad (4.10)$$

as we have seen in Example 2.6. Thus the representation of (4.6) by the differential equation

$$dx_t = x_t\,d\beta_t$$

involves errors of the order of dt and is consequently meaningless (truncation errors add linearly). Apparently the $\delta\beta_t^2$ term in (4.7) cannot be omitted.

Compute[2]

$$\mathscr{E}\left\{\delta x_t - x_t\,\delta\beta_t - \frac{x_t}{2}\,\delta\beta_t^2\right\} = o(\delta t),$$

$$\mathscr{E}\left\{\left|\,\delta x_t - x_t\,\delta\beta_t - \frac{x_t}{2}\,\delta\beta_t^2\,\right|^2\right\} = o(\delta t^2). \qquad (4.11)$$

Therefore, in the mean square sense, we have the differential equation representation

$$dx_t = x_t\,d\beta_t + \tfrac{1}{2}x_t\,d\beta_t^2, \qquad x_0 = 1 \text{ wp } 1. \qquad (4.12)$$

Our manipulations with differentials have been purely formal. We have not defined these differentials, and, as a result, we have not given a meaning to (4.12). Integrating (4.12) formally,

$$x_t - 1 = \int_0^t x_\tau\,d\beta_\tau + \frac{1}{2}\int_0^t x_\tau\,d\beta_\tau^2. \qquad (4.13)$$

---

[1] $\mathscr{E}\{|\,x_t\,|^n\}$ exists for all n. See Example 3.11. $O(\delta t)$ means that $O(\delta t)/\delta t \to M$ ($\delta t \to 0$), $0 < M < \infty$.

[2] $o(\delta t)$ means that $o(\delta t)/\delta t \to 0$ ($\delta t \to 0$).

We shall define the integrals in (4.13). Then the differential dx_t in (4.12) will be considered as a formal notation for (4.13). That is, the differential will be defined in terms of the integral.

Obviously, (4.13) [and consequently (4.12)] should have (4.6) as its solution. Having defined the integrals in (4.13), we shall return to this problem to see that this is so. See Examples 4.3, 4.7, and 4.10.

Example 4.1. We compute the solution of (4.12) by formal manipulations with differentials. Multiply (4.12) by the integrating factor $e^{-\beta t}$:

$$e^{-\beta t}\, dx_t = x_t e^{-\beta t}\, d\beta_t + \tfrac{1}{2} x_t e^{-\beta t}\, d\beta_t{}^2.$$

Now by an expansion procedure similar to (4.7) it is easy to compute the differential

$$d(x_t e^{-\beta t}) = e^{-\beta t}\, dx_t - x_t e^{-\beta t}\, d\beta_t - \tfrac{1}{2} x_t e^{-\beta t}\, d\beta_t{}^2$$

in the mean square sense, where we have used (4.12). Combining these two equations, we get

$$d(x_t e^{-\beta t}) = 0,$$

so that

$$x_t = c e^{\beta t}.$$

In view of the initial condition, $c = 1$, and we get (4.6) as the solution. ▲

3. ITÔ STOCHASTIC INTEGRAL

The integral

$$\int_a^b g(\tau)\, d\beta_\tau \,,$$

where $g(t)$ is a fixed (deterministic) real function and $\{\beta_t\}$ is a scalar Brownian motion process, was introduced by Wiener [18] and is called a *Wiener integral*. This was subsequently generalized by Itô [9] to the case where the function $g_t(\omega)$ is random, and the integral

$$\int_a^b g_\tau(\omega)\, d\beta_\tau$$

will be called the *Itô (stochastic) integral*. In this section, we define the Itô integral and determine some of its properties. Its existence is not rigorously proved, but a plausibility argument is given. The general mathematical treatment of the Itô integral may be found in Itô [9–11], Doob [6, Chapter IX], Skorokhod [15], or Dynkin [7].

Let $T = [a, b]$, and let $\{\beta_t, t \in T\}$ be a scalar Brownian motion process with variance parameter σ^2. Suppose that the random function $g_t(\omega)$ is defined on T. Partition T:

$$a = t_0 < t_1 < \cdots < t_n = b.$$

First consider *step* functions

$$g_t(\omega) = \begin{cases} 0, & t < t_0 \\ g_i(\omega), & t_i \leqslant t < t_{i+1} \\ 0, & t \geqslant t_n, \end{cases} \tag{4.14}$$

where $g_i(\omega)$ is independent of $\{\beta_{t_k} - \beta_{t_l} : t_i \leqslant t_l \leqslant t_k \leqslant b\}$, and $\mathscr{E}\{|g_i(\omega)|^2\} < \infty$. For such step functions, the Itô integral is defined by

$$\int_T g_t(\omega)\, d\beta_t \triangleq \sum_{i=0}^{n-1} g_i(\omega)(\beta_{t_{i+1}} - \beta_{t_i}). \tag{4.15}$$

We easily see that

$$\mathscr{E}\left\{\int_T g_t(\omega)\, d\beta_t\right\} = 0, \tag{4.16}$$

in view of the independence assumption. Let $f_t(\omega)$ be another step function similarly defined. Then $(j > i)$

$$\mathscr{E}\{g_j(\omega) f_i(\omega)(\beta_{t_{j+1}} - \beta_{t_j})(\beta_{t_{i+1}} - \beta_{t_i})\} = 0,$$

since $(\beta_{t_{j+1}} - \beta_{t_j})$ is independent of the other random variables, and

$$\mathscr{E}\{\beta_{t_{j+1}} - \beta_{t_j}\} = 0.$$

Therefore,

$$\mathscr{E}\left\{\left[\sum_{i=0}^{n-1} g_i(\beta_{t_{i+1}} - \beta_{t_i})\right]\left[\sum_{i=0}^{n-1} f_i(\beta_{t_{i+1}} - \beta_{t_i})\right]\right\}$$

$$= \sigma^2 \sum_{i=0}^{n-1} \mathscr{E}\{g_i f_i\}(t_{i+1} - t_i) = \sigma^2 \int_T \mathscr{E}\{g_t f_t\}\, dt.$$

Thus we have

$$\mathscr{E}\left\{\int_T g_t(\omega)\, d\beta_t \cdot \int_T f_t(\omega)\, d\beta_t\right\} = \sigma^2 \int_T \mathscr{E}\{g_t f_t\}\, dt. \tag{4.17}$$

Now suppose $\{g_t{}^n\}$ is a sequence of step functions converging to the random function $g_t(\omega)$ in the sense that

$$\int_T \mathscr{E}\{|\, g_t - g_t{}^n\,|^2\}\, dt \to 0 \qquad (n \to \infty). \qquad (4.18)$$

Then, in view of (4.17),

$$\mathscr{E}\left\{\left|\int_T g_t\, d\beta_t - \int_T g_t{}^n\, d\beta_t\right|^2\right\} = \mathscr{E}\left\{\left|\int_T (g_t - g_t{}^n)\, d\beta_t\right|^2\right\}$$

$$= \sigma^2 \int_T \mathscr{E}\{|\, g_t - g_t{}^n\,|^2\}\, dt \to 0 \qquad (n \to \infty).$$

That is,

$$\int_T g_t(\omega)\, d\beta_t = \underset{n \to \infty}{\text{l.i.m.}} \int_T g_t{}^n\, d\beta_t\,. \qquad (4.19)$$

Thus, for all functions $g_t(\omega)$ that can be approximated by step functions in the sense of (4.18), the mean square limit in (4.19) exists. It can be shown (Doob [6, p. 439]) that the class of $g_t(\omega)$ functions satisfying

$$g_t(\omega) \text{ independent of } \{\beta_{t_k} - \beta_{t_l} : t \leqslant t_l \leqslant t_k \leqslant b\}, \qquad \text{for all} \quad t \in T, \qquad (4.20)$$

and

$$\int_T \mathscr{E}\{|\, g_t(\omega)|^2\}\, dt < \infty \qquad (4.21)$$

can be so approximated. Therefore, for all $g_t(\omega)$ functions satisfying (4.20) and (4.21), the mean square limit in (4.19) exists and defines the **Itô stochastic integral**. In view of Theorem 3.1 (iv), the Itô integral is linear and additive in the domain of integration.

We now show that (4.16) and (4.17) hold in general, since they hold for step function integrands.

Theorem 4.1. *Let the random functions $g_t(\omega)$ and $f_t(\omega)$ satisfy (4.20) and (4.21). Then their Itô integrals are well defined [by (4.19)], and*

$$\mathscr{E}\left\{\int_T g_t(\omega)\, d\beta_t\right\} = 0, \qquad (4.22)$$

$$\mathscr{E}\left\{\int_T g_t(\omega)\, d\beta_t \cdot \int_T f_t(\omega)\, d\beta_t\right\} = \sigma^2 \int_T \mathscr{E}\{g_t f_t\}\, dt. \qquad (4.23)$$

PROOF: (4.22) holds for step function integrands; that is,

$$\mathscr{E}\left\{\int_T g_t^n \, d\beta_t\right\} = 0, \qquad \text{all } n,$$

in view of (4.16). Therefore,

$$\lim_{n\to\infty} \mathscr{E}\left\{\int_T g_t^n \, d\beta_t\right\} = 0.$$

But then, by Theorem 3.1 (v),

$$\mathscr{E}\left\{\operatorname*{l.i.m.}_{n\to\infty} \int_T g_t^n \, d\beta_t\right\} = 0,$$

which is (4.22). To prove (4.23), we shall prove that

$$\lim_{n\to\infty} \int_T \mathscr{E}\{g_t^n f_t^n\} \, dt = \int_T \mathscr{E}\{g_t f_t\} \, dt, \qquad (4.24)$$

where the sequences of step functions converge according to (4.18). (4.23) will then follow by Theorem 3.1 (vii). Compute

$$\left| \int_T \mathscr{E}\{g_t^n f_t^n - g_t f_t\} \, dt \right|$$

$$= \left| \int_T \mathscr{E}\{g_t^n (f_t^n - f_t) + (g_t^n - g_t) f_t\} \, dt \right|$$

$$\leqslant \int_T |\mathscr{E}\{g_t^n (f_t^n - f_t)\}| \, dt + \int_T |\mathscr{E}\{(g_t^n - g_t) f_t\}| \, dt$$

$$\leqslant \int_T \mathscr{E}^{1/2}\{|g_t^n|^2\} \mathscr{E}^{1/2}\{|f_t^n - f_t|^2\} \, dt + \int_T \mathscr{E}^{1/2}\{|g_t^n - g_t|^2\} \mathscr{E}^{1/2}\{|f_t|^2\} \, dt$$

$$\leqslant \left(\int_T \mathscr{E}\{|g_t^n|^2\} \, dt\right)^{1/2} \left(\int_T \mathscr{E}\{|f_t^n - f_t|^2\} \, dt\right)^{1/2}$$

$$+ \left(\int_T \mathscr{E}\{|g_t^n - g_t|^2\} \, dt\right)^{1/2} \left(\int_T \mathscr{E}\{|f_t|^2\} \, dt\right)^{1/2} \to 0 \qquad (n \to \infty)$$

in view of (4.18).   ∎

Although the definition of the Itô integral is in terms of step functions, we show that under certain conditions it is the mean square limit of Riemann–Stieltjes sums.

Theorem 4.2.   Let $\mathscr{E}\{|g_t(\omega)|^2\} < \infty$ for all $t \in T$. Suppose $g_t(\omega)$ is *mean square continuous on T and satisfies the independence condition (4.20). Partition T: $a = t_0 < t_1 < \cdots < t_n = b$, and let $\rho = \max_i(t_{i+1} - t_i)$. Then the Itô stochastic integral equals the following mean square limit:*

$$\underset{\rho \to 0}{\text{l.i.m.}} \sum_{i=0}^{n-1} g_{t_i}(\omega)(\beta_{t_{i+1}} - \beta_{t_i}) = \int_T g_t(\omega) \, d\beta_t. \tag{4.25}$$

PROOF: By Theorem 3.2, $\mathscr{E}\{|g_t(\omega)|^2\}$ is continuous on T. Therefore, condition (4.21) is satisfied, and the Itô integral is well defined. Let

$$\varDelta = \int_T g_t(\omega) \, d\beta_t - \sum_{i=0}^{n-1} g_{t_i}(\omega)(\beta_{t_{i+1}} - \beta_{t_i}).$$

Let

$$\tau(t) = t_i \, ; \quad t_i \leqslant t < t_{i+1}.$$

Then

$$\varDelta = \int_T [g_t(\omega) - g_{\tau(t)}(\omega)] \, d\beta_t.$$

By Theorem 4.1,

$$\mathscr{E}\{|\varDelta|^2\} = \sigma^2 \int_T \mathscr{E}\{|g_t(\omega) - g_{\tau(t)}(\omega)|^2\} \, dt.$$

Now $\mathscr{E}\{|g_t|^2\}$ is continuous on T and therefore uniformly bounded on T. Therefore, $\mathscr{E}\{|g_t - g_\tau|^2\}$ is uniformly bounded on T. Furthermore, as $\rho \to 0$, $\tau \to t$, and therefore

$$\mathscr{E}\{|g_t - g_\tau|^2\} \to 0 \quad (\rho \to 0).$$

Therefore, by bounded convergence,

$$\mathscr{E}\{|\varDelta|^2\} \to 0 \quad (\rho \to 0). \quad \blacksquare$$

In studying stochastic differential equations, we need to consider the Itô integral as a function of its upper limit. Let

$$x_t = \int_a^t g_\tau(\omega) \, d\beta_\tau, \quad t \in T. \tag{4.26}$$

Theorem 4.3. *Conditions of Theorem 4.2. Then x_t is mean square continuous on T.*

PROOF: In view of Theorems 4.1 and 4.2, $\mathscr{E}\{|\,x_t\,|^2\} < \infty$, $t \in T$. Now, by Theorem 4.1,

$$\mathscr{E}\left\{\left|\int_t^{t+h} g_\tau(\omega)\,d\beta_\tau\right|^2\right\} = \sigma^2 \int_t^{t+h} \mathscr{E}\{|\,g_\tau\,|^2\}\,d\tau \to 0 \qquad (h \to 0),$$

since the (last) Riemann integral is a continuous function of its upper limit. ∎

Remark.   It can also be shown (Doob [6, p. 445]) that the sample functions of the $\{x_t\}$ process are almost all continuous on T. That is, x_t is continuous wp 1 on T.

In Section 2 we encountered the integral

$$\int_T g_t(\omega)\,d\beta_t{}^2.$$

We now characterize this integral. For our usual partition of the interval T, we define the **second-order stochastic integral**:

$$\int_T g_t(\omega)\,d\beta_t{}^2 \triangleq \underset{\rho\to 0}{\mathrm{l.i.m.}} \sum_{i=0}^{n-1} g_{t_i}(\beta_{t_{i+1}} - \beta_{t_i})^2. \tag{4.27}$$

Theorem 4.4. *Conditions of Theorem 4.2. Then*

$$\int_T g_t(\omega)\,d\beta_t{}^2 = \sigma^2 \int_T g_t(\omega)\,dt. \tag{4.28}$$

PROOF: Note that the second integral in (4.28) is a well-defined mean square Riemann integral (Theorem 3.7, Corollary 2). In view of the definitions of the integrals in (4.28), we have to prove that

$$\mathscr{E}\{|\,\Delta\,|^2\} = \mathscr{E}\left\{\left|\sum_{i=0}^{n-1} g_{t_i}[(\beta_{t_{i+1}} - \beta_{t_i})^2 - \sigma^2(t_{i+1} - t_i)]\right|^2\right\}$$

goes to zero with ρ. Now

$$\mathscr{E}\{|\,\Delta\,|^2\} = \mathscr{E}\left\{\sum_{i=0}^{n-1}\sum_{j=0}^{n-1} g_{t_i}g_{t_j}[(\beta_{t_{i+1}} - \beta_{t_i})^2 - \sigma^2(t_{i+1} - t_i)]\right.$$
$$\left. \times\,[(\beta_{t_{j+1}} - \beta_{t_j})^2 - \sigma^2(t_{j+1} - t_j)]\right\}.$$

Consider an $i \neq j$ term in the double sum. Suppose $j > i$. Then

$$[(\beta_{t_{j+1}} - \beta_{t_j})^2 - \sigma^2(t_{j+1} - t_j)] \qquad (4.29)$$

is independent of the other rv's in that term. Since the expectation of (4.29) is zero, that term is zero. Therefore,

$$\mathscr{E}\{|\Delta|^2\} = \mathscr{E}\left\{\sum_{i=0}^{n-1} |g_{t_i}|^2 |(\beta_{t_{i+1}} - \beta_{t_i})^2 - \sigma^2(t_{i+1} - t_i)|^2\right\}$$

$$= \sum_{i=0}^{n-1} \mathscr{E}\{|g_{t_i}|^2\} \mathscr{E}\{|(\beta_{t_{i+1}} - \beta_{t_i})^2 - \sigma^2(t_{i+1} - t_i)|^2\},$$

in view of the independence condition (4.20). But

$$\mathscr{E}\{|(\beta_{t_{i+1}} - \beta_{t_i})^2 - \sigma^2(t_{i+1} - t_i)|^2\} = 2\sigma^4(t_{i+1} - t_i)^2, \qquad (4.30)$$

in view of (4.10). Therefore,

$$\mathscr{E}\{|\Delta|^2\} = 2\sigma^4 \sum_{i=0}^{n-1} \mathscr{E}\{|g_{t_i}|^2\}(t_{i+1} - t_i)^2$$

$$\leqslant 2\sigma^4 \rho \sum_{i=0}^{n-1} \mathscr{E}\{|g_{t_i}|^2\}(t_{i+1} - t_i) \to 0 \qquad (\rho \to 0),$$

since

$$\int_T \mathscr{E}\{|g_t|^2\} \, dt < \infty. \quad \blacksquare$$

Example 4.2. One might similarly define a third-order stochastic integral. Let the reader show that

$$\int_T g_t(\omega) \, d\beta_t^3 = 0. \quad \blacktriangle$$

Example 4.3. Let us for a moment return to the stochastic differential equation (4.12), which we derived for the $e^{\beta t}$ process in Section 2. In view of Theorem 4.4, (4.13) becomes

$$x_t - 1 = \int_0^t x_\tau \, d\beta_\tau + \sigma^2/2 \int_0^t x_\tau \, d\tau. \qquad (4.31)$$

Since, as we have said, the integral defines the differential, (4.12) becomes

$$dx_t = x_t \, d\beta_t + \sigma^2/2 \, x_t \, dt, \qquad x_0 = 1 \quad \text{wp 1}. \qquad (4.32)$$

Note that $d\beta_t^2$ *is replaced by its expectation*, which is $\sigma^2\, dt$! Note also that the Itô integral in (4.31) apparently cannot be evaluated by the formal rules of integration. For if it could be so evaluated, then $x_t = e^{\beta_t}$ would satisfy

$$x_t - 1 = \int_0^t x_\tau\, d\beta_\tau\,,$$

but, in fact, e^{β_t} satisfies (4.31).   ▲

Example 4.4.[3]   We evaluate a simple Itô integral. Let $g_t(\omega)$ be the explicit function of β_t,

$$g_t(\omega) = \beta_t - \beta_a\,, \tag{4.33}$$

and evaluate

$$\int_a^b (\beta_t - \beta_a)\, d\beta_t\,.$$

According to Theorem 4.2,

$$\int_a^b (\beta_t - \beta_a)\, d\beta_t = \operatorname*{l.i.m.}_{\rho \to 0} \sum_{i=0}^{n-1} (\beta_{t_i} - \beta_a)(\beta_{t_{i+1}} - \beta_{t_i}),$$

since $g_t(\omega)$ (4.33) satisfies the conditions of that theorem (Section 3.5). Now

$$\sum_{i=0}^{n-1} (\beta_{t_i} - \beta_a)(\beta_{t_{i+1}} - \beta_{t_i}) = \frac{1}{2}(\beta_b - \beta_a)^2 - \frac{1}{2}\sum_{i=0}^{n-1} (\beta_{t_{i+1}} - \beta_{t_i})^2$$

and

$$\mathscr{E}\left\{\left|\sum_{i=0}^{n-1} (\beta_{t_{i+1}} - \beta_{t_i})^2 - \sigma^2(b - a)\right|^2\right\}$$

$$= \mathscr{E}\left\{\left|\sum_{i=0}^{n-1} [(\beta_{t_{i+1}} - \beta_{t_i})^2 - \sigma^2(t_{i+1} - t_i)]\right|^2\right\}$$

$$= \mathscr{E}\left\{\sum_{i=0}^{n-1} [(\beta_{t_{i+1}} - \beta_{t_i})^2 - \sigma^2(t_{i+1} - t_i)]^2\right\}$$

$$= 2\sigma^4 \sum_{i=0}^{n-1} (t_{i+1} - t_i)^2$$

$$\leqslant 2\sigma^4\rho(b - a) \to 0 \qquad (\rho \to 0).$$

[3] This example is due to Doob [6, p. 443].

We used the independence condition (4.20), and (4.30). Therefore,

$$\int_a^b (\beta_t - \beta_a) \, d\beta_t = (1/2)(\beta_b - \beta_a)^2 - (\sigma^2/2)(b - a) \qquad (4.34)$$

(see Theorem 3.1). Note that formal rules of integration would give only the first term on the right-hand side of (4.34). ▲

Finally, we note that, if $g_t(\omega)$ is a random n-vector function, then

$$\int_a^b g_t(\omega) \, d\beta_t = \begin{bmatrix} \int_a^b [g_t(\omega)]_1 \, d\beta_t \\ \vdots \\ \int_a^b [g_t(\omega)]_n \, d\beta_t \end{bmatrix}.$$

The integrals of the elements of $g_t(\omega)$ have been defined.

4. STOCHASTIC DIFFERENTIAL EQUATIONS

Having defined the Itô stochastic integral, we are in a position to study the stochastic differential equation (4.4):

$$dx_t = f(x_t, t) \, dt + g(x_t, t) \, d\beta_t, \qquad t \in [t_0, T]. \qquad (4.35)$$

We suppose for simplicity that (4.35) is a scalar equation and $\{\beta_t, t \geq t_0\}$ is a scalar Brownian motion process with unit variance parameter. The assumption of a unit variance parameter is no loss of generality, since $\sigma^2 \neq 1$ can be absorbed in the function g (replace g by σg). By (4.35), we mean the integral equation

$$x_t - x_{t_0} = \int_{t_0}^t f(x_\tau, \tau) \, d\tau + \int_{t_0}^t g(x_\tau, \tau) \, d\beta_\tau, \qquad t \in [t_0, T]. \qquad (4.36)$$

The first integral will be a mean square Riemann integral, whereas the second one will be the Itô stochastic integral. Equation (4.35) is therefore called an *Itô stochastic differential equation.*

Theorem 4.5. *Suppose the real functions f and g, and initial condition x_{t_0}, satisfy the hypotheses:*

H_1. *There is a $K > 0$ such that*

$$|f(x, t)| \leq K(1 + |x|^2)^{1/2},$$
$$|g(x, t)| \leq K(1 + |x|^2)^{1/2}.$$

H_2.[4] *f and g satisfy uniform Lipschitz conditions in x:*

$$|f(x_2, t) - f(x_1, t)| \leqslant K |x_2 - x_1|,$$
$$|g(x_2, t) - g(x_1, t)| \leqslant K |x_2 - x_1|.$$

H_3.[5] *f and g satisfy Lipschitz conditions in t on* $[t_0, T]$:

$$|f(x, t_2) - f(x, t_1)| \leqslant K |t_2 - t_1|,$$
$$|g(x, t_2) - g(x, t_1)| \leqslant K |t_2 - t_1|.$$

H_4. x_{t_0} *is any random variable with* $\mathscr{E}\{|x_{t_0}|^2\} < \infty$, *independent of* $\{d\beta_t, t \in [t_0, T]\}$.

Then (4.36) has a solution on $[t_0, T]$ *in the mean square sense. The solution* $\{x_t\}$ *process has the following properties:*

P_1. x_t *is mean square continuous in* $[t_0, T]$.
P_2. $\mathscr{E}\{|x_t|^2\} < \infty$, *all* $t \in [t_0, T]$.
P_2'. $\mathscr{E}\{|x_t|^2\} < M$ *on* $[t_0, T]$ $(P_2' \Rightarrow P_2)$.
P_3. $\int_{t_0}^{T} \mathscr{E}\{|x_t|^2\} \, dt < \infty$.
P_4. $x_t - x_{t_0}$ *is independent of* $\{d\beta_\tau, \tau \geqslant t\}$ *for every* $t \in [t_0, T]$.

Furthermore, the $\{x_t\}$ *process is a Markov process, and, in the mean square sense, is uniquely determined by the initial condition* x_{t_0}.

The existence proof will be by successive approximations. The following lemma defines the successive approximations.

Lemma 4.1. *Assume hypotheses* H_1, H_2, H_3. *If an* $\{x_t\}$ *process has properties* P_1, P_2, P_3, *and* P_4, *then the* $\{y_t\}$ *process defined by*

$$y_t = \int_{t_0}^{t} f(x_\tau, \tau) \, d\tau + \int_{t_0}^{t} g(x_\tau, \tau) \, d\beta_\tau, \qquad t \in [t_0, T], \qquad (4.37)$$

has properties P_1, P_2', P_3, *and* P_4.

PROOF OF LEMMA: By H_1 and P_2,

$$\mathscr{E}\{|f(x_t, t)|^2\} \leqslant K^2(1 + \mathscr{E}\{|x_t|^2\}) < \infty, \qquad \text{all} \quad t \in [t_0, T],$$

---

[4] Actually, $H_2 \Rightarrow H_1$. See Wong and Zakai [21, p. 7].
[5] H_3 can be weakened to continuity of f and g in t on $[t_0, T]$. See Lemma 3 in Wong and Zakai [21]. However, this complicates the proof.

and similarly for g. By H_2, H_3, and P_1,

$$\mathscr{E}\{|f(x_{t+h}, t+h) - f(x_t, t)|^2\}$$
$$= \mathscr{E}\{|f(x_{t+h}, t+h) - f(x_t, t+h) + f(x_t, t+h) - f(x_t, t)|^2\}$$
$$\leqslant 2\mathscr{E}\{|f(x_{t+h}, t+h) - f(x_t, t+h)|^2\} + 2\mathscr{E}\{|f(x_t, t+h) - f(x_t, t)|^2\}$$
$$\leqslant 2K^2\mathscr{E}\{|x_{t+h} - x_t|^2\} + 2K^2h^2 \to 0 \qquad (h \to 0),$$

and similarly for g. Therefore f and g are ms continuous. By Corollary 2 of Theorem 3.7, f is ms Riemann integrable, and, by Theorem 3.9, the first integral in (4.37) is an ms continuous function of its upper limit. Also, the hypotheses of Theorem 4.2 are satisfied in view of what has been said and P_4, and therefore, by Theorem 4.3, the Itô integral in (4.37) is an ms continuous function of its upper limit. Therefore, $\{y_t\}$ has property P_1. It clearly has property P_4. Furthermore, $P_1 \Rightarrow P_3$ (for y_t) by Theorem 3.2. It remains to demonstrate property P_2'.

$$\mathscr{E}\{|y_t|^2\} \leqslant 2\mathscr{E}\left\{\left|\int_{t_0}^t f(x_\tau, \tau)\, d\tau\right|^2\right\} + 2\mathscr{E}\left\{\left|\int_{t_0}^t g(x_\tau, \tau)\, d\beta_\tau\right|^2\right\}$$

$$= 2\int_{t_0}^t \int_{t_0}^t \mathscr{E}\{f_\tau f_s\}\, d\tau\, ds + 2\int_{t_0}^t \mathscr{E}\{|g_\tau|^2\}\, d\tau$$

$$\leqslant 2\left|\int_{t_0}^T \mathscr{E}^{1/2}\{|f_\tau|^2\}\, d\tau\right|^2 + 2\int_{t_0}^T \mathscr{E}\{|g_\tau|^2\}\, d\tau$$

$$\leqslant 2(T - t_0)\int_{t_0}^T \mathscr{E}\{|f_\tau|^2\}\, d\tau + 2\int_{t_0}^T \mathscr{E}\{|g_\tau|^2\}\, d\tau$$

$$\leqslant 2(T - t_0)K^2 \int_{t_0}^T (1 + \mathscr{E}\{|x_\tau|^2\})\, d\tau$$

$$+ 2K^2 \int_{t_0}^T (1 + \mathscr{E}\{|x_\tau|^2\})\, d\tau < M < \infty,$$

where the last inequality follows from P_3. ∎

PROOF OF THEOREM 4.5: Let $\{x_t^0\}$ be any process with properties P_1, P_2, P_3, and P_4 (for example, $x_t^0 \equiv 0$). We define the successive approximations

$$x_t^n = x_{t_0} + \int_{t_0}^t f(x_\tau^{n-1}, \tau)\, d\tau + \int_{t_0}^t g(x_\tau^{n-1}, \tau)\, d\beta_\tau, \qquad n \geqslant 1. \quad (4.38)$$

According to Lemma 4.1, the $\{x_t^n\}$ process has properties P_1, P_2', P_3, and P_4 for all n. We shall first prove that

$$\underset{n \to \infty}{\text{l.i.m.}} \, x_t^n = x_t, \qquad t \in [t_0, T], \tag{4.39}$$

uniformly in t, and that the $\{x_t\}$ process in (4.39) has properties P_1, P_2', P_3, and P_4. We then show that

$$\underset{n \to \infty}{\text{l.i.m.}} \int_{t_0}^{t} f(x_\tau^n, \tau) \, d\tau = \int_{t_0}^{t} f(x_\tau, \tau) \, d\tau, \qquad t \in [t_0, T], \tag{4.40}$$

and

$$\underset{n \to \infty}{\text{l.i.m.}} \int_{t_0}^{t} g(x_\tau^n, \tau) \, d\beta_\tau = \int_{t_0}^{t} g(x_\tau, \tau) \, d\beta_\tau, \qquad t \in [t_0, T], \tag{4.41}$$

uniformly in t. (4.40) and (4.41) show that the ms limit in (4.39) is a solution of (4.36). This will prove existence.

We first develop an inequality for later use. Let

$$\Delta_n x_t = x_t^n - x_t^{n-1},$$
$$\Delta_n f_t = f(x_t^n, t) - f(x_t^{n-1}, t),$$
$$\Delta_n g_t = g(x_t^n, t) - g(x_t^{n-1}, t).$$

Then

$$\mathscr{E}\{|\Delta_n x_t|^2\} \leqslant 2\mathscr{E}\left\{\left|\int_{t_0}^{t} \Delta_{n-1} f_\tau \, d\tau\right|^2\right\} + 2\mathscr{E}\left\{\left|\int_{t_0}^{t} \Delta_{n-1} g_\tau \, d\beta_\tau\right|^2\right\}$$

$$\leqslant 2(T - t_0) \int_{t_0}^{t} \mathscr{E}\{|\Delta_{n-1} f_\tau|^2\} \, d\tau + 2 \int_{t_0}^{t} \mathscr{E}\{|\Delta_{n-1} g_\tau|^2\} \, d\tau$$

$$\leqslant 2(T - t_0) K^2 \int_{t_0}^{t} \mathscr{E}\{|\Delta_{n-1} x_\tau|^2\} \, d\tau + 2K^2 \int_{t_0}^{t} \mathscr{E}\{|\Delta_{n-1} x_\tau|^2\} \, d\tau$$

$$= 2K^2(T - t_0 + 1) \int_{t_0}^{t} \mathscr{E}\{|\Delta_{n-1} x_\tau|^2\} \, d\tau. \tag{4.42}$$

Hence,

$$\mathscr{E}\{|\Delta_2 x_t|^2\} \leqslant 2K^2(T - t_0 + 1) \int_{t_0}^{t} \mathscr{E}\{|\Delta_1 x_\tau|^2\} \, d\tau,$$

$$\mathscr{E}\{|\Delta_3 x_t|^2\} \leqslant [2K^2(T - t_0 + 1)]^2 \int_{t_0}^{t} d\tau \int_{t_0}^{\tau} \mathscr{E}\{|\Delta_1 x_s|^2\} \, ds$$

$$= [2K^2(T - t_0 + 1)]^2 \int_{t_0}^{t} (t - \tau) \, \mathscr{E}\{|\Delta_1 x_\tau|^2\} \, d\tau,$$

where the last equality is obtained by integration by parts. Therefore, by induction,

$$\mathscr{E}\{|\, \varDelta_n x_t\,|^2\} \leqslant [2K^2(T - t_0 + 1)]^{n-1} \int_{t_0}^t \frac{(t - \tau)^{n-2}}{(n - 2)!}\, \mathscr{E}\{|\, \varDelta_1 x_\tau\,|^2\}\, d\tau$$

$$\leqslant \frac{C^n}{n!}, \qquad \text{some} \quad C > 0, \qquad t \in [t_0\,, T]. \tag{4.43}$$

We now prove (4.39). For each t and $n > m$,

$$\mathscr{E}\{|\, x_t{}^n - x_t{}^m\,|^2\} = \mathscr{E}\left\{\left|\, \sum_{i=m+1}^n \varDelta_i x_t\,\right|^2\right\} \leqslant 2^{-m} \sum_{i=m+1}^n 2^i \mathscr{E}\{|\, \varDelta_i x_t\,|^2\},$$

and by (4.43),

$$\mathscr{E}\{|\, x_t{}^n - x_t{}^m\,|^2\} \leqslant 2^{-m} \sum_{i=m+1}^n \frac{(2C)^i}{i!} \leqslant 2^{-m} \sum_1^\infty \frac{(2C)^i}{i!} \to 0 \qquad (m \to \infty). \tag{4.44}$$

Thus by the Cauchy criterion (3.44), the mean square limit in (4.39) exists uniformly in t. By Theorem 3.3, the $\{x_t\}$ process has property P_1. By Theorem 3.2 it has property P_3. It clearly has property P_4. We show that it has property $P_2{}'$. When $n \to \infty$ in (4.44), we get

$$\mathscr{E}\{|\, x_t - x_t{}^m\,|^2\} \leqslant 2^{-m} \sum_1^\infty \frac{(2C)^i}{i!}\,. \tag{4.45}$$

But

$$|\, x_t - x_t{}^m\,| \geqslant |\, x_t\,| - |\, x_t{}^m\,|,$$

so that

$$\mathscr{E}\{|\, x_t\,|^2\} \leqslant 2\mathscr{E}\{|\, x_t{}^m\,|^2\} + 2^{1-m} \sum_1^\infty \frac{(2C)^i}{i!}\,,$$

and $\{x_t\}$ has property $P_2{}'$, since $\{x_t{}^m\}$ does.

We next prove (4.40) and (4.41):

$$\mathscr{E}\left\{\left|\int_{t_0}^t [f(x_\tau{}^n, \tau) - f(x_\tau\,, \tau)]\, d\tau\,\right|^2\right\} \leqslant (T - t_0) \int_{t_0}^T \mathscr{E}\{|\, f(x_\tau{}^n, \tau) - f(x_\tau\,, \tau)|^2\}\, d\tau$$

$$\leqslant (T - t_0)\, K^2 \int_{t_0}^T \mathscr{E}\{|\, x_\tau{}^n - x_\tau\,|^2\}\, d\tau$$

$$\leqslant (T - t_0)^2 K^2 2^{-n} \sum_1^\infty \frac{(2C)^i}{i!} \to 0 \qquad (n \to \infty).$$

This proves (4.40). (4.41) is proved similarly, using Theorem 4.1

instead of Theorem 3.8 and Schwarz's inequality. This completes the proof of existence.

Let Δ_t be the difference between two solutions of (4.36) with the same initial condition. Then, in the same way we obtained (4.42), we get

$$\mathscr{E}\{|\Delta_t|^2\} \leqslant 2K^2(T - t_0 + 1) \int_{t_0}^{t} \mathscr{E}\{|\Delta_\tau|^2\}\, d\tau. \tag{4.46}$$

Applying (4.46) to itself, we have

$$\mathscr{E}\{|\Delta_t|^2\} \leqslant C^n/n!, \qquad \text{all } n.$$

This proves uniqueness.

Suppose $t, s \in T$, $t \geqslant s$. Then

$$x_t = x_s + \int_s^t f(x_\tau, \tau)\, d\tau + \int_s^t g(x_\tau, \tau)\, d\beta_\tau$$

by the additive property of the integrals. Thus x_t depends on x_s and $\{d\beta_\tau, s \leqslant \tau \leqslant t\}$. By property P_4 and hypothesis H_4, these increments are independent of x_τ, $\tau \leqslant s$. Therefore,

$$p(x_t \mid x_\tau, \tau \leqslant s) = p(x_t \mid x_s),$$

and the $\{x_t\}$ process is Markov. ∎

Theorem 4.5 can readily be generalized to the vector case

$$dx_t = f(x_t, t)\, dt + G(x_t, t)\, d\beta_t, \tag{4.47}$$

where x_t and f are n-vectors, G is an $n \times m$ matrix, and β_t is a vector of m independent Brownian motions, each with unit variance parameter. Use the norms

$$|x| = \left(\sum_{i=1}^{n} x_i^2\right)^{1/2} = (x^{\mathrm{T}}x)^{1/2}$$

and

$$|G| = \left(\sum_{i=1}^{n}\sum_{j=1}^{m} G_{ij}^2\right)^{1/2} = [\operatorname{tr}(GG^{\mathrm{T}})]^{1/2}.$$

See Section 3.3 for mean square convergence of random vector sequences. We leave this as an exercise for the reader.

It can be shown (Doob [6, p. 277]) that all the conclusions of Theorem 4.5 also hold with probability 1. That is (4.39),

$$\lim_{n \to \infty} x_t^n = x_t \quad \text{wp 1} \tag{4.48}$$

uniformly in t, and [(4.40) and (4.41)]

$$\lim_{n \to \infty} \int_{t_0}^{t} f(x_\tau^n, \tau) \, d\tau = \int_{t_0}^{t} f(x_\tau, \tau) \, d\tau, \quad \text{wp 1}, \tag{4.49}$$

$$\lim_{n \to \infty} \int_{t_0}^{t} g(x_\tau^n, \tau) \, d\beta_\tau = \int_{t_0}^{t} g(x_\tau, \tau) \, d\beta_\tau, \quad \text{wp 1}, \tag{4.50}$$

uniformly in t. Also (uniqueness),

$$\Pr\{\Delta_t = 0, \, t \in [t_0, T]\} = 1. \tag{4.51}$$

That is, (4.36) has a unique solution with probability 1. Furthermore, the x_t sample functions are almost all continuous, and properties P_2', P_3, and P_4 hold. We may note that x_t *does not* have a derivative since the Brownian motion process is not differentiable.

Since, in view of hypotheses H_2 and H_3, $f(\cdot, \cdot)$ is a continuous function of both arguments and x_t is continuous wp 1, f is a continuous function of t wp 1. Therefore $f(x_t, t)$ is Riemann integrable on $[t_0, T]$ wp 1. Thus the first integral in (4.36) is also a well-defined Riemann integral for the sample functions.

Consider a special case of (4.47), the *linear* stochastic differential equation

$$dx_t = F(t)x_t \, dt + G(t) \, d\beta_t, \tag{4.52}$$

where F and G are, respectively, $n \times n$ and $n \times m$ nonrandom matrix functions. In integral form,

$$x_t - x_{t_0} = \int_{t_0}^{t} F(\tau)x_\tau \, d\tau + \int_{t_0}^{t} G(\tau) \, d\beta_\tau. \tag{4.53}$$

Let the initial condition x_{t_0} be Gaussian. Suppose we solve (4.53) by successive approximations (4.38). Let x_t^0 be Gaussian (or $x_t^0 \equiv 0$). Since the integrals in (4.53) are mean square limits, it follows by Theorem 3.11 that x_t^1 is Gaussian. By induction, x_t^n is Gaussian for all n. Since

$$x_t = \text{l.i.m. } x_t^n,$$

x_t is also Gaussian (Theorem 3.11 again). It therefore follows that the process generated by the linear stochastic differential equation (4.52) is a *Gauss–Markov* process.

5. ITÔ STOCHASTIC CALCULUS

We have already seen in Section 2 and in Examples 4.3 and 4.4 that formal integration and differentiation rules cannot be applied to the Itô stochastic integral. In the present section we develop some rules of stochastic calculus, including the stochastic differential and its relationship with the stochastic integral.

Lemma 4.2 (Itô). *Let* x_t *be the unique solution of the vector Itô stochastic differential equation*

$$dx_t = f(x_t, t)\, dt + G(x_t, t)\, d\beta_t, \qquad t \geqslant t_0, \tag{4.54}$$

where x and f are n-vectors, G is $n \times m$, and $\{\beta_t, t \geqslant t_0\}$ is an m-vector Brownian motion process with $\mathscr{E}\{d\beta_t\, d\beta_t^{\mathrm{T}}\} = Q(t)\, dt$. Let $\varphi(x_t, t)$ be a scalar-valued real function, continuously differentiable in t and having continuous second mixed partial derivatives with respect to the elements of x. Then the (stochastic) differential $d\varphi$ of φ is

$$d\varphi = \varphi_t\, dt + \varphi_x^{\mathrm{T}}\, dx_t + \tfrac{1}{2} \operatorname{tr} GQG^{\mathrm{T}}\varphi_{xx}\, dt, \tag{4.55}$$

where

$$\varphi_t = \frac{\partial \varphi}{\partial t}; \qquad \varphi_x^{\mathrm{T}} = \left[\frac{\partial \varphi}{\partial x_1}, \; \ldots, \; \frac{\partial \varphi}{\partial x_n} \right];$$

$$\varphi_{xx} = \begin{bmatrix} \dfrac{\partial^2 \varphi}{\partial x_1{}^2}, & \dfrac{\partial^2 \varphi}{\partial x_1\, \partial x_2}, & \cdots, & \dfrac{\partial^2 \varphi}{\partial x_1\, \partial x_n} \\ \vdots & \vdots & & \vdots \\ \dfrac{\partial^2 \varphi}{\partial x_n\, \partial x_1}, & \dfrac{\partial^2 \varphi}{\partial x_n\, \partial x_2}, & \cdots, & \dfrac{\partial^2 \varphi}{\partial x_n{}^2} \end{bmatrix}.$$

Equation (4.55) is to be understood in the sense that the integral of its right side over $[t_0, T]$ (where the stochastic integral is as defined in Section 3) is $\varphi(T) - \varphi(t_0)$. That is, the differential is defined in terms of the integral. Equation (4.55) tells us what stochastic differential equation a function of the solution of another stochastic differential equation satisfies.

Lemma 4.2 is proved for the scalar case in Itô [11] and Skorokhod [15, p. 24], and the extension to the vector case is straightforward. We shall not prove it here, but give a formal derivation of the result (4.55).

As we did in Section 2, we expand φ in a Taylor's series, retaining second-order terms in $d\beta_t$:

$$
\begin{aligned}
d\varphi &= \varphi_t \, dt + \varphi_x{}^{\mathrm{T}} \, dx_t + \tfrac{1}{2} \, dx_t{}^{\mathrm{T}} \, \varphi_{xx} \, dx_t + \cdots \\
&= \varphi_t \, dt + \varphi_x{}^{\mathrm{T}}(f \, dt + G \, d\beta_t) + \tfrac{1}{2} \, d\beta_t{}^{\mathrm{T}} \, G^{\mathrm{T}}\varphi_{xx}G \, d\beta_t + \cdots \\
&= \varphi_t \, dt + \varphi_x{}^{\mathrm{T}} \, dx_t + \tfrac{1}{2} \operatorname{tr} G \, d\beta_t \, d\beta_t{}^{\mathrm{T}} \, G^{\mathrm{T}}\varphi_{xx} + \cdots.
\end{aligned}
$$

As we saw in Example 4.3, $d\beta_t \, d\beta_t{}^{\mathrm{T}}$ has to be replaced by its expectation, and so we get

$$
d\varphi = \varphi_t \, dt + \varphi_x{}^{\mathrm{T}} \, dx_t + \tfrac{1}{2} \operatorname{tr} GQG^{\mathrm{T}}\varphi_{xx} \, dt + o(dt),
$$

which is the desired result. Note that (4.56) can be written as

$$
d\varphi = (\varphi_t + \varphi_x{}^{\mathrm{T}}f + \tfrac{1}{2} \operatorname{tr} GQG^{\mathrm{T}}\varphi_{xx}) \, dt + \varphi_x{}^{\mathrm{T}}G \, d\beta_t . \tag{4.56}
$$

Example 4.5. In Example 4.4, we saw that

$$
\int_a^t (\beta_\tau - \beta_a) \, d\beta_\tau = (1/2)(\beta_t - \beta_a)^2 - (\sigma^2/2)(t - a),
$$

where $\{\beta_t , t \geqslant a\}$ has variance parameter σ^2. Let

$$
\varphi = \tfrac{1}{2}(x_t - \beta_a)^2 - (\sigma^2/2)(t - a),
$$

and let

$$
dx_t = d\beta_t .
$$

Applying Lemma 4.2, we get

$$
\begin{aligned}
d\varphi &= -(\sigma^2/2) \, dt + (x_t - \beta_a) \, dx_t + (\sigma^2/2) \, dt \\
&= (\beta_t - \beta_a) \, d\beta_t .
\end{aligned}
$$

Thus, we see that the differential (4.55) is consistent with the definition of the Itô integral. ▲

Corollary 1 (Fundamental Theorem of Itô Stochastic Calculus). *Let $\varphi(x)$ be a twice continuously differentiable real scalar function of the real variable x. Let $\psi = \varphi_x$. Then ($b > a$)*

$$
\int_a^b \psi(\beta_t) \, d\beta_t = \varphi(\beta_b) - \varphi(\beta_a) - (\sigma^2/2) \int_a^b \varphi_{xx}(\beta_t) \, dt. \tag{4.57}
$$

PROOF: Apply Lemma 4.2 to $\varphi(x_t)$ with $dx_t = d\beta_t$, $\mathscr{E}\{d\beta_t^2\} = \sigma^2\, dt$. We get

$$d\varphi = \varphi_x\, d\beta_t + (\sigma^2/2)\varphi_{xx}\, dt,$$

and integration produces (4.57). ∎

If, given ψ, we can find φ such that $\varphi_x = \psi$, we may use (4.57) to evaluate the stochastic integral (in terms of a Riemann integral).

Example 4.6.   Evaluate

$$\int_a^b \beta_t^n\, d\beta_t\,.$$

We have $\psi = \beta_t^n$, $\varphi = (n+1)^{-1}\beta_t^{n+1}$, and $\varphi_{\beta\beta} = n\beta_t^{n-1}$. From (4.57),

$$\int_a^b \beta_t^n\, d\beta_t = (n+1)^{-1}(\beta_b^{n+1} - \beta_a^{n+1}) - (n\sigma^2/2)\int_a^b \beta_t^{n-1}\, dt. \qquad (4.58)$$

Thus,

$$\int_a^b \beta_t\, d\beta_t = \tfrac{1}{2}(\beta_b^2 - \beta_a^2) - (\sigma^2/2)(b - a),$$

$$\int_a^b \beta_t^2\, d\beta_t = \tfrac{1}{3}(\beta_b^3 - \beta_a^3) - (\sigma^2/2)(\beta_b^2 - \beta_a^2) + 2(\sigma^2/2)^2(b - a),$$

and by induction

$$\int_a^b \beta_t^n\, d\beta_t = (n+1)^{-1}(\beta_b^{n+1} - \beta_a^{n+1}) + \sum_{i=2}^n (-1)^{n-i+1}(n!/i!)(\sigma^2/2)^{n-i+1}(\beta_b^i - \beta_a^i)$$

$$+ (-1)^n n!(\sigma^2/2)^n(b - a). \qquad \blacktriangle \qquad (4.59)$$

Example 4.7.   In Section 2 and Example 4.3 we saw that

$$dx_t = x_t\, d\beta_t + (\sigma^2/2)x_t\, dt, \qquad t \geqslant 0, \qquad x_0 = 1,$$

or equivalently,

$$x_t - 1 = \int_0^t x_\tau\, d\beta_\tau + (\sigma^2/2)\int_0^t x_\tau\, d\tau, \qquad (4.60)$$

should have the solution

$$x_t = e^{\beta_t}.$$

Let $\psi = e^{\beta_t}$. Then $\varphi = \varphi_{\beta\beta} = e^{\beta_t}$. Then with $b = t$, $a = 0$, Eq. (4.57) becomes

$$\int_0^t e^{\beta_\tau}\, d\beta_\tau = e^{\beta_t} - 1 - (\sigma^2/2) \int_0^t e^{\beta_\tau}\, d\tau \qquad (4.61)$$

($\beta_0 = 0$ by definition). Comparing (4.60) with (4.61), we see that e^{β_t} is indeed the solution. ▲

Example 4.8. We saw in Example 4.7 that the solution of

$$dx_t = x_t\, d\beta_t + (\sigma^2/2)x_t\, dt, \qquad t \geqslant 0, \qquad x_0 = 1,$$

is

$$x_t = e^{\beta_t}.$$

What is the solution of

$$dx_t = x_t\, d\beta_t, \qquad t \geqslant 0, \qquad x_0 = 1\, ?$$

It is easy to verify, with the aid of Lemma 4.2, that the solution is

$$x_t = \exp[\beta_t - (\sigma^2/2)t]. ▲$$

Let x_t be the unique solution of the Itô equation (4.54), and let y_t be the unique solution of the Itô equation

$$dy_t = \bar{f}(y_t, t)\, dt + \bar{G}(y_t, t)\, d\beta_t, \qquad t \geqslant t_0,$$

where $\{\beta_t, t \geqslant t_0\}$, here and in (4.54), is the same Brownian motion process. Then, by considering the joint process

$$dz_t = \begin{bmatrix} f \\ \bar{f} \end{bmatrix} dt + \begin{bmatrix} G \\ \bar{G} \end{bmatrix} d\beta_t, \qquad z_t = [x_t, y_t]^{\mathrm{T}},$$

it is easy to prove the following corollaries of Lemma 4.2.

Corollary 2. *Let the product $\varphi\psi$ of the scalar functions $\varphi(x_t, t)$ and $\psi(y_t, t)$ satisfy the hypotheses of Lemma 4.2. Then*

$$d(\varphi\psi) = \psi\, d\varphi + \varphi\, d\psi + \operatorname{tr} G Q \bar{G}^{\mathrm{T}} \psi_y \varphi_x^{\mathrm{T}}\, dt.$$

Corollary 3. *Let $\psi(y_t, t) > 0$, and let the quotient $\varphi\psi^{-1}$ of the scalar functions $\varphi(x_t, t)$ and $\psi(y_t, t)$ satisfy the hypotheses of Lemma 4.2. Then*

$$d(\varphi\psi^{-1}) = \psi^{-1}\, d\varphi - \varphi\psi^{-2}\, d\psi - \psi^{-2}\operatorname{tr} G Q \bar{G}^{\mathrm{T}} \psi_y \varphi_x^{\mathrm{T}}\, dt$$
$$+ \varphi\psi^{-3}\operatorname{tr} \bar{G} Q \bar{G}^{\mathrm{T}} \psi_y \psi_y^{\mathrm{T}}\, dt.$$

6. STOCHASTIC INTEGRAL OF STRATONOVICH

Recently Stratonovich [17] proposed a new "symmetric" definition of a stochastic integral. Let $T = [a, b]$, partition T:

$$a = t_0 < t_1 < \cdots < t_n = b,$$

and let $\rho = \max_i(t_{i+1} - t_i)$. Let the scalar random function $g_t(\omega)$, defined on T, be an explicit function of β_t, and denote it by $g(\beta_t, t)$, where $\{\beta_t, t \in T\}$ is a scalar Brownian motion process with variance parameter σ^2. Then the *Stratonovich stochastic integral* is defined by

$$\oint_T g(\beta_t, t)\, d\beta_t \triangleq \underset{\rho \to 0}{\text{l.i.m.}} \sum_{i=0}^{n-1} g\left(\frac{\beta_{t_i} + \beta_{t_{i+1}}}{2}, t_i\right)(\beta_{t_{i+1}} - \beta_{t_i}). \qquad (4.62)$$

Equivalently, $g[(\beta_{t_i} + \beta_{t_{i+1}})/2, (t_i + t_{i+1})/2]$ can be used instead of $g[(\beta_{t_i} + \beta_{t_{i+1}})/2, t_i]$ in Eq. (4.62). Gray and Caughey [8] noted that $[g(\beta_{t_i}, t_i) + g(\beta_{t_{i+1}}, t_{i+1})]/2$ is also equivalent. Compare (4.62) with (4.25) of Theorem 4.2.

Stratonovich showed that, if $g(\beta, t)$ is continuous in t and has a continuous partial derivative $\partial g / \partial \beta = g_\beta(\beta, t)$, and further satisfies

$$\int_T \mathscr{E}\{|g(\beta_t, t)|^2\}\, dt < \infty,$$

then the mean square limit in (4.62) exists and is related to the Itô integral by

$$\oint_T g(\beta_t, t)\, d\beta_t = \int_T g(\beta_t, t)\, d\beta_t + (\sigma^2/2) \int_T g_\beta(\beta_t, t)\, dt, \quad \text{wp 1.} \qquad (4.63)$$

The last integral in (4.63) is a well-defined Riemann integral for the sample functions in view of the conditions imposed on g.

We note that the Stratonovich integral is only defined for *explicit* functions of β_t, whereas [see (4.20) and (4.21)] the Itô integral is, for example, defined for functionals on $\{\beta_\tau, \tau \leqslant t\}$. Thus the Stratonovich integral is defined for a much more restricted class of functions than is the Itô integral.

Let us evaluate

$$\oint_a^b (\beta_t - \beta_a)\, d\beta_t.$$

In view of (4.62), we have to compute the ms limit of

$$\sum_{i=0}^{n-1} [(\beta_{t_i} + \beta_{t_{i+1}})/2 - \beta_a](\beta_{t_{i+1}} - \beta_{t_i}) = \sum_{i=0}^{n-1} (\beta_{t_i} - \beta_a)(\beta_{t_{i+1}} - \beta_{t_i})$$

$$+ \frac{1}{2} \sum_{i=0}^{n-1} (\beta_{t_{i+1}} - \beta_{t_i})^2.$$

Therefore, from Example 4.4,

$$\oint_a^b (\beta_t - \beta_a)\, d\beta_t = \tfrac{1}{2}(\beta_b - \beta_a)^2. \tag{4.64}$$

Thus we see that this Stratonovich integral can be evaluated by formal rules.

In Example 4.7 we saw that $x_t = e^{\beta_t}$ is the solution of the Itô equation

$$dx_t = x_t\, d\beta_t + (\sigma^2/2)x_t\, dt, \qquad t \geqslant 0, \qquad x_0 = 1,$$

or

$$\int_0^t dx_\tau = \int_0^t x_\tau\, d\beta_\tau + (\sigma^2/2) \int_0^t x_\tau\, d\tau. \tag{4.65}$$

Substituting for the Itô integral in (4.65) the Stratonovich integral (4.63), and noting that $x_t = e^{\beta_t}$, we get

$$\int_0^t dx_\tau = \oint_0^t x_\tau\, d\beta_\tau - (\sigma^2/2) \int_0^t x_\tau\, d\tau + (\sigma^2/2) \int_0^t x_\tau\, d\tau,$$

so that

$$\int_0^t dx_\tau = \oint_0^t x_\tau\, d\beta_\tau.$$

Therefore, $x_t = e^{\beta_t}$ is the solution of the Stratonovich equation (with the stochastic integral understood in the Stratonovich sense)

$$dx_t = x_t\, d\beta_t, \qquad t \geqslant 0, \qquad x_0 = 1. \tag{4.66}$$

We could have obtained (4.66) by differentiating $x_t = e^{\beta_t}$ formally as if β_t were continuously differentiable. See Example 4.9.

Example 4.9.[6] In Section 2 we gave a motivation for the Itô stochastic differential equation by modeling the scalar process

$$x_t = e^{\beta_t}, \qquad t \geqslant 0, \qquad \beta_0 = 0 \quad \text{wp } 1, \tag{4.6}$$

[6] This example is due to Clark [4].

via a *forward* difference equation, and thus we obtained (4.12) in the mean square sense. In view of the symmetric definition of the Stratonovich integral (4.62), let us model (4.6) via a *central* difference equation. In fact, let us consider the general difference operator

$$\delta^\lambda x_t \triangleq x_{t+\lambda\delta t} - x_{t-(1-\lambda)\delta t}, \qquad \delta t > 0.$$

With $\lambda = 1$, this becomes a forward difference; with $\lambda = \frac{1}{2}$, a central difference; and with $\lambda = 0$, a backward difference. Using the usual power series expansion, we get from (4.6)

$$\delta^\lambda x_t = e^{\beta_{t+\lambda\delta t}} - e^{\beta_{t-(1-\lambda)\delta t}}$$

$$= x_t[\delta\beta_t + (\lambda - \tfrac{1}{2})\,\delta\beta_t{}^2 + O(\delta\beta_t{}^3)],$$

where

$$\delta\beta_t = \beta_{t+\delta t} - \beta_t.$$

Thus for a *central* difference ($\lambda = \frac{1}{2}$), the $\delta\beta_t{}^2$ term is absent, and we have the Stratonovich equation [for (4.6)]

$$dx_t = x_t\,d\beta_t$$

in the mean square sense. Note, however, that this happy· circumstance occurs only for $\lambda = \frac{1}{2}$. ▲

It can, in fact, be shown that the Stratonovich integral and the differential that it defines satisfy all the formal rules of calculus, including integration by parts, variable substitution, the chain rule of differentiation, and so on. In the Stratonovich sense, β_t can be treated as if it were continuously differentiable. As a final example, we note that, in view of (4.63), Eq. (4.57) of Corollary 1 (of Lemma 4.2) becomes

$$\int_a^b \psi(\beta_t)\,d\beta_t = \varphi(\beta_b) - \varphi(\beta_a).$$

At this point the reader might ask the following question. Two different stochastic integrals have been defined. Which one is "right?" The answer is that neither one is right, or wrong. They are merely different definitions. If one wants to model the process e^{β_t} via a stochastic differential equation, then one can equivalently use the Itô equation

$$dx_t = x_t\,d\beta_t + (\sigma^2/2)x_t\,dt \qquad \text{(stochastic integral in Itô sense),}$$

or the Stratonovich equation

$$dx_t = x_t \, d\beta_t \quad \text{(stochastic integral in Stratonovich sense).}$$

The key question is not which equation (integral) to use, but what process is being modeled. The modeling and interpretation of the resultant differential equation have to be consistent. We shall have more to say on the question of modeling in Section 8. In this book we use the Itô definition of the stochastic integral. Our reasons for doing so are outlined in Section 10.

For the Itô equation (4.35),

$$(I) \qquad dx_t = f(x_t, t) \, dt + g(x_t, t) \, d\beta_t, \qquad t \in [t_0, T], \qquad (4.67)$$

there is an equivalent Stratonovich equation. We develop it here formally; the proof can be found in Stratonovich [17]. Let $g(\beta_t, t)$ in (4.63) be $g(x(\beta_t), t)$. Then

$$g_\beta(x(\beta_t), t) = g_x(x_t, t) \, (dx_t/d\beta_t) = g_x(x_t, t) \, g(x_t, t),$$

in view of (4.67). Therefore (4.63) becomes

$$\oint_{t_0}^{t} g(x_\tau, \tau) \, d\beta_\tau = \int_{t_0}^{t} g(x_\tau, \tau) \, d\beta_\tau + \tfrac{1}{2} \int_{t_0}^{t} g(x_\tau, \tau) \, g_x(x_\tau, \tau) \, d\tau. \quad (4.68)$$

As a result, the equivalent Stratonovich equation is

$$(S) \quad dx_t = [f(x_t, t) - \tfrac{1}{2} g(x_t, t) \, g_x(x_t, t)] \, dt + g(x_t, t) \, d\beta_t, \qquad t \in [t_0, T].$$
$$(4.69)$$

(β_t is assumed to have unit variance parameter.) Conversely, in view of (4.68), if

$$(S) \qquad dx_t = f(x_t, t) \, dt + g(x_t, t) \, d\beta_t \qquad (4.70)$$

is a Stratonovich equation, the equivalent Itô equation is

$$(I) \qquad dx_t = [f(x_t, t) + \tfrac{1}{2} g(x_t, t) \, g_x(x_t, t)] \, dt + g(x_t, t) \, d\beta_t. \quad (4.71)$$

For the vector equation (4.47), the transformation $(I \leftrightarrow S)$ is as follows. The ith component of the vector f is modified by

$$\frac{1}{2} \sum_{k=1}^{n} \sum_{j=1}^{m} G_{kj} \frac{\partial G_{ij}}{\partial x_k}. \qquad (4.72)$$

Finally, we note that if the noise coefficient g is a nonrandom time function, the Itô and Stratonovich integrals are equivalent. Thus the linear equation (4.52) is the same in the Itô and Stratonovich sense. Furthermore, in this case, the stochastic integral can be manipulated by formal rules.

7. EVALUATION OF STOCHASTIC (ITÔ) INTEGRALS

We have seen in Example 4.6 how the fundamental theorem of Itô stochastic calculus (Lemma 4.2, Corollary 1) can be used to evaluate stochastic integrals. The relationship between Itô and Stratonovich integrals (4.63) is also useful since the Stratonovich integral can be manipulated by formal rules. Here we present an important result due to Wong and Zakai [19, 21] which relates the stochastic integral to its polygonal approximations. We shall state their result without proof.

Let $T = [a, b]$, and let $a = t_0 < t_1 < \cdots < t_n = b$ be the usual partition of T, with $\rho = \max_i(t_{i+1} - t_i)$. Let $\{\beta_t, t \in T\}$ be a scalar Brownian motion process with variance parameter σ^2. Consider the polygonal approximations to β_t:

$$\beta_t^\rho = \beta_{t_i}^\rho + \frac{\beta_{t_{i+1}}^\rho - \beta_{t_i}^\rho}{t_{i+1} - t_i}(t - t_i), \qquad t_i \leqslant t \leqslant t_{i+1}. \tag{4.73}$$

Theorem 4.6 (Wong and Zakai). *Let $\mathscr{E}\{| g(\beta_t, t)|^2\} \leqslant M$ on T, and suppose $g(\beta_t, t)$ is mean square continuous on T. Furthermore, suppose $g(\beta, t)$ is continuous in t and $g_\beta(\beta, t)$ is continuous in β and t, and $| g_\beta(\beta, t)| \leqslant K$ on T. Then*

$$\operatorname*{l.i.m.}_{\rho \to 0} \int_a^b g(\beta_t^\rho, t)\, d\beta_t^\rho = \int_a^b g(\beta_t, t)\, d\beta_t + (\sigma^2/2) \int_a^b g_\beta(\beta_t, t)\, dt, \tag{4.74}$$

and, as a consequence of (4.63),

$$\operatorname*{l.i.m.}_{\rho \to 0} \int_a^b g(\beta_t^\rho, t)\, d\beta_t^\rho = \oint_a^b g(\beta_t, t)\, d\beta_t. \tag{4.75}$$

Now the polygonal approximations (4.73) are piecewise differentiable, and, as a result of this and the conditions imposed on g, the sequence

$$\left\{ \int_a^b g(\beta_t^\rho, t)\, d\beta_t^\rho \right\}$$

is a sequence of ordinary integrals for the sample functions. Therefore, Theorem 4.6 enables us to evaluate the stochastic integral [Itô integral in (4.74) or Stratonovich integral in (4.75)] in terms of ordinary integrals.

Such polygonal approximations are natural physical approximations to the Brownian motion process, which is a mathematical abstraction. (The Brownian motion process is differentiable nowhere, whereas physical Brownian motion is.) It is therefore interesting to note that these approximations converge to the Stratonovich integral, which indicates that this integral may better model a physical process than the Itô integral. We deal with this subject in more detail in the next section.

Example 4.10. For each ρ, the integral

$$\int_a^b \beta_t^{\rho} \, d\beta_t^{\rho} = \tfrac{1}{2}(\beta_b^2 - \beta_a^2),$$

and, as a result, so does its ms limit. Compare this with Example 4.6. ▲

Example 4.11. For each ρ, the integral

$$\int_a^b \exp(\beta_t^{\rho}) \, d\beta_t^{\rho} = e^{\beta_b} - e^{\beta_a},$$

so that

$$\text{l.i.m.}_{\rho \to 0} \int_a^b \exp(\beta_t^{\rho}) \, d\beta_t^{\rho} = e^{\beta_b} - e^{\beta_a} = \oint_a^b e^{\beta_t} \, d\beta_t$$

$$= \int_a^b e^{\beta_t} \, d\beta_t + (\sigma^2/2) \int_a^b e^{\beta_t} \, dt, \qquad (4.76)$$

in view of (4.74) and (4.75). Therefore, with $\beta_0 = 0$ wp 1,

$$e^{\beta_t} - 1 = \int_0^t e^{\beta_\tau} \, d\beta_\tau + (\sigma^2/2) \int_0^t e^{\beta_\tau} \, d\tau = \oint_0^t e^{\beta_\tau} \, d\beta_\tau, \qquad (4.77)$$

and again we see that the solution of the Itô equation (4.32) is e^{β_t}, which is also the solution of the Stratonovich equation

$$\text{(S)} \qquad dx_t = x_t \, d\beta_t. \quad ▲$$

8. MORE ON MODELING

We have already noted in Sections 4 and 7 that solutions of stochastic differential equations (be they of the Itô or Stratonovich type) have no derivatives. This is, of course, a result of the erratic properties of the

Brownian motion process. Thus stochastic differential equations of the type we are studying are merely idealizations of physical processes, and can only serve to, in some sense, approximate a physical process. We recall (Section 3.8) that white noise approximates a physical process with short correlation times. The question arises: In what sense, if any, does the solution of a stochastic differential equation approximate a physical process?

To be specific, consider a physical process that is modeled by the scalar differential equation

$$\dot{y}_t = f(y_t, t) + g(y_t, t)m_t,$$ (4.78)

where m_t is a random Gaussian disturbance. Let $\{m_t\}$ be a zero-mean, exponentially correlated [see (3.107)] stationary process with correlation function

$$\gamma^\rho(t + \tau, t) = \mathscr{E}\{m_{t+\tau}m_t\} = \sigma^2(\rho/2)\,e^{-\rho|\tau|}.$$ (4.79)

We saw (Section 3.8) that, for ρ large, $\{m_t\}$ is a wide-band process that approximates white noise. We note that if $\{m_t\}$ is correlated, then $\{y_t\}$ is not a Markov process. The $\{y_t\}$ process is, however, smooth. The correlation function (4.79) is integrable, so that $\{y_t\}$ is differentiable; Eq. (4.78) is a well-defined differential equation for the sample functions.

Now the correlated process $\{m_t\}$, sometimes called *colored* noise, can be generated by a linear stochastic differential equation forced by white noise (see Example 4.12),

$$dm_t = -\rho m_t\,dt + \sigma\rho\,d\beta_t,$$ (4.80)

where $\{\beta_t\}$ is a Brownian motion process with unit variance parameter, and $\rho, \sigma > 0$ are fixed constants. Taking (4.78) and (4.80) together,

$$dy_t = [f(y_t, t) + g(y_t, t)m_t]\,dt \equiv h(y_t, m_t, t)\,dt,$$
$$dm_t = -\rho m_t\,dt + \sigma\rho\,d\beta_t,$$ (4.81)

we have a vector stochastic differential equation, with state vector

$$x_t^{\mathrm{T}} = [y_t, m_t],$$

of the type studied in Section 4. $\{x_t\}$ is a Markov process. Thus (4.78) can be modeled by a stochastic differential equation.[7] We have, however,

---

[7] In this section, we reserve the term *stochastic differential equation* to mean the equation of Section 4, be it interpreted in the Itô or Stratonovich sense. Equations such as (4.78) are models for physical processes.

increased the dimension of the problem. Instead of a scalar equation (4.78), we now have a two-dimensional system (4.81).

Example 4.12. Consider the linear stochastic differential equation

$$dm_t = -\rho m_t \, dt + \sigma\rho \, d\beta_t, \quad t \geqslant 0, \tag{4.82}$$

where $\{\beta_t, t \geqslant 0\}$ is a Brownian motion process with unit variance parameter, and ρ, $\sigma > 0$ are fixed constants. Let $m_0 \sim N(0, \sigma^2\rho/2)$, independent of $\{\beta_t\}$. We saw in Section 6 that (4.82) is the same in the Itô and Stratonovich sense (since the coefficient of $d\beta_t$ is nonrandom), and can be manipulated by formal rules as if β_t were continuously differentiable. We therefore consider the equivalent equation

$$\dot{m}_t = -\rho m_t + \sigma\rho w_t, \quad t \geqslant 0, \tag{4.83}$$

where $\{w_t\}$ is zero-mean, white Gaussian noise with

$$\mathscr{E}\{w_t w_\tau\} = \delta(t - \tau).$$

The solution of (4.83) is

$$m_t = m_0 e^{-\rho t} + \sigma\rho \int_0^t e^{-\rho(t-s)} w_s \, ds. \tag{4.84}$$

Clearly,

$$\mathscr{E}\{m_t\} \equiv 0.$$

Compute $(t > \tau)$

$$\mathscr{E}\{m_t m_\tau\} = \mathscr{E}\left\{\left[m_0 e^{-\rho t} + \sigma\rho \int_0^t e^{-\rho(t-s)} w_s \, ds\right]\right.$$

$$\left. \times \left[m_0 e^{-\rho\tau} + \sigma\rho \int_0^\tau e^{-\rho(\tau-\xi)} w_\xi \, d\xi\right]\right\}$$

$$= \sigma^2(\rho/2)e^{-\rho(t+\tau)} + \sigma^2\rho^2 \int_0^t \int_0^\tau e^{-\rho(t-s)} e^{-\rho(\tau-\xi)} \, \delta(s - \xi) \, ds \, d\xi$$

$$= \sigma^2(\rho/2)e^{-\rho(t+\tau)} + \sigma^2\rho^2 \int_0^\tau e^{-\rho(t-\xi)} e^{-\rho(\tau-\xi)} \, d\xi$$

$$= \sigma^2(\rho/2)e^{-\rho(t-\tau)}. \tag{4.85}$$

Thus (4.82) generates a stationary, exponentially correlated, Gaussian process. $\{m_t\}$ is sometimes called *colored* noise. ▲

It is clear that if $\{y_t\}$ is a vector physical process, say of dimension n, then we need at least a $2n$-dimensional system of stochastic differential equations to model that physical process by the method we have just described. We are therefore forced to deal with a process of high dimension.

Is it possible, in some sense, to model the physical process $\{y_t\}$ by a stochastic differential equation of the same dimension? This is an area in which much theoretical and experimental research remains to be done. Some interesting results have been obtained by Clark [4] and Wong and Zakai [19–21]. A detailed treatment of this subject is outside the scope of this book, and we present only a simple example due to W. M. Wonham (see Kushner [13]).

Consider a special case of the physical process (4.78), namely,

$$\dot{y}_t = y_t m_t, \qquad t \geqslant 0, \qquad y_0 = 1. \tag{4.86}$$

Equation (4.86) has the solution

$$y_t^\rho = \exp\left(\int_0^t m_\tau \, d\tau\right), \tag{4.87}$$

where the superscript ρ indicates that m_t comes from (4.80) for a fixed ρ. Compute

$$\mathscr{E}\left\{\left|\int_0^t m_\tau \, d\tau - \sigma\beta_t\right|^2\right\} = \int_0^t \int_0^t \mathscr{E}\{m_\tau m_s\} \, d\tau \, ds - 2\sigma\mathscr{E}\left\{\beta_t \int_0^t m_\tau \, d\tau\right\} + \sigma^2 t.$$

With the aid of (4.85), it is easy to compute

$$\int_0^t \int_0^t \mathscr{E}\{m_\tau m_s\} \, d\tau \, ds = \sigma^2 t + (\sigma^2/\rho)(e^{-\rho t} - 1).$$

From (4.80),

$$-2\sigma\mathscr{E}\left\{\beta_t \int_0^t m_\tau \, d\tau\right\} = (2t/\rho) \, \mathscr{E}\{(m_t - m_0)\} - 2\sigma^2 t = -2\sigma^2 t.$$

Therefore,

$$\mathscr{E}\left\{\left|\int_0^t m_\tau \, d\tau - \sigma\beta_t\right|^2\right\} = (\sigma^2/\rho)(e^{-\rho t} - 1) \to 0 \qquad (\rho \to \infty).$$

In other words,

$$\operatorname*{l.i.m.}_{\rho \to \infty} \int_0^t m_\tau \, d\tau = \sigma\beta_t \, . \tag{4.88}$$

Therefore, by Example 3.11,

$$\underset{\rho\to\infty}{\text{l.i.m.}}\, y_t{}^\rho = e^{\sigma\beta_t}. \tag{4.89}$$

But $e^{\sigma\beta_t}$ is the solution of the Stratonovich equation

$$(S) \qquad dx_t = x_t\, d\bar{\beta}_t \tag{4.90}$$

and the Itô equation

$$(I) \qquad dx_t = x_t\, d\bar{\beta}_t + (\sigma^2/2)x_t\, dt, \tag{4.91}$$

where $\{\bar{\beta}_t\}$ has variance parameter σ^2 (see Section 6 and Example 4.11). Therefore, for ρ large, (4.86) can be modeled by (4.90) or (4.91).

We present, without proof, an important related result due to Wong and Zakai [19]. Consider a sequence of differential equations

$$dx_t{}^\rho = f(x_t{}^\rho, t)\, dt + g(x_t{}^\rho, t)\, d\beta_t{}^\rho, \qquad t \in [a, b] = T, \tag{4.92}$$

where $\beta_t{}^\rho$ is a sequence of polygonal approximations to β_t, given in (4.73). Let $x_a{}^\rho = x_a$. Since the polygonal approximations are piecewise differentiable wp 1, Eqs. (4.92) are ordinary differential equations for the sample functions.

Theorem 4.7 (Wong and Zakai).[8] *Let f and g satisfy the hypotheses of Theorem 4.5, and in addition, let $g_x(x, t)$ be continuous in x and t, and let $g(x, t)\, g_x(x, t)$ satisfy a uniform Lipschitz condition in x and be continuous in t. Furthermore, suppose $\mathscr{E}\{|\, x_a\,|^4\} < \infty$. Then*

$$\underset{\rho\to 0}{\text{l.i.m.}}\, x_t{}^\rho = x_t, \tag{4.93}$$

where x_t is the unique solution of the Stratonovich equation

$$(S) \qquad dx_t = f(x_t, t)\, dt + g(x_t, t)\, d\beta_t \tag{4.94}$$

and the Itô equation

$$(I) \qquad dx_t = [f(x_t, t) + \tfrac{1}{2}g(x_t, t)\, g_x(x_t, t)]\, dt + g(x_t, t)\, d\beta_t. \tag{4.95}$$

This theorem is useful in the numerical simulation of stochastic differential equations. In this connection, see also Example 4.9.

---

[8] A related result for more general approximations to the Brownian motion process is given by Wong and Zakai [20].

In general, consider the vector physical process

$$\dot{y}_t = f(y_t, t) + G(y_t, t)m_t, \qquad (4.96)$$

where the components of m_t are *independent*. Let m_t be a wide-band or rapidly fluctuating but still continuous Gaussian process and let

$$\text{l.i.m.} \int_0^t m_\tau \, d\tau = \beta_t, \qquad (4.97)$$

where β_t is a vector process of independent Brownian motions with unit variance parameters. Then (see Clark [4]) in the mean square sense, (4.96) is modeled by the Stratonovich equation

$$(\text{S}) \qquad dx_t = f(x_t, t) \, dt + G(x_t, t) \, d\beta_t \qquad (4.98)$$

or the equivalent Itô equation

$$(\text{I}) \qquad (dx_t)_i = \left[f_i(x_t, t) + \frac{1}{2} \sum_{j,k} G_{kj} \frac{\partial G_{ij}}{\partial x_k} \right] dt + \sum_j G_{ij} (d\beta_t)_j. \qquad (4.99)$$

These results are in general not valid if the components of m_t are not independent (see Clark [4] for explicit conditions and assumptions).

In connection with modeling, the reader might also consult Stratonovich [16], Kulman [12], Astrom [1], and Gray and Caughey [8].

9. KOLMOGOROV'S EQUATIONS

As was proved in Theorem 4.5, the $\{x_t, t \in [t_0, T]\}$ process generated by the Itô stochastic differential equation (4.35),

$$dx_t = f(x_t, t) \, dt + g(x_t, t) \, d\beta_t, \qquad t \in [t_0, T], \qquad (4.100)$$

is a Markov process. [As in Section 4, we assume without loss of generality that $\{\beta_t\}$ has unit variance parameter. Unless otherwise specified, (4.100) is understood in the Itô sense.] The $\{x_t\}$ process is therefore characterized by the density function

$$p(x_t) = p(x, t), \qquad \text{all} \quad t \in [t_0, T], \qquad (4.101)$$

and the transition probability density function

$$p(x_t \mid x_\tau) = p_{x_t \mid x_\tau}(x \mid y) = p(x, t; y, \tau), \qquad \text{all} \quad t > \tau \in [t_0, T]. \qquad (4.102)$$

See Section 3.7 on Markov processes, and Section 3.2 for the notation.

In this section, we formally derive the laws of evolution of the densities in (4.101) and (4.102), assuming the existence of the continuous partial derivatives

$$\frac{\partial p}{\partial t}, \quad \frac{\partial[pf(x, t)]}{\partial x}, \quad \frac{\partial^2[pg^2(x, t)]}{\partial x^2},$$

where p stands for $p(x, t)$ and $p(x, t; y, \tau)$. In the following, $\delta(\cdot)$ stands for a forward difference

$$\delta\xi = \xi_{t+\delta t} - \xi_t, \qquad \delta t > 0.$$

We consider the transition probability density function first.

Let $R(x)$ be a nonnegative, twice continuously differentiable real function such that $(x_1 < x_2)$

$$R(x) = 0, \qquad x < x_1 \quad \text{and} \quad x > x_2,$$

$$R(x_1) = R(x_2) = R'(x_1) = R'(x_2) = R''(x_1) = R''(x_2) = 0,$$

where the prime denotes the derivative. Then

$$\int_{x_1}^{x_2} \left[\frac{\partial p(x, t; y, \tau)}{\partial t} \delta t + o(\delta t) \right] R(x)\, dx$$

$$= \int_{-\infty}^{\infty} [p(x, t + \delta t; y, \tau) - p(x, t; y, \tau)]\, R(x)\, dx. \qquad (4.103)$$

Now according to the Chapman–Kolmogorov equation (3.98),

$$p(x, t + \delta t; y, \tau) = \int_{-\infty}^{\infty} p(x, t + \delta t; z, t)\, p(z, t; y, \tau)\, dz. \qquad (4.104)$$

Using (4.104), the right-hand side of (4.103) becomes

$$\int_{-\infty}^{\infty} \int_{-\infty}^{\infty} p(x, t + \delta t; z, t)\, p(z, t; y, \tau)\, R(x)\, dz\, dx - \int_{-\infty}^{\infty} p(x, t; y, \tau)\, R(x)\, dx$$

$$= \int_{-\infty}^{\infty} \int_{-\infty}^{\infty} p(z, t + \delta t; x, t)\, p(x, t; y, \tau)\, R(z)\, dx\, dz$$

$$- \int_{-\infty}^{\infty} p(x, t; y, \tau)\, R(x)\, dx$$

$$= \int_{-\infty}^{\infty} p(x, t; y, \tau) \left\{ \int_{-\infty}^{\infty} p(z, t + \delta t; x, t)\, R(z)\, dz - R(x) \right\} dx, \qquad (4.105)$$

where we replaced the dummy variable of integration x by z, and z by x.

We next reduce the term in braces in (4.105). Expand $R(z)$ in Taylor's series:

$$R(z) = R(x) + (z - x) R'(x) + \tfrac{1}{2}(z - x)^2 R''(x) + o(z - x)^2.$$

Then

$$\int_{-\infty}^{\infty} p(z, t + \delta t; x, t) R(z) \, dz - R(x)$$

$$= R(x) \int_{-\infty}^{\infty} p(z, t + \delta t; x, t) \, dz + R'(x) \int_{-\infty}^{\infty} (z - x) p(z, t + \delta t; x, t) \, dz$$

$$+ \tfrac{1}{2} R''(x) \int_{-\infty}^{\infty} (z - x)^2 p(z, t + \delta t; x, t) \, dz$$

$$+ \int_{-\infty}^{\infty} o(z - x)^2 p(z, t + \delta t; x, t) \, dz - R(x). \tag{4.106}$$

Now

$$\int_{-\infty}^{\infty} p(z, t + \delta t; x, t) \, dz = 1, \tag{4.107}$$

since the integrand is a density function. Notice that, in (4.106), z is a realization of $x_{t+\delta t}$ and x is a realization of x_t. Therefore, $z - x = \delta x$ is a realization of δx_t. As a result,

$$p(z, t + \delta t; x, t) = p_{x_{t+\delta t}|x_t}(z \mid x) = p_{x_{t+\delta t}|x_t}(x + \delta x \mid x)$$

$$= p_{\delta x_t|x_t}(\delta x \mid x). \tag{4.108}$$

Consequently,

$$\int_{-\infty}^{\infty} (z - x)^k p(z, t + \delta t; x, t) \, dz = \mathcal{E}\{\delta x_t^k \mid x_t\}, \qquad k = 1, 2, \ldots . \tag{4.109}$$

But, in view of (4.100) and (4.10),

$$\mathcal{E}\{\delta x_t \mid x_t\} = f(x, t) \, \delta t,$$

$$\mathcal{E}\{\delta x_t^2 \mid x_t\} = g^2(x, t) \, \delta t + o(\delta t), \tag{4.110}$$

$$\mathcal{E}\{\delta x_t^k \mid x_t\} = o(\delta t), \qquad k > 2.$$

Combining (4.109) with (4.110) and using (4.107), (4.106) becomes

$$\int_{-\infty}^{\infty} p(z, t + \delta t; x, t) R(z) \, dz - R(x)$$

$$= R'(x) f(x, t) \, \delta t + \tfrac{1}{2} R''(x) g^2(x, t) \, \delta t + o(\delta t). \tag{4.111}$$

Therefore, in view of (4.105), (4.103) becomes

$$\int_{x_1}^{x_2} \left[\frac{\partial p(x, t; y, \tau)}{\partial t} \, \delta t + o(\delta t) \right] R(x) \, dx$$

$$= \int_{-\infty}^{\infty} p(x, t; y, \tau)[R'(x) f(x, t) \, \delta t + \tfrac{1}{2} R''(x) \, g^2(x, t) \, \delta t + o(\delta t)] \, dx,$$

and, dividing by δt and taking the $\delta t = 0$ limit, we finally get

$$\int_{x_1}^{x_2} \frac{\partial p(x, t; y, \tau)}{\partial t} R(x) \, dx = \int_{x_1}^{x_2} p(x, t; y, \tau) f(x, t) \, R'(x) \, dx$$

$$+ \frac{1}{2} \int_{x_1}^{x_2} p(x, t; y, \tau) \, g^2(x, t) \, R''(x) \, dx. \quad (4.112)$$

Our derivation is almost complete. We evaluate the integrals on the right-hand side of (4.112) by parts and, using the properties of $R(x)$, obtain

$$\int_{x_1}^{x_2} \left\{ \frac{\partial p}{\partial t} + \frac{\partial (pf)}{\partial x} - \frac{1}{2} \frac{\partial^2 (pg^2)}{\partial x^2} \right\} R(x) \, dx = 0. \quad (4.113)$$

Since $R(x)$ is arbitrary, the term in braces must vanish:

$$\frac{\partial p(x, t; y, \tau)}{\partial t} = - \frac{\partial [p(x, t; y, \tau) f(x, t)]}{\partial x}$$

$$+ \frac{1}{2} \frac{\partial^2 [p(x, t; y, \tau) g^2(x, t)]}{\partial x^2}. \quad (4.114)$$

The partial differential equation (4.114) is known as **Kolmogorov's forward equation** or the **Fokker–Planck** equation. It describes the evolution of the transition probability density of the Markov process generated by the Itô equation (4.100). A Markov process whose transition probability density satisfies Kolmogorov's equation is called a **diffusion** process.

The initial condition for (4.114) is clearly

$$\lim_{t \to \tau} p_{x_t | x_\tau}(x \mid y) = \delta(x - y). \quad (4.115)$$

If $p(x, t; y, \tau)$ is assumed "well behaved" at infinity, the boundary conditions are

$$p_{x_t | x_\tau}(\infty \mid y) = p_{x_t | x_\tau}(-\infty \mid y) = 0. \quad (4.116)$$

One can also similarly derive the so-called **Kolmogorov's backward equation**, which is

$$-\frac{\partial p(x, t; y, \tau)}{\partial \tau} = f(y, \tau) \frac{\partial p(x, t; y, \tau)}{\partial y}$$

$$+ \tfrac{1}{2} g^2(y, \tau) \frac{\partial^2 p(x, t; y, \tau)}{\partial y^2}. \qquad (4.117)$$

The backward equation is the formal adjoint of the forward equation, but it will be of little interest in this book.

In the vector case, that is, for the vector Itô stochastic differential equation (4.47),

$$dx_t = f(x_t, t) \, dt + G(x_t, t) \, d\beta_t \qquad (4.118)$$

[x, f are n-vectors, G is $n \times m$, $\{\beta_t\}$ is an m-vector Brownian motion process with $\mathscr{E}\{d\beta_t \, d\beta_t^T\} = Q(t) \, dt$], Kolmogorov's forward equation becomes

$$\frac{\partial p}{\partial t} = -\sum_{i=1}^{n} \frac{\partial (p f_i)}{\partial x_i} + \frac{1}{2} \sum_{i,j=1}^{n} \frac{\partial^2 [p(GQG^T)_{ij}]}{\partial x_i \, \partial x_j}. \qquad (4.119)$$

Its derivation is completely analogous to that of the scalar equation and is left for the reader.

Recall that $p(x, t; y, \tau)$ is a random variable since it depends on the values y taken on by x_τ. With this in mind, we take the expectation of (4.114) and, interchanging expectation with differentiation, obtain

$$\frac{\partial p(x, t)}{\partial t} = -\frac{\partial [p(x, t) f(x, t)]}{\partial x} + \frac{1}{2} \frac{\partial^2 [p(x, t) g^2(x, t)]}{\partial x^2}. \qquad (4.120)$$

Thus we see that $p(x_t)$ also satisfies Kolmogorov's forward equation. The initial condition for (4.120) is clearly $p(x_{t_0})$, which is assumed given.

Kolmogorov's forward equation (4.119) can be written as

$$dp = \mathscr{L}(p) \, dt, \qquad (4.121)$$

were $\mathscr{L}$ is called the *forward diffusion operator* of the $\{x_t\}$ process generated by the Itô equation (4.118):

$$\mathscr{L}(\cdot) = -\sum_{i=1}^{n} \frac{\partial (\cdot f_i)}{\partial x_i} + \frac{1}{2} \sum_{i,j=1}^{n} \frac{\partial^2 [\cdot (GQG^T)_{ij}]}{\partial x_i \, \partial x_j}. \qquad (4.122)$$

Example 4.13. Kolmogorov's forward equation has been solved only in a few simple cases. Some of these solutions may be found in Bharucha–Reid [2]. Using Laplace transform techniques, let the reader solve the Kolmogorov's forward equation:

$$\frac{\partial p(x, t)}{\partial t} = \frac{1}{2} \frac{\partial^2 p(x, t)}{\partial x^2}, \qquad t \geqslant 0, \qquad p(x, 0) = \delta(x).$$

This is the Kolmogorov's equation for the Itô equation

$$dx_t = d\beta_t, \qquad t \geqslant 0, \qquad \beta_0 = 0 \quad \text{wp } 1,$$

and thus clearly must have the solution

$$p(x, t) = (1/2\pi t)^{1/2} \exp(-x^2/2t). \quad \blacktriangle$$

Example 4.14. Suppose the stochastic differential equation (4.118) is interpreted in the Stratonovich sense. Then (see Sect. 6) the equivalent Itô equation is

$$d(x_t)_i = \left[f_i(x_t, t) + \frac{1}{2} \sum_{k=1}^{n} \sum_{j=1}^{m} \sum_{l=1}^{m} G_{kj} Q_{jl} \frac{\partial G_{il}}{\partial x_k} \right] dt$$

$$+ \sum_{j=1}^{m} G_{ij} \, d(\beta_t)_j. \qquad (4.123)$$

Substituting the quantity in brackets for f_i in (4.119), we get

$$\frac{\partial p}{\partial t} = - \sum_{i=1}^{n} \frac{\partial (p f_i)}{\partial x_i} + \frac{1}{2} \sum_{i,k=1}^{n} \sum_{j=1}^{m} \frac{\partial}{\partial x_i} \left[(GQ)_{ij} \frac{\partial}{\partial x_k} (p G_{kj}) \right]. \qquad (4.124)$$

This is Kolmogorov's forward equation for the *Stratonovich* equation

$$(\text{S}) \qquad dx_t = f(x_t, t) \, dt + G(x_t, t) \, d\beta_t. \qquad (4.125)$$

One can similarly obtain Kolmogorov's backward equation for (4.125). $\blacktriangle$

Consider the *linear* Itô stochastic differential equation

$$dx_t = F(t) x_t \, dt + G(t) \, d\beta_t, \qquad t \geqslant t_0, \qquad (4.126)$$

where x_t is an n-vector, F and G are, respectively, $n \times n$ and $n \times m$ time matrix functions, and $\{\beta_t, t \geqslant t_0\}$ is an m-vector Brownian motion process with

$$\mathscr{E}\{d\beta_t \, d\beta_t^{\mathrm{T}}\} = Q(t) \, dt. \qquad (4.127)$$

Let $x_{t_0} \sim N(\hat{x}_{t_0}, P_{t_0})$, independent of $\{\beta_t, t \geqslant t_0\}$. We saw in Section 4 that the $\{x_t, t \geqslant t_0\}$ process generated by (4.126) is Gauss–Markov. As a result, $p(x_t)$ is completely characterized by its mean and covariance matrix

$$\hat{x}_t = \mathscr{E}\{x_t\}, \tag{4.128}$$

$$P_t = \mathscr{E}\{(x_t - \hat{x}_t)(x_t - \hat{x}_t)^\mathsf{T}\}. \tag{4.129}$$

In the next four examples we derive the equations of evolution for $\hat{x}_t$ and P_t by four different methods.

Example 4.15. For the linear system (4.126), Kolmogorov's forward equation (4.119) becomes

$$\partial p / \partial t = -p \operatorname{tr} F - p_x^\mathsf{T} Fx + \tfrac{1}{2} \operatorname{tr} GQG^\mathsf{T} p_{xx}. \tag{4.130}$$

Note that in this linear case (4.124) also reduces to (4.130). This is consistent with the fact that (Section 6) the linear system (4.126) is the same in the Itô and Stratonovich sense.

Recall that the characteristic function of x_t is just the Fourier transform of $p(x, t)$ (2.71):

$$\varphi(u, t) = \int_{-\infty}^{\infty} e^{iu^\mathsf{T}x} p(x, t) \, dx. \tag{4.131}$$

Taking the Fourier transform in (4.130),

$$\partial \varphi / \partial t = \int e^{iu^\mathsf{T}x} (\partial p / \partial t) \, dx = -\varphi \operatorname{tr} F - \int e^{iu^\mathsf{T}x} p_x^\mathsf{T} Fx \, dx$$

$$+ \tfrac{1}{2} \operatorname{tr} GQG^\mathsf{T} \int e^{iu^\mathsf{T}x} p_{xx} \, dx$$

$$= -\varphi \operatorname{tr} F - \operatorname{tr} F \int x p_x^\mathsf{T} e^{iu^\mathsf{T}x} \, dx + \tfrac{1}{2} \operatorname{tr} GQG^\mathsf{T} \int p_{xx} e^{iu^\mathsf{T}x} \, dx. \tag{4.132}$$

Evaluating the last two integrals by parts and assuming that p is well behaved at infinity (p and its partial derivatives with respect to x vanish at infinity), we get

$$\partial \varphi / \partial t = u^\mathsf{T} F \varphi_u - \tfrac{1}{2} \varphi u^\mathsf{T} GQG^\mathsf{T} u, \tag{4.133}$$

which is a partial differential equation for the characteristic function.

Since x_t is Gaussian, we know that its characteristic function is simply (2.130)

$$\varphi(u, t) = \exp(iu^\mathsf{T}\hat{x}_t - \tfrac{1}{2}u^\mathsf{T} P_t u). \tag{4.134}$$

Differentiating (4.134),

$$\partial \varphi / \partial t = \varphi[i u^{\mathrm{T}}(d\hat{x}_t/dt) - \tfrac{1}{2} u^{\mathrm{T}}(dP_t/dt)u], \tag{4.135}$$

$$\varphi_u = \varphi[i\hat{x}_t - P_t u], \tag{4.136}$$

and substituting in (4.133),

$$\varphi i u^{\mathrm{T}}(d\hat{x}_t/dt) - \tfrac{1}{2}\varphi u^{\mathrm{T}}(dP_t/dt)u$$
$$= \varphi i u^{\mathrm{T}} F \hat{x}_t - \tfrac{1}{2}\varphi u^{\mathrm{T}}(FP_t + P_t F^{\mathrm{T}})u - \tfrac{1}{2}\varphi u^{\mathrm{T}} GQG^{\mathrm{T}} u. \tag{4.137}$$

(We symmetrized FP_t, since its skew-symmetric part vanishes in the quadratic form.)

Now equating the real and imaginary parts in (4.137), we get

$$d\hat{x}_t/dt = F(t)\hat{x}_t,$$
$$dP_t/dt = F(t)P_t + P_t F^{\mathrm{T}}(t) + G(t)Q(t)G^{\mathrm{T}}(t), \tag{4.138}$$

which are ordinary differential equations describing the evolution of $\hat{x}_t$ and P_t. Initial conditions for (4.138) are given, namely, $\hat{x}_{t_0}$ and P_{t_0}.   ▲

Example 4.16.   We now obtain (4.138) directly from (4.126). Taking the expectation in (4.126), and interchanging d and $\mathscr{E}$, we get

$$d\hat{x}_t = F(t)\hat{x}_t \, dt, \tag{4.139}$$

which is the first of (4.138) ($\mathscr{E}\{d\beta_t\} \equiv 0$). Now

$$dP_t = d[\mathscr{E}\{x_t x_t^{\mathrm{T}}\} - \hat{x}_t \hat{x}_t^{\mathrm{T}}] = \mathscr{E}\{d(x_t x_t^{\mathrm{T}})\} - d(\hat{x}_t \hat{x}_t^{\mathrm{T}}). \tag{4.140}$$

Applying Itô's lemma 4.2 to the elements of $x_t x_t^{\mathrm{T}}$,

$$d(x_t x_t^{\mathrm{T}}) = x_t \, dx_t^{\mathrm{T}} + (dx_t)x_t^{\mathrm{T}} + GQG^{\mathrm{T}} \, dt, \tag{4.141}$$

and a trivial application of Itô's lemma gives

$$d(\hat{x}_t \hat{x}_t^{\mathrm{T}}) = \hat{x}_t \, d\hat{x}_t^{\mathrm{T}} + (d\hat{x}_t)\hat{x}_t^{\mathrm{T}}, \tag{4.142}$$

since the differential equation for $\hat{x}_t$ (4.139) is not stochastic. Now using (4.126) and (4.139),

$$dP_t = \mathscr{E}\{x_t(x_t^{\mathrm{T}} F^{\mathrm{T}} \, dt + d\beta_t^{\mathrm{T}} G^{\mathrm{T}}) + (Fx_t \, dt + G \, d\beta_t)x_t^{\mathrm{T}} + GQG^{\mathrm{T}} \, dt\}$$
$$\quad - \hat{x}_t \hat{x}_t^{\mathrm{T}} F^{\mathrm{T}} \, dt - F\hat{x}_t \hat{x}_t^{\mathrm{T}} \, dt$$
$$= (\mathscr{E}\{x_t x_t^{\mathrm{T}}\} - \hat{x}_t \hat{x}_t^{\mathrm{T}})F^{\mathrm{T}} \, dt + F(\mathscr{E}\{x_t x_t^{\mathrm{T}}\} - \hat{x}_t \hat{x}_t^{\mathrm{T}}) \, dt$$
$$\quad + GQG^{\mathrm{T}} \, dt, \tag{4.143}$$

since $\mathscr{E}\{x_t \, d\beta_t^{\mathrm{T}}\} = 0$. (4.143) is just the second of (4.138).   ▲

We used Itô's stochastic calculus in Example 4.16, namely, Lemma 4.2, to derive (4.138). Since (4.126) is linear, and the stochastic integral in that equation is therefore the same in the Itô and Stratonovich sense, we could use formal rules of differentiation to obtain the same result. The GQG^T term in (4.141) would then be absent, but the differential operator d in that equation would have to be defined as a limit of a *central* difference, rather than the usual limit of a forward difference (see Example 4.9). Defined this way, the first two terms on the right-hand side of (4.141) themselves contribute a GQG^T term since

$$\mathscr{E}\{x_t \, d^{1/2}\beta_t^T\} \neq 0$$

if $d^{1/2}$ is a "central difference" differential. We leave the details as an exercise for the reader.

Example 4.17.   We saw in Section 6 that the linear equation (4.126) can be manipulated by formal rules. Consider, therefore, the equivalent equation

$$\dot{x}_t = F(t)x_t + G(t)w_t, \qquad t \geq t_0, \tag{4.144}$$

where $\{w_t, t \geq t_0\}$ is a zero-mean white Gaussian noise process with

$$\mathscr{E}\{w_t w_\tau^T\} = Q(t)\, \delta(t - \tau).$$

The solution of (4.144) is

$$x_t = \Phi(t, t_0)x_{t_0} + \int_{t_0}^t \Phi(t, \tau)\, G(\tau)w_\tau \, d\tau, \tag{4.145}$$

where $\Phi(t, \tau)$ is the fundamental matrix (state transition matrix) of the homogeneous system

$$\dot{x}_t = F(t)x_t,$$

satisfying the matrix differential equation

$$\dot{\Phi} = F(t)\Phi, \qquad \Phi(t, t) = I. \tag{4.146}$$

Taking the expectation in (4.145), we get

$$\hat{x}_t = \Phi(t, t_0)\hat{x}_{t_0},$$

so that

$$\dot{\hat{x}}_t = F(t)\hat{x}_t, \tag{4.147}$$

which is the first equation in (4.138).

Now

$$\frac{dP_t}{dt} = \frac{d}{dt} \mathscr{E}\{(x_t - \hat{x}_t)(x_t - \hat{x}_t)^{\mathrm{T}}\} = \mathscr{E}\left\{\left(\frac{d}{dt}(x_t - \hat{x}_t)\right)(x_t - \hat{x}_t)^{\mathrm{T}}\right\}$$

$$+ \mathscr{E}\left\{(x_t - \hat{x}_t)\frac{d}{dt}(x_t - \hat{x}_t)^{\mathrm{T}}\right\}. \tag{4.148}$$

From (4.144) and (4.147), we have

$$\frac{d}{dt}(x_t - \hat{x}_t) = F(x_t - \hat{x}_t) + Gw_t ,$$

so that

$$dP_t/dt = FP_t + \mathscr{E}\{Gw_t(x_t - \hat{x}_t)^{\mathrm{T}}\} + P_tF^{\mathrm{T}} + \mathscr{E}\{(x_t - \hat{x}_t)w_t^{\mathrm{T}}G^{\mathrm{T}}\}. \tag{4.149}$$

Now

$$\mathscr{E}\{(x_t - \hat{x}_t)w_t^{\mathrm{T}}G^{\mathrm{T}}\} = \mathscr{E}\left\{\left[\Phi(t, t_0)(x_{t_0} - \hat{x}_{t_0}) + \int_{t_0}^t \Phi(t, \tau)G(\tau)w_\tau\, d\tau\right]w_t^{\mathrm{T}}G^{\mathrm{T}}\right\}$$

$$= \int_{t_0}^t \Phi(t, \tau)G(\tau)\,\delta(\tau - t)\,G^{\mathrm{T}}(t)\,d\tau$$

$$= \tfrac{1}{2}G(t)Q(t)\,G^{\mathrm{T}}(t),$$

and, similarly,

$$\mathscr{E}\{Gw_t(x_t - \hat{x}_t)^{\mathrm{T}}\} = \tfrac{1}{2}G(t)Q(t)\,G^{\mathrm{T}}(t).$$

Equation (4.149) therefore becomes

$$dP_t/dt = FP_t + P_tF^{\mathrm{T}} + GQG^{\mathrm{T}},$$

which is (4.138). ▲

Example 4.18. In Section 3.9, Eqs. (3.128)–(3.132), we developed equations of evolution (difference equations) for $\hat{x}$ and P for a discrete linear stochastic system. Using that development, we obtain here differential equations for $\hat{x}_t$ and P_t for the continuous system (4.126) by a formal limiting operation. We consider the white noise equivalent of (4.126), namely, (4.144). For $\Delta = t_{k+1} - t_k$ small, we write (4.144)

$$x_{t_{k+1}} = x_{t_k} + F(t_k)x_{t_k}\Delta + G(t_k)w_{t_k}\Delta + o(\Delta), \tag{4.150}$$

where, according to Example 3.20,

$$\mathscr{E}\{w_{t_k}w_{t_l}^{\mathrm{T}}\} = [Q(t_k)/\Delta]\,\delta_{kl}. \tag{4.151}$$

According to the results in Section 3.9,

$$\hat{x}_{t_{k+1}} = [I + F(t_k)\, \Delta]\hat{x}_{t_k},$$

$$P_{t_{k+1}} = [I + F(t_k)\, \Delta]\, P_{t_k}[I + F(t_k)\, \Delta]^{\mathrm{T}} + \Delta^2 G(t_k)[Q(t_k)/\Delta]\, G^{\mathrm{T}}(t_k).$$

These equations are rewritten as

$$(\hat{x}_{t_{k+1}} - \hat{x}_{t_k})/\Delta = F(t_k)\hat{x}_{t_k}, \tag{4.152}$$

$$(P_{t_{k+1}} - P_{t_k})/\Delta = F(t_k)P_{t_k} + P_{t_k}F^{\mathrm{T}}(t_k) + \Delta F(t_k)\, P_{t_k}F^{\mathrm{T}}(t_k)$$

$$+ G(t_k)\, Q(t_k)\, G^{\mathrm{T}}(t_k), \tag{4.153}$$

and taking the $\Delta = 0$ limit we get (4.138). ▲

Consider the scalar *nonlinear* Itô stochastic differential equation,

$$dx_t = f(x_t, t)\, dt + g(x_t, t)\, d\beta_t, \qquad t \geq t_0, \tag{4.154}$$

where $\{\beta_t, t \geq t_0\}$ has variance parameter $q(t)$. The density $p(x_{t_0})$ is given and x_{t_0} is independent of $\{\beta_t\}$. Since (4.154) is nonlinear, $\hat{x}_t$(4.128) and P_t (4.129) *do not*, in general, characterize $p(x, t)$. They do, however, determine the mean path of (4.154) and the dispersion about that path, and are therefore of great interest. In the next two examples, we develop equations of evolution for $\hat{x}_t$ and P_t by two different methods. In fact, we first develop the equation of evolution for $\mathscr{E}\{\varphi(x_t)\}$, where $\varphi(x_t)$ is an arbitrary function of x_t. Thus we can write down the equation of evolution for any moment of $p(x, t)$. We shall see that these equations are integral-differential equations. The methods we use have apparently been used by others. See Bogdanoff and Kozin [3] and Cummings [5]. Our analysis readily generalizes to the vector case.

Example 4.19.   Let $\varphi(x)$ satisfy the hypotheses of Lemma 4.2. Then, by that lemma,

$$d\varphi(x_t) = \varphi_x(x_t)\, dx_t + \tfrac{1}{2}q(t)\, g^2(x_t, t)\, \varphi_{xx}(x_t)\, dt,$$

and substituting for dx_t from (4.154),

$$d\varphi(x_t) = [\varphi_x(x_t)f(x_t, t) + \tfrac{1}{2}q(t)\, g^2(x_t, t)\, \varphi_{xx}(x_t)]\, dt$$

$$+ \varphi_x(x_t)\, g(x_t, t)\, d\beta_t. \tag{4.155}$$

Since

$$\mathscr{E}\{\varphi_x(x_t)\,g(x_t\,,t)\,d\beta_t\} = \mathscr{E}\{\mathscr{E}\{\varphi_x(x_t)\,g(x_t\,,t)\,d\beta_t \mid x_t\}\}$$
$$= \mathscr{E}\{\varphi_x(x_t)\,g(x_t\,,t)\,\mathscr{E}\{d\beta_t\}\}$$
$$= 0,$$

upon taking the expectation in (4.155), we get

$$d\mathscr{E}\{\varphi(x_t)\} = \mathscr{E}\{\varphi_x(x_t)\,f(x_t\,,t)\}\,dt + \tfrac{1}{2}q(t)\,\mathscr{E}\{g^2(x_t\,,t)\,\varphi_{xx}(x_t)\}\,dt. \quad (4.156)$$

Now let $\varphi = x_t$. Then we immediately get

$$d\hat{x}_t/dt = \hat{f}(x_t\,,t), \quad (4.157)$$

where

$$\widehat{(\cdot)} \triangleq \mathscr{E}\{(\cdot)\}.$$

Now letting $\varphi = x_t{}^2$,

$$d\mathscr{E}\{x_t{}^2\} = 2\widehat{x_t f}(x_t\,,t)\,dt + q(t)\,\widehat{g^2}(x_t\,,t)\,dt. \quad (4.158)$$

But, using (4.157),

$$d(\hat{x}_t{}^2) = 2\hat{x}_t\hat{f}(x_t\,,t)\,dt,$$

and combining this with (4.158),

$$dP_t/dt = d\mathscr{E}\{x_t{}^2\}/dt - d(\hat{x}_t{}^2)/dt$$
$$= 2[\widehat{x_t f}(x_t\,,t) - \hat{x}_t\hat{f}(x_t\,,t)] + q(t)\,\widehat{g^2}(x_t\,,t).$$

To summarize, $\hat{x}_t$ and P_t satisfy

$$d\hat{x}_t/dt = \hat{f}(x_t\,,t),$$
$$dP_t/dt = 2[\widehat{x_t f}(x_t\,,t) - \hat{x}_t\hat{f}(x_t\,,t)] + q(t)\,\widehat{g^2}(x_t\,,t). \quad (4.159)$$

Equations (4.159) are not ordinary differential equations. To emphasize this, we write

$$d\hat{x}_t/dt = \int f(x,t)\,p(x,t)\,dx,$$

which in general depends on all the moments of x_t. We note that, if

(4.154) is linear, $f = F(t)x_t$, and g a function of t alone, then (4.159) reduce to

$$d\hat{x}_t/dt = F(t)\hat{x}_t,$$

$$dP_t/dt = 2F(t)[\mathscr{E}\{x_t^2\} - \hat{x}_t^2] + q(t)g^2(t),$$

which is just the result in the linear case (4.138). ▲

Example 4.20. We now derive (4.159) using Kolmogorov's forward equation (4.119). Actually, we derive (4.156), and then (4.159) follow as in Example 4.19.

By definition,

$$\mathscr{E}\{\varphi(x_t)\} = \int \varphi(x)\, p(x, t)\, dx,$$

and therefore

$$d\mathscr{E}\{\varphi(x_t)\}/dt = \int \varphi(x)\frac{\partial p}{\partial t}\, dx. \qquad (4.160)$$

Using Kolmogorov's forward equation, (4.160) becomes

$$d\mathscr{E}\{\varphi(x_t)\}/dt = -\int \varphi(x)\frac{\partial(pf)}{\partial x}\, dx$$

$$+ \tfrac{1}{2}q(t)\int \varphi(x)\frac{\partial^2(pg^2)}{\partial x^2}\, dx. \qquad (4.161)$$

Evaluating these integrals by parts, and assuming p is well-behaved at infinity, we get

$$d\mathscr{E}\{\varphi(x_t)\}/dt = \int \varphi_x(x)f(x, t)p(x, t)\, dx + \tfrac{1}{2}q(t)\int \varphi_{xx}(x)g^2(x, t)p(x, t)\, dx,$$

which is just (4.156). ▲

Just as we did in Example 4.18 in the linear case, we may derive (4.159) by a limiting operation from the analogous result for the discrete nonlinear system that we derived in Section 3.9. The details of the development are left for the reader.

Example 4.21. If (4.154) is interpreted in the Stratonovich sense, then the analysis in Example 4.20, but using (4.124) instead of (4.119) as Kolmogorov's equation, leads to

$$d\mathscr{E}\{\varphi(x_t)\}/dt = \mathscr{E}\{\varphi_x f\} + \tfrac{1}{2}q[\mathscr{E}\{\varphi_{xx}g^2\} + \mathscr{E}\{\varphi_x g g_x\}].$$ (4.162)

Then, instead of (4.159), we get

$$d\hat{x}_t/dt = \hat{f} + \tfrac{1}{2}q(t)\widehat{gg_x},$$

$$dP_t/dt = 2(\widehat{x_t f} - \hat{x}_t \hat{f}) + q(t)(\widehat{g^2} + [x_t g g_x]\hat{}). \quad \blacktriangle$$ (4.163)

As was already noted in Example 4.19, the equations we derived for $\hat{x}_t$ and P_t (4.159) require the knowledge of $p(x, t)$ for their evaluation. We may obtain *approximate* ordinary differential equations for $\hat{x}_t$ and P_t by making certain assumptions about $p(x, t)$. As in Section 3.9, we assume that $p(x, t)$ is symmetric and "close to the mean" (P_t small). Thus, we might neglect third- and higher-order central moments of $p(x, t)$ in evaluating the expectations on the right-hand sides of Eqs. (4.159). The right-hand sides of (4.159) will then depend only on $\hat{x}_t$ and P_t, and these equations can then be solved. As in Section 3.9, we expand in Taylor's series the functions on the right-hand sides of (4.159) and then take their expectation. Using our assumptions about $p(x, t)$, we get the following approximate equations for $\hat{x}_t$ and P_t:

$$d\hat{x}_t/dt = f(\hat{x}_t, t) + (P_t/2)f_{xx}(\hat{x}_t, t),$$

$$dP_t/dt = 2P_t f_x(\hat{x}_t, t)$$
$$+ q(t)[g^2(\hat{x}_t, t) + P_t g_x^2(\hat{x}_t, t) + P_t g(\hat{x}_t, t) g_{xx}(\hat{x}_t, t)].$$ (4.164)

The comments in footnote 17 of Section 3.9 and the comments at the end of that section with respect to our approximation apply here as well. We also note that equations of evolution for higher-order moments can easily be obtained from (4.156). These higher-order moments can then be retained in the approximate equations.

10. DISCUSSION

In this book we shall use Itô's stochastic integral, and all stochastic differential equations (except for isolated examples) will be understood in the Itô sense. Our reasons for using the Itô integral instead of the Stratonovich integral are the following.

It is much easier to compute expectations of the Itô integral than the Stratonovich integral. Theorem 4.1 does not hold for the Stratonovich integral. The Itô integral (see Doob [6, Chapter IX]) has other nice mathematical properties not possessed by the Stratonovich integral. As a result, most theoretical work is more conveniently done in the Itô framework. In fact, stochastic stability theory (see Kushner [14]) owes its existence to these properties of the Itô integral. Virtually all the theoretical work in stability and control (see Kushner [13]) for stochastic systems is done using the Itô integral, which provides us with another important reason for using the Itô integral here. The background provided the readers of this book will make this theoretical work more accessible for them. Most important of all is the fact that the Itô integral is defined for a much broader class of functions than the Stratonovich integral. The Stratonovich integral is so restrictive that *it is not applicable to nonlinear filtering theory* (see footnote 3 in Chapter 6).

One might argue that the Stratonovich integral should be used because it is simpler, since it can be manipulated by the formal rules. This is intuitively appealing but beside the point. The Stratonovich integral does not offer any new mathematical insight or content. As a matter of fact, all the results concerning the Stratonovich integral are proved using Itô's theory. Of itself, the Stratonovich integral does not offer any additional physical insight. It does, under certain conditions, more directly model a physical process (Section 8).

We have given (in Sections 6 and 9) transformations from Itô to Stratonovich. For every Itô equation, there is an equivalent Stratonovich equation, and conversely. Thus one and the same process can be modeled via either equation. The transformations we have given enable the interested reader to translate our results to the Stratonovich framework.

REFERENCES

1. K. J. Astrom, On a First Order Stochastic Differential Equation, *Intern. J. Control* 1, 301–326 (1965).
2. A. T. Bharucha-Reid, "Elements of the Theory of Markov Processes and Their Applications." McGraw-Hill, New York, 1960.
3. J. L. Bogdanoff and F. Kozin, Moments of the Output of Linear Random Systems, *J. Acoust. Soc. Am.* 34, 1063–1066 (1962).
4. J. M. C. Clark, The Representation of Nonlinear Stochastic Systems with Applications to Filtering. Ph. D. Thesis, Electrical Engr. Dept., Imperial College, London, England, 1966.
5. I. G. Cummings, Derivation of the Moments of a Continuous Stochastic System, *Intern. J. Control* 5, 85–90 (1967).
6. J. L. Doob, "Stochastic Processes." Wiley, New York, 1953.
7. E. B. Dynkin, "Markov Processes." Academic Press, New York, 1965.
8. A. H. Gray, Jr. and T. K. Caughey, A Controversy in Problems Involving Random Parameter Excitation, *J. Math. and Phys.* 44, 288–296 (1965).

9. K. Itô, Stochastic Integral, *Proc. Imp. Acad. Tokyo* **20**, 519–524 (1944).
10. K. Itô, On Stochastic Differential Equations, *Mem. Amer. Math. Soc.* **4** (1951).
11. K. Itô, Lectures on Stochastic Processes (mimeographed notes), Tata Inst. Fundamental Research, Bombay, India, 1961.
12. N. K. Kulman, A Note on the Differential Equations of Conditional Probability Density Functions. *J. Math. Anal. Appl.* **14**, 301–308 (1966).
13. H. J. Kushner, On the Status of Optimal Control and Stability for Stochastic Systems, *IEEE Intern. Conv. Record, Part 6* **14**, 143–151 (1966).
14. H. J. Kushner, "Stochastic Stability and Control." Academic Press, New York, 1967.
15. A. V. Skorokhod, "Studies in the Theory of Random Processes." Addison-Wesley, Reading, Massachusetts, 1965.
16. R. L. Stratonovich, "Topics in the Theory of Random Noise." Gordon and Breach, New York, 1963.
17. R. L. Stratonovich, A New Form of Representing Stochastic Integrals and Equations, *J. SIAM Control* **4**, 362–371 (1966).
18. N. Wiener, Generalized Harmonic Analysis, *Acta Math.* **55**, 117–258 (1930).
19. E. Wong and M. Zakai, On the Relation Between Ordinary and Stochastic Differential Equations, *Internat. J. Engrg. Sci.* **3**, 213–229 (1965).
20. E. Wong and M. Zakai, On the Convergence of Ordinary Integrals to Stochastic Integrals, *Ann. Math. Stat.* **36**, 1560–1564 (1965).
21. E. Wong and M. Zakai, On Convergence of the Solutions of Differential Equations Involving Brownian Motion. Rept. No. 65-5, AF-AFOSR-139-64, Electronics Research Lab, Univ. of California, Berkeley, 1965.

5

Introduction to Filtering Theory

1. INTRODUCTION

The problem of estimating the state of a stochastic dynamical system from noisy observations taken on the state is of central importance in engineering. Interest in this problem dates back almost two centuries to the work of Gauss. Gauss was interested in determining the orbital elements of a celestial body from (many) observations and developed the technique that is known today as *least squares*. More recently, the names of Wiener and Kalman are associated with advances in estimation theory. Estimation theory had its beginnings in problems of space (astronomy, celestial mechanics), and its most recent advances are associated with modern aerospace problems.

It is not our purpose to give a historical account of the development of estimation theory. We shall outline the major mathematical methods used in estimation and credit (hopefully accurately) the development of these methods. First, however, we define the estimation problem in the context of the mathematical model we use in this book.

Consider the discrete stochastic dynamical system (Section 3.9) described by the stochastic vector difference equation

$$x_{k+1} = \varphi(x_k, t_{k+1}, t_k) + \Gamma(x_k, t_k)\, w_{k+1}, \qquad k = 0, 1, \ldots, \qquad (5.1)$$

where the state at t_k is x_k, an n-vector, φ is an n-vector function, Γ is

142

$n \times r$, and $\{w_k, k = 1,...\}$ is an r-vector, white Gaussian sequence, $w_k \sim N(0, Q_k)$. The distribution of the initial condition x_0 is assumed given, and x_0 is independent of $\{w_k\}$. We saw in Section 3.9 that the random sequence generated by (5.1) is Markov. Let discrete, noisy, m-vector observations (measurements) y_k be given by

$$y_k = h(x_k, t_k) + v_k, \qquad k = 1, 2,..., \qquad (5.2)$$

where h is an m-vector function and $\{v_k, k = 1,...\}$ is an m-vector, white Gaussian sequence, $v_k \sim N(0, R_k)$, $R_k > 0$. For simplicity, $\{w_k\}$ and $\{v_k\}$ are assumed independent, and $\{w_k\}$ is independent of x_0. We note that the joint $\{x_k, y_k\}$ process is Markov.

Now let Y_l be the sequence of observations

$$Y_l = \{y_1 ,..., y_l\}. \qquad (5.3)$$

Given a realization of the sequence of observations $\{y_1 ,..., y_l\}$, that is, given Y_l, the **discrete estimation** problem consists of computing an estimate of x_k based on Y_l. If $k < l$, the problem is called the **discrete smoothing** problem; if $k = l$, it is called the **discrete filtering** problem; and if $k > l$, it is called the **discrete prediction** problem.

The filtering and prediction problems are usually associated with real-time operations, in which estimates are required on the basis of observations or data available now. In a *post mortem* (after the fact) analysis, it is possible to wait for more observations to accumulate. In that case, the estimate can be improved by smoothing.

Now consider the continuous stochastic dynamical system (Section 4.1) described by the vector (Itô) stochastic differential equation (Section 4.4),

$$dx_t = f(x_t, t)\, dt + G(x_t, t)\, d\beta_t, \qquad t \geqslant t_0, \qquad (5.4)$$

where x_t and f are n-vectors, G is $n \times r$, and $\{\beta_t, t \geqslant t_0\}$ is an r-vector Brownian motion process with $\mathscr{E}\{d\beta_t\, d\beta_t^{\mathrm{T}}\} = Q(t)\, dt$. [Equation (5.4) is formally equivalent to

$$dx_t/dt = f(x_t, t) + G(x_t, t)\, w_t,$$

where $\{w_t, t \geqslant t_0\}$ is a white Gaussian process, $w_t \sim N(0, Q(t))$. See Chapter 4, particularly Sections 4.1, 4.8, and 4.10.] The properties of solutions of (5.4) are summarized in Theorem 4.5. Suppose continuous observations are taken on system (5.4), of the form

$$dz_t = h(x_t, t)\, dt + d\eta_t, \qquad t \geqslant t_0, \qquad (5.5)$$

where z_t and h are m-vectors and $\{\eta_t, t \geqslant t_0\}$ is an m-vector Brownian motion process with $\mathscr{E}\{d\eta_t \, d\eta_t^T\} = R(t) \, dt, R(t) > 0$. We suppose that $\{\beta_t\}$, $\{\eta_t\}$, and x_{t_0} are independent. [Equation (5.5) is formally equivalent to

$$y_t = h(x_t, t) + v_t,$$

with the identifications

$$y_t \sim dz_t/dt, \qquad v_t \sim d\eta_t/dt,$$

and with $\{v_t, t \geqslant t_0\}$ a white Gaussian noise process, $v_t \sim N(0, R(t))$.] Equations (5.4) and (5.5), taken together, constitute a vector (Itô) stochastic differential equation.

Suppose a realization

$$Y_\tau = \{z_s, t_0 \leqslant s \leqslant \tau\} \tag{5.6}$$

is given. The problem of estimating x_t, given Y_τ (based on Y_τ), is the **continuous estimation** problem. If $t < \tau$, the problem is called the **continuous smoothing** problem; if $t = \tau$, it is the **continuous filtering** problem; and if $t > \tau$, it is called the **continuous prediction** problem.

In addition to the discrete and continuous problems, we shall also study the mixed **continuous-discrete estimation** problem. In this problem, the state x_t evolves according to the stochastic differential equation (5.4), and observations (5.2) are taken at discrete time instants. The smoothing, filtering, and prediction problems are defined in an analogous way.

Important special cases of the problems we have defined result when the dynamics and observations are *linear*. In the discrete problem, this means that the function φ is linear in x_k :

$$\varphi(x_k, t_{k+1}, t_k) = \Phi(t_{k+1}, t_k) x_k$$

(Φ an $n \times n$ matrix), the matrix function Γ is independent of x_k, and the measurement function h is linear in x_k :

$$h(x_k, t_k) = M(t_k) x_k$$

(M an $m \times n$ matrix). Similarly, in the continuous problem,

$$f(x_t, t) = F(t) x_t$$

(F an $n \times n$ matrix), G is independent of x_t, and

$$h(x_t, t) = M(t) x_t.$$

In the linear problems, the stochastic processes involved are Gaussian (see Sections 3.9 and 4.4) as well as Markov, and as a result the linear problems admit a particularly tractable solution.

Discrete (and continuous-discrete) estimation theory has its beginnings in the work of Gauss, which we have already mentioned. Differential correction techniques, based on the idea of least squares, have been used widely in orbit determination. The interested reader can find an account of this aspect of estimation theory in Deutsch [16]. This earlier work does not place the estimation problem in a probabilistic framework as we have done. Rather, the problem is looked upon as a deterministic problem of minimizing errors. This will be further pursued in Section 3.

Continuous estimation theory begins with Wiener [42]. Wiener solves the linear, stationary problem (F, G, M, Q, and R constant) using frequency domain techniques. His model is a probabilistic one, like ours, but the stochastic processes involved are not modeled by stochastic differential equations. The continuous filtering problem is reduced to the solution of an integral equation (the Wiener–Hopf equation). Wiener [43] gives a general, practical method of solution in the linear-stationary case, but the nonlinear and nonstationary problems remain essentially unsolved. See the text by Laning and Battin [24], for example, for further details and additional bibliography. See also Parzen [29] for an excellent summary and a new approach to Wiener's problem.

The general linear (nonstationary) filtering and prediction problem is essentially completely solved in the pioneering work of Kalman [19], Kalman and Bucy [20], and Kalman [21]. The parallel work of Stratonovich [37, 38] and Kushner [22, 23] provides the basis for subsequent developments in nonlinear filtering and prediction theory. These authors adopt the probabilistic approach that is introduced in Section 2, in which the problem is modeled via stochastic difference and differential equations, as we have done in the foregoing. We pursue this approach in Section 2.

This book deals almost exclusively with the problems of filtering and prediction. Smoothing, for linear problems, is briefly treated in Examples 7.8 and 7.16. Additional material on smoothing may be found in Bryson and Frazier [7], Rauch [31], Rauch et al. [32], Cox [12, 13], Friedland and Bernstein [17], and Meditch [26].

2. PROBABILISTIC APPROACH

Consider, for concreteness, the continuous filtering problem defined in the preceding section. It is clear that the conditional probability density function of x_t given Y_t, which we write as

$$p(x, t \mid Y_t), \tag{5.7}$$

is the complete solution of the filtering problem. This is simply because $p(x, t \mid Y_t)$ embodies all the statistical information about x_t which is contained in the available observations *and* in the initial condition $p(x, t_0)$. We are here adopting the so-called *Bayesian* point of view in that we are taking into account *a priori* (initial) data or information about x_t which is contained in $p(x, t_0)$. (See the definition of a conditional density function in Section 2.5, and also Section 3.5. The conditional density is defined in terms of the *joint* density.) Similarly, the complete solution of the continuous prediction problem is $(t > \tau)$

$$p(x, t \mid Y_\tau). \tag{5.8}$$

Analogous statements hold for the discrete and continuous-discrete filtering and prediction problems.

In the case of linear filtering and prediction problems, the densities in (5.7) and (5.8) are Gaussian (see Sections 3.9 and 4.4 and Theorem 2.13). As a result, they are characterized by their (respective) mean vectors and covariance matrices. It is then a relatively simple matter to compute these conditional densities. In the nonlinear case, the situation is vastly more difficult. In general, there does not exist a finite set of parameters which characterizes these densities. We might say that in the linear case the filter state is finite, consisting of the mean and covariance matrix. In the nonlinear case, the filter state is infinite—the whole function $p(x, t \mid Y_t)$.

With the densities (5.7) and (5.8) in hand, there still remains the question of what the estimate of the state should be. Several obvious possibilities suggest themselves, as for example, the mean, the mode (peak), the median, and so on. We need a criterion that will permit us to compare various possible estimates, and to choose the best (optimal) one.

Let $\hat{x}_t$ be an estimate of x_t given Y_τ, $t \geqslant \tau$. Then the error in the estimate is simply

$$\tilde{x}_t \equiv x_t - \hat{x}_t. \tag{5.9}$$

A good estimate should clearly have the property of producing small errors, or rather a small statistical measure of the errors. We proceed to define such a measure.

Let $\rho(\xi)$ be a real-valued, nonnegative, convex[1] function of the real

---

[1] $\rho(\xi)$ is *convex* if "straight-line interpolation" overestimates ρ. More precisely, at every ξ^0,

$$\lambda\rho(\xi) + (1 - \lambda)\rho(\xi^0) \geqslant \rho(\lambda\xi + (1 - \lambda)\xi^0),$$

$0 \leqslant \lambda \leqslant 1$, for all ξ.

n-vector ξ. We define a *loss* or criterion function $L(\tilde{x}_t)$ as any real-valued function with the properties

$$L(0) = 0, \qquad \rho(\xi^2) \geqslant \rho(\xi^1) \geqslant 0 \Rightarrow L(\xi^2) \geqslant L(\xi^1) \geqslant 0. \qquad (5.10)$$

As an example, an admissible function ρ is

$$\rho(\xi) = |\xi| = (\xi^T\xi)^{1/2}. \qquad (5.11)$$

ρ clearly measures the distance from the origin, and, in view of (5.10), the loss function is nondecreasing with this distance. Note that L itself can, but need not be, convex. In the scalar case, a subclass of loss functions with properties (5.10) is the class of symmetric functions

$$L(\xi) = L(-\xi) \geqslant 0, \qquad \xi^2 \geqslant \xi^1 \geqslant 0 \Rightarrow L(\xi^2) \geqslant L(\xi^1). \qquad (5.12)$$

Take the ρ in (5.11). Clearly,

$$L(\xi) = \xi^T S \xi, \qquad (5.13)$$

where the $n \times n$ (symmetric) matrix S is positive semidefinite, has properties (5.10).

Having defined a reasonable class of loss functions, we seek that estimate $\hat{x}_t$ of x_t which minimizes the average or *expected loss*

$$\mathscr{E}\{L(\tilde{x}_t)\}.$$

For a certain class of density functions $p(x, t \mid Y_\tau)$, the answer is contained in the following theorem due to Sherman [34], which we state here without proof. See also Sherman [35] for a discussion of loss functions (5.12).

Theorem 5.1 (Sherman). *Let x be a random vector with mean μ and density function $p(x)$. Let $L(\tilde{x})$, $\tilde{x} = x - \hat{x}$, be of class (5.10). If $p(x)$ is symmetric about μ, and unimodal,*[2] *then $\hat{x} = \mu$ minimizes $\mathscr{E}\{L(\tilde{x})\}$.*

Now the expected loss can be written ($t \geqslant \tau$) as

$$\mathscr{E}\{L(\tilde{x}_t)\} = \mathscr{E}\{\mathscr{E}\{L(\tilde{x}_t)\mid Y_\tau\}\}. \qquad (5.14)$$

Since $\mathscr{E}\{L(\tilde{x}_t)\mid Y_\tau\}$ depends only on Y_τ (is independent of $\hat{x}_t$), $\mathscr{E}\{L(\tilde{x}_t)\}$

---

[2] $p(x)$ is *unimodal* if it has only one "peak." More precisely, if $p(x)$ is symmetric about μ, then $p(x)$ is unimodal if the distribution function $F(x)$ is convex for $x_i \leqslant \mu_i$.

is minimized by minimizing $\mathscr{E}\{L(\tilde{x}_t)|\ Y_\tau\}$. Applying Theorem 5.1 with $p(x,\ t\ |\ Y_\tau)$, we have

Theorem 5.2. *Let $p(x,\ t\ |\ Y_\tau)$ have mean $\hat{x}_t{}^\tau \triangleq \mathscr{E}\{x_t\ |\ Y_\tau\}$. Suppose $p(x,\ t\ |\ Y_\tau)$ is symmetric about $\hat{x}_t{}^\tau$ and unimodal. Let $L(\tilde{x}_t)$ be a class (5.10). Then the optimal estimate of x_t in the continuous, discrete, and continuous-discrete filtering and prediction problems, in the sense of minimizing $\mathscr{E}\{L(\tilde{x}_t)\}$, is the conditional mean $\hat{x}_t{}^\tau$.*

The hypotheses of Theorem 5.2 are clearly satisfied if $p(x,\ t\ |\ Y_\tau)$ is Gaussian. We therefore have

Corollary. *Let $L(\tilde{x}_t)$ be of class (5.10). Then the optimal estimate of x_t in linear filtering and prediction problems (in the sense of minimizing $\mathscr{E}\{L(\tilde{x}_t)\}$) is the conditional mean $\hat{x}_t{}^\tau \triangleq \mathscr{E}\{x_t\ |\ Y_\tau\}$.*

Examples of loss functions of class (5.10) include

$$L_1(\xi) = \xi^{\mathsf{T}}S\xi, \tag{5.13}$$

$$L_2(\xi) = \sum_i c_i\ |\ \xi_i\ |, \qquad c_i > 0,[3] \tag{5.15}$$

$$L_3(\xi) = \begin{cases} \xi^{\mathsf{T}}S\xi & \xi^{\mathsf{T}}S\xi \leqslant c \\ c, & \xi^{\mathsf{T}}S\xi > c,[3] \end{cases} \tag{5.16}$$

and, in the scalar case [class (5.12)],

$$L_4(\xi) = c\xi^4, \qquad c > 0,[4] \tag{5.17}$$

$$L_5(\xi) = c(1 - e^{-\xi^2}), \qquad c > 0,[4] \tag{5.18}$$

and so on. Additional loss functions are sketched in Fig. 5.1. Three of them are of class (5.12) (symmetric), whereas one is asymmetric.[5]

We see that the conditional mean provides the minimum expected loss for a large class of filtering and prediction problems, and for many loss functions. Clearly, if $p(x,\ t\ |\ Y_\tau)$ is available, then the optimal estimate can be computed for any loss function.

A particularly tractable loss function is the quadratic form (5.13)

$$L_1(\xi) = \xi^{\mathsf{T}}S\xi, \tag{5.19}$$

---

[3] From Cox [12].

[4] From Kalman [19].

[5] Brown [6] shows that, in the case of scalar, Gaussian processes, the conditional mean is also optimal for asymmetric loss functions that are (i) nonnegative, (ii) nondecreasing for $\xi \geqslant 0$, and (iii) nonincreasing for $\xi \leqslant 0$.

where $S \geqslant 0$. The estimate that minimizes $\mathscr{E}\{L_1(\tilde{x}_t)\}$ is called (for obvious reasons) the **minimum variance** or **minimum mean square error** estimate. It turns out that the conditional mean is the minimum variance estimate for *all* filtering and prediction problems, regardless of the properties of the conditional density function.

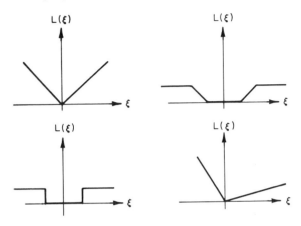

FIG. 5.1.   Loss functions of class (5.10).

Theorem 5.3. *Let the estimate be a functional on Y_τ. Then the minimum variance estimate is the conditional mean.*

PROOF: Write

$$\mathscr{E}\{L_1\} = \mathscr{E}\{\mathscr{E}\{(x_t - \mu_t + \mu_t - \hat{x}_t)^{\mathrm{T}} S(x_t - \mu_t + \mu_t - \hat{x}_t)|\ Y_\tau\}\}, \quad (5.20)$$

and let

$$\mu_t = \mathscr{E}\{x_t \mid Y_\tau\}.$$

Since

$$\mathscr{E}\{(x_t - \mu_t)^{\mathrm{T}} S(\mu_t - \hat{x}_t)|\ Y_\tau\} = (\mu_t - \hat{x}_t)^{\mathrm{T}} S\mathscr{E}\{(x_t - \mu_t)|\ Y_\tau\}$$

$$= (\mu_t - \hat{x}_t)^{\mathrm{T}} S(\mu_t - \mu_t) = 0$$

(μ_t and the estimate $\hat{x}_t$ are constant with respect to expectation conditioned on Y_τ), and since

$$\mathscr{E}\{(x_t - \mu_t)^{\mathrm{T}} S(x_t - \mu_t)\}$$

is independent of $\hat{x}_t$, (5.20) becomes

$$\mathscr{E}\{L_1\} = \mathrm{const} + \mathscr{E}\{(\mu_t - \hat{x}_t)^{\mathrm{T}} S(\mu_t - \hat{x}_t)\}. \quad (5.21)$$

Since S is positive semidefinite, (5.21) is clearly minimized by setting $\hat{x}_t = \mu_t$. ■

We note that the conditional mean is an *unbiased estimate*, that is,

$$\mathscr{E}\{(x_t - \hat{x}_t^\tau)\} = \mathscr{E}\{x_t\} - \mathscr{E}\{\mathscr{E}\{x_t \mid Y_\tau\}\} = 0.$$

From the results of this section, it is clear that the first order of business in solving filtering and prediction problems is the determination of the conditional probability density function $p(x, t \mid Y_\tau)$. This is essentially what we mean by the probabilistic approach. With the conditional density in hand, the optimal estimate can be computed for any loss function. However, in view of Theorem 5.2, and especially Theorem 5.3, our second order of business will be the determination of the conditional mean. This is the minimum variance estimate, and our concern will be primarily with this estimate. What we seek are equations of evolution of the conditional density and the conditional mean. We have already developed such equations in the absence of observations. (See Section 3.9 in the discrete case and Section 4.9 in the continuous case.) These equations will constitute the optimal filter and/or predictor.

We will find that the conditional mean, which is the first moment of the conditional density function, in general depends on the higher-order moments. Of course, in linear problems only second moments are involved. These are contained in the *conditional covariance matrix*

$$P_t^\tau \triangleq \mathscr{E}\{(x_t - \hat{x}_t^\tau)(x_t - \hat{x}_t^\tau)^{\mathrm{T}} \mid Y_\tau\},$$

and we will have to develop an equation of evolution for this matrix. Since

$$\mathscr{E}\{(x_t - \hat{x}_t^\tau)^{\mathrm{T}} (x_t - \hat{x}_t^\tau)\} = \mathrm{tr}\, \mathscr{E}\{P_t^\tau\},$$

P_t^τ provides a measure of the goodness of the estimate. It can be argued that knowledge of P_t^τ is just as important as knowing the estimate $\hat{x}_t^\tau$ itself. An estimate is meaningless unless one knows how good it is.

3. STATISTICAL METHODS

Consider the discrete estimation problem defined in Section 1. Instead of regarding w_k and v_k in (5.1) and (5.2) as random inputs with well-defined statistics, consider them simply as errors of unknown character. Equation (5.1) is no longer a stochastic difference equation, but rather an ordinary difference equation that could be solved if the errors were known. w_k might represent a deterministic error in modeling the dynamical system via the difference equation (5.1). Suppose some *a priori* estimate of x_0, call it $\bar{x}_0$, is available. Suppose noisy observations $\{y_1, ..., y_N\}$ are given by (5.2). We wish to determine (estimate) $\{x_0, ..., x_N\}$ so that, simultaneously, the errors in the dynamical system and in the observations are small. Roughly speaking, we want to pass the solution of (5.1), as closely as possible, through the observations.

Taking the classical *least squares* approach, it is appropriate to minimize, with respect to $\{x_0, ..., x_N ; w_1, ..., w_N\}$, the function

$$J_N = \frac{1}{2}(x_0 - \bar{x}_0)^T P_0^{-1}(x_0 - \bar{x}_0) + \frac{1}{2}\sum_{k=1}^{N}(y_k - h(x_k, t_k))^T R_k^{-1}(y_k - h(x_k, t_k))$$

$$+ \frac{1}{2}\sum_{k=1}^{N} w_k^T Q_k^{-1} w_k, \tag{5.22}$$

subject to the constraints

$$x_{k+1} = \varphi(x_k, t_{k+1}, t_k) + \Gamma(x_k, t_k) w_{k+1}, \qquad k = 0, 1, ..., N - 1. \tag{5.23}$$

The appearance of the first term in J_N reflects our belief in the prior estimate. The positive definite matrices P_0^{-1}, R_k^{-1}, and Q_k^{-1} are here to be regarded as *weighting* matrices, which are quantitative measures of our belief in the prior estimate, the observation model, and the dynamical system model, respectively. In practice, they are "fiddling parameters" that help us obtain a good least squares fit.

Similar considerations in the continuous estimation problem lead to the minimization, with respect to $\{x_\tau, t_0 \leqslant \tau \leqslant t\}$ and $\{w_\tau, t_0 \leqslant \tau \leqslant t\}$, of the functional

$$J_t = \frac{1}{2}(x_{t_0} - \bar{x}_{t_0})^T P_{t_0}^{-1}(x_{t_0} - \bar{x}_{t_0})$$

$$+ \frac{1}{2}\int_{t_0}^{t}(y_\tau - h(x_\tau, \tau))^T R^{-1}(\tau)(y_\tau - h(x_\tau, \tau))\, d\tau$$

$$+ \frac{1}{2}\int_{t_0}^{t} w_\tau^T Q^{-1}(\tau)\, w_\tau\, d\tau, \tag{5.24}$$

subject to the constraint

$$dx_\tau/d\tau = f(x_\tau, \tau) + G(x_\tau, \tau) w_\tau, \qquad t_0 \leqslant \tau \leqslant t. \tag{5.25}$$

To be specific, consider the discrete problem in (5.22) and (5.23). The solution of that problem, the minimizing sequence $\{\bar{x}_0, ..., \bar{x}_N\}$, is the desired smoothing solution. $\bar{x}_N$ is the filtering solution. Prediction (to t_M, say) can be accomplished by adding

$$\frac{1}{2} \sum_{k=N+1}^{M-1} w_k^T Q_k^{-1} w_k \qquad (5.26)$$

to J_N. It is easy to see that the minimizing sequence $\{w_{N+1}, ..., w_{M-1}\}$ is

$$w_k \equiv 0, \qquad k = N+1, ..., M-1.$$

As a result, prediction is accomplished by assuming an errorless dynamical system:

$$y_{k+1} = \varphi(y_k, t_{k+1}, t_k); \qquad k = N, N+1, ...; \qquad y_N = \bar{x}_N. \qquad (5.27)$$

Now the minimization of J_N obviously cannot be accomplished until all observations $\{y_1, ..., y_N\}$ have been collected. When the additional observation y_{N+1} is obtained, a new problem, J_{N+1}, has to be solved. Proceeding this way, it soon becomes painfully clear that observations accumulate *ad infinitum*. Furthermore, the computations involved in solving the least squares minimization problem grow with N.

One way to avoid this growth in memory and computations is to separate the observations into batches,

$$\{y_1, ..., y_N\}, \{y_{N+1}, ..., y_{2N}\}, ...,$$

and to solve the least squares problem separately for each batch of observations. This procedure is sometimes called *batch processing*. Information from one batch can be carried over to the next via the first term in J_N. That is, information about x_0 from previous batches can be considered as *a priori* information for the current batch. This is clearly, in general, an *ad hoc* procedure and is not equivalent to processing all the data simultaneously.

Another procedure, called *recursive least squares*, consists of solving the problem J_{N+1} in terms of the observation y_{N+1} and $\bar{x}_N$ (the filtering solution for problem J_N). The goal is to obtain a difference equation for the estimate $\bar{x}_N$ with the current observation as a forcing term. This would lead to a filter similar in structure to that described in Section 2. It turns out that a recursive least squares solution can be obtained for linear problems. Approximate recursive solutions are available in the nonlinear case. The recursive least squares approach was actually inspired by probabilistic results that automatically produce an equation of evolution for the estimate (the conditional mean). In fact, much of the recent least squares work did nothing more than rederive the probabilistic results (perhaps in an attempt to understand them). As a result, much of the least squares work contributes very little to estimation theory. Notable exceptions include work in smoothing. Smoothing solutions were first obtained via the least squares approach.

The least squares approach that we developed above has no probabilistic meaning. We simply set out to minimize errors in a deterministic problem. By reinterpreting the errors $\{w_k\}$ and $\{v_k\}$ as random sequences with statistics defined in Section 1, we shall see that the discrete least squares problem in (5.22), (5.23) is equivalent to maximizing the conditional probability density function

$$p(x_0, ..., x_N \mid y_1, ..., y_N) \tag{5.28}$$

with respect to $\{x_0, ..., x_N\}$, provided Γ in (5.1) is a time function $\Gamma(t_k)$, independent of x_k.[6] A similar interpretation will be made in the continuous problem. Our development follows Cox [12, 13].

Using Bayes' rule, we write the density in (5.28) as

$$p(x_0, ..., x_N \mid y_1, ..., y_N) = \frac{p(y_1, ..., y_N \mid x_0, ..., x_N)\, p(x_0, ..., x_N)}{p(y_1, ..., y_N)}. \tag{5.29}$$

Note that only the numerator in (5.29) depends on $\{x_0, ..., x_N\}$. Now with $\{x_0, ..., x_N\}$ given, $\{y_1, ..., y_N\}$ [see (5.2)] are independent, since the $\{v_k\}$ sequence is white. Therefore, applying Theorem 2.7, we get

$$p(y_1, ..., y_N \mid x_0, ..., x_N) = \prod_{k=1}^{N} p_{v_k}(y_k - h(x_k, t_k)). \tag{5.30}$$

Since the sequence generated by (5.1) is Markov, we have

$$p(x_0, ..., x_N) = p(x_0) \prod_{k=1}^{N} p(x_k \mid x_{k-1})$$

$$= p(x_0) \prod_{k=1}^{N} p_{\Gamma w_k}(x_k - \varphi(x_{k-1}, t_k, t_{k-1})). \tag{5.31}$$

---

[6] See Friedland and Bernstein [17] for the case in which Γ is state dependent. We note that Bryson and Frazier [7] and Cox [13] contain an error in that the state dependence of Γ (G in the continuous problem) is there neglected.

The last equality in (5.31) follows by Theorem 2.7 in view of the fact that the $\{w_k\}$ sequence is white. Collecting these results, we have

$$p(x_0,...,x_N \mid y_1,...,y_N) = cp(x_0) \prod_{k=1}^{N} p_{v_k}(y_k - h(x_k, t_k))$$

$$\times \prod_{k=1}^{N} p_{\Gamma w_k}(x_k - \varphi(x_{k-1}, t_k, t_{k-1})), \qquad (5.32)$$

where c is a constant independent of $\{x_0,...,x_N\}$.

Let us assume that the initial condition $x_0 \sim N(\bar{x}_0, P_0)$, $P_0 > 0$. Recall that $v_k \sim N(0, R_k)$, $R_k > 0$, and $w_k \sim N(0, Q_k)$, $Q_k \geqslant 0$, so that

$$\Gamma w_k \sim N(0, \Gamma Q_k \Gamma^{\mathrm{T}}).$$

We assume for simplicity (see Section 3.9) that $\Gamma Q_k \Gamma^{\mathrm{T}} > 0$ and $Q_k > 0$; both Γ and Q_k are $n \times n$ and nonsingular. Then (5.32) becomes

$$p(x_0,...,x_N \mid y_1,...,y_N)$$

$$= c' \exp\left\{-\tfrac{1}{2}(x_0 - \bar{x}_0)^{\mathrm{T}} P_0^{-1}(x_0 - \bar{x}_0)\right.$$

$$-\frac{1}{2}\sum_{k=1}^{N}(y_k - h(x_k, t_k))^{\mathrm{T}} R_k^{-1}(y_k - h(x_k, t_k))$$

$$\left.-\frac{1}{2}\sum_{k=1}^{N}(x_k - \varphi(x_{k-1}, t_k, t_{k-1}))^{\mathrm{T}} (\Gamma Q_k \Gamma^{\mathrm{T}})^{-1} (x_k - \varphi(x_{k-1}, t_k, t_{k-1}))\right\},$$

$$(5.33)$$

where c' is a constant independent of $\{x_0,...,x_N\}$. Maximizing (5.33) with respect to $\{x_0,...,x_N\}$ is equivalent to minimizing

$$\frac{1}{2}(x_0 - \bar{x}_0)^{\mathrm{T}} P_0^{-1}(x_0 - \bar{x}_0) + \frac{1}{2}\sum_{k=1}^{N}(y_k - h(x_k, t_k))^{\mathrm{T}} R_k^{-1}(y_k - h(x_k, t_k))$$

$$+\frac{1}{2}\sum_{k=1}^{N}(x_k - \varphi(x_{k-1}, t_k, t_{k-1}))^{\mathrm{T}} (\Gamma Q_k \Gamma^{\mathrm{T}})^{-1} (x_k - \varphi(x_{k-1}, t_k, t_{k-1})),$$

$$(5.34)$$

which, in turn, is equivalent to minimizing J_N (5.22) with respect to $\{x_0,...,x_N; w_1,...,w_N\}$, subject to constraints (5.23).

We now consider the continuous least squares problem (5.24), (5.25).

Our approach is to construct an analog of the conditional probability density function (5.28) in the continuous case. We first discretize (5.25), construct the required probability density functions, and then pass to the limit. Our development is formal.

Assume, for simplicity, that $t_0 = 0$, $Q(t) > 0$, $G(t) Q(t) G^T(t) > 0$; $\underset{\sim}{Q}$ and G are $n \times n$ and nonsingular. Partition $[0, t]$

$$0, \delta\tau, ..., (N-1)\, \delta\tau = t,$$

with $\delta\tau > 0$ and small. Then the discretized dynamics and observations are

$$x_{(k+1)\delta\tau} = x_{k\delta\tau} + f(x_{k\delta\tau}, k\, \delta\tau)\, \delta\tau + G(k\, \delta\tau)\, w_{k\delta\tau}\, \delta\tau,$$

$$y_{k\delta\tau} = h(x_{k\delta\tau}, k\, \delta\tau) + v_{k\delta\tau}, \tag{5.35}$$

where now $w_{k\delta\tau} \sim N(0, Q(k\, \delta\tau)/\delta\tau)$, $v_{k\delta\tau} \sim N(0, R(k\, \delta\tau)/\delta\tau)$ (see Section 3.8, Example 3.20).

In view of (5.29), we have

$$p(x_0, x_{\delta\tau}, ..., x_t \mid y_0, y_{\delta\tau}, ..., y_t) = \frac{p(y_0, ..., y_t \mid x_0, ..., x_t)\, p(x_0, ..., x_t)}{p(y_0, ..., y_t)}. \tag{5.36}$$

Since only the numerator in (5.36) depends on x_τ, we shall be concerned only with maximizing

$$p(y_0, ..., y_t \mid x_0, ..., x_t)\, p(x_0, ..., x_t). \tag{5.37}$$

Now, in view of (5.30),

$$p(y_0, ..., y_t \mid x_0, ..., x_t) = \prod_{k=0}^{N-1} p_{v_{k\delta\tau}}(y_{k\delta\tau} - h(x_{k\delta\tau}, k\, \delta\tau))$$

$$= c_1 \exp\left\{ -\frac{1}{2} \sum_{k=0}^{N-1} (y_{k\delta\tau} - h)^T R^{-1}(k\, \delta\tau)(y_{k\delta\tau} - h)\, \delta\tau \right\}. \tag{5.38}$$

If we attempt to pass to the $\delta\tau = 0 \,(N = \infty)$ limit in (5.38), the normalizing constant c_1 will go to zero. To avoid this difficulty, we utilize the concept of a *probability density functional*.[7] We divide (5.38) by

$$\prod_{k=0}^{N-1} p_{v_{k\delta\tau}}(0) = c_1,$$

---

[7] For a rigorous treatment of this concept, see Parzen [30].

which is independent of x_τ, and *define* the probability density functional

$$p[y_\tau, \tau \in [0, t] | x_\tau, \tau \in [0, t]]$$

$$\triangleq \lim_{\substack{\delta\tau \to 0 \\ N \to \infty}} \frac{p(y_0, ..., y_t | x_0, ..., x_t)}{c_1}$$

$$= \exp\left\{-\frac{1}{2} \int_0^t (y_\tau - h(x_\tau, \tau))^{\mathrm{T}} R^{-1}(\tau)(y_\tau - h(x_\tau, \tau))\, d\tau\right\}. \quad (5.39)$$

In a completely analogous way, we get

$$p[x_\tau, \tau \in [0, t]] = p(x_0) \exp\left\{-\frac{1}{2} \int_0^t (dx_\tau/d\tau - f(x_\tau, \tau))^{\mathrm{T}} (GQG^{\mathrm{T}})^{-1}\right.$$

$$\left. \times (dx_\tau/d\tau - f(x_\tau, \tau))\, d\tau\right\}. \quad (5.40)$$

It is easy to see that maximization of the product

$$p[y_\tau, \tau \in [0, t] | x_\tau, \tau \in [0, t]]\, p[x_\tau, \tau \in [0, t]]$$

with respect to $x_\tau, 0 \leqslant \tau \leqslant t$ is equivalent to the continuous least squares problem (5.24), (5.25).

Estimation based on the maximization of the conditional probability density function (5.28) may be called *joint maximum likelihood (Bayesian) estimation*. The estimate is the peak or *mode* of the joint conditional density (5.28), and as such is called the maximum likelihood or most probable estimate. It is obtained by maximizing the *joint* density function (5.28) rather than the marginal density

$$p(x_N | y_1, ..., y_N). \quad (5.41)$$

The *maximum likelihood (Bayesian) estimate* x_N^* obtained by maximizing the marginal density (5.41) is, in general, not the same as the *joint* maximum likelihood (Bayesian) estimate $\bar{x}_N$ obtained by maximizing (5.28). The optimizing sequence $\{\bar{x}_0, ..., \bar{x}_N\}$ is sometimes called the *modal trajectory*. x_N^* does not, in general, lie on the modal trajectory. (It does in linear problems.) The estimate is *Bayesian*, since it is obtained by maximizing an *a posteriori* density that includes the prior data $p(x_0)$ via Bayes' rule. A non-Bayesian joint maximum likelihood estimate may be obtained by maximizing

$$p(y_1, ..., y_N | x_0, ..., x_N). \quad (5.42)$$

Compare this with (5.29).

Consider the maximum likelihood (Bayesian) estimate obtained by maximizing (5.41) or, in the continuous case, by maximizing

$$p(x, t \mid Y_t). \tag{5.43}$$

If these densities are Gaussian (linear problems), or just symmetric and unimodal, then their (respective) modes coincide with their (respective) means. In that case, the *maximum likelihood (Bayesian) estimate is the same as the minimum variance estimate* defined in Section 2. We note that maximum likelihood estimation is of questionable value unless the density function is unimodal and concentrated near the mode.

Many investigators have studied the least squares estimation problem in the form in which we define it, and in other related and equivalent forms. The objective of most of these studies is the development of recursive least squares filters and predictors. Some of these authors give the probabilistic interpretation of least squares which we have outlined. It is impossible to develop all these works here. Furthermore, they are not central to the theme of this book, which is the probabilistic approach to filtering and prediction. As was already noted, much of this work simply rederives probabilistic results. In the next paragraphs, we cite some of the recent least squares work. The interested reader may go to the references.

Perhaps the earliest work in discrete linear recursive least squares is that of Swerling [39]. His work predates Kalman's [19] probabilistic results, but considers only the case in which the dynamical system is noise free ($w_k \equiv 0$). Following the publication of Kalman's work, a number of authors rederive his (linear) results. These include Ho [18], Battin [2], Schmidt [33], and Mowery [28]. Swerling [40, 41] gives a modern treatment of least squares estimation. A theoretical treatment of recursive least squares is available in Albert and Sittler [1]. Cox [11, 12] studies the nonlinear discrete problem via dynamic programming and variational techniques, rederives the linear results, and gives an approximate solution for the nonlinear problem. He also gives the linear smoothing solution. Rauch *et al.* [32] rederive Kalman's result by maximizing (5.41). They also give the linear smoothing solution. Cox [14] proposes a dynamic programming formulation for a minimax problem (minimizing maximum errors, rather than their squares). Larson and Peschon [25] formulate a general nonlinear problem with non-Gaussian errors via dynamic programming. The last problem is also studied by Friedland and Bernstein [17] via variational techniques.

Kalman and Bucy's [20] continuous (linear) filtering and prediction results are rederived by Bryson and Frazier [7] using variational techniques. They also give the linear smoothing solution. Cox [13] studies the nonlinear continuous problem via dynamic programming, rederives the linear results, and gives an approximate solution for the nonlinear problem. He also gives the linear smoothing solution, as do Rauch *et al.* [32]. The continuous nonlinear least squares estimation problem is also studied by Bellman *et al.* [3, 4] using the quasi-linearization technique. That same problem is also studied by Bellman *et al.* [5], and by Detchmendy and Sridhar [15], who, using the technique of invariant imbedding, give an approximate solution for the filtering problem. Such an approximate solution is also obtained by Friedland and Bernstein [17] using variational techniques.

The statistical methods, without a probabilistic interpretation, do not involve stochastic differential equations, and as a result are conceptually and theoretically simpler than the probabilistic approach. There is no need to make any assumptions about the noise; the problem is one of minimizing deterministic errors. On the other hand, results obtained via statistical methods are more difficult to interpret. Weighting matrices (in the least squares problems) are arbitrary, and the meaning and significance of the results are not clear. Once a probabilistic interpretation of the statistical methods is made, these methods are formal approaches to the probabilistic problem, without a rigorous mathematical basis.

4. FOREWORD AND DISCUSSION

This book develops the probabilistic approach to filtering and prediction problems. Some of the statistical methods will be pursued in examples, but are of secondary interest here. The smoothing problem is not treated in detail. Only the linear smoothers are derived in Examples 7.8 and 7.16, and the interested reader can consult the references cited in Section 1 for additional material on smoothing.

There are several ways in which our definitions of the estimation problems can be generalized. We have assumed that the system noise (w) and measurement noise (v) are independent. More generally, these could be assumed correlated. This case is treated by Kalman [21], Cox [13], Friedland and Bernstein [17], and Kushner [23].[8] We have also assumed that all the noise inputs are white (uncorrelated in time). The case of colored (time-correlated) system noise can often be reduced to the white noise case by state-augmentation (Section 4.8), and therefore our white noise assumption represents little loss of generality. See

---

[8] Kushner's paper contains a small algebraic error in terms involving this correlation.

Section 4.8 on modeling, and also Section 4.10. A similar procedure in case of colored measurement noise leads to a singular covariance matrix (R), which is also the case when some of the observations are noise-free. This presents no particular theoretical difficulties in discrete problems (see Kalman [21]), but is essential in the continuous case. In this connection, additional results for discrete problems are contained in Cox [12] and Bryson and Henrikson [9]. For the continuous case, see Bryson and Johansen [8], Cox [13], Mehra and Bryson [27], and Bucy [10]. Some of the generalizations we have noted, although they do not appear in the main body of the book, will be treated in examples.

Finally, we note that our work is restricted to dynamical systems in which the state is a continuous or discrete parameter, continuous state space (Markov) process. Nonlinear filtering results for continuous parameter Markov chains are available in Wonham [44] and Shiryaev [36].

REFERENCES

1. A. Albert and R. W. Sittler, A Method for Computing Least Squares Estimators that Keep up with the Data, *SIAM J. Control* 3, 384–417 (1966).
2. R. H. Battin, "Astronautical Guidance," pp. 303–340. McGraw-Hill, New York, 1964.
3. R. Bellman, H. Kagiwada, and R. Kalaba, Orbit Determination as a Multi-Point Boundary-Value Problem and Quasilinearization, *Proc. Natl. Acad. Sci.* 48, 1327–1329 (1962).
4. R. Bellman, H. Kagiwada, and R. Kalaba, Quasilinearization, System Identification and Prediction, The RAND Corporation, Santa Monica, California, Memorandum RM-3812-PR, August 1963.
5. R. E. Bellman, H. H. Kagiwada, R. E. Kalaba, and R. Sridhar, Invariant Imbedding and Nonlinear Filtering Theory, *J. Astronautical Sci.* 13, 110–115 (1966).
6. J. L. Brown, Asymmetric Non-Mean-Square Error Criteria, *IRE Trans. Automatic Control* 7, 64–66 (1962).
7. A. E. Bryson and M. Frazier, Smoothing for Linear and Nonlinear Dynamic Systems, *Proc. Optimum Systems Synthesis Conf.*, U.S. Air Force Tech. Rept. ASD-TDR-063-119, February 1963.
8. A. E. Bryson and D. E. Johansen, Linear Filtering for Time-Varying Systems Using Measurements Containing Colored Noise, *IEEE Trans. Automatic Control* 10, 4–10 (1965).
9. A. E. Bryson and L. J. Henrikson, Estimation Using Sampled-Data Containing Sequentially Correlated Noise, *AIAA Guidance, Control, Flight Dynamics Conf.*, Huntsville, Alabama. Paper 67-541, 1967.
10. R. S. Bucy, Optimal Filtering for Correlated Noise, *J. Math. Anal. Appl.* 20, 1–8 (1967).
11. H. Cox, On Estimation of State Variables and Parameters, *Joint AIAA-IMS-SIAM-ONR Symp.*, *Control and System Optimization*, Monterey, California, January 1964.

12. H. Cox, On the Estimation of State Variables and Parameters for Noisy Dynamic Systems, *IEEE Trans. Automatic Control* 9, 5–12 (1964).

13. H. Cox, Estimation of State Variables via Dynamic Programming, *Proc. 1964 Joint Automatic Control Conf.* pp. 376–381, Stanford, California, 1964.

14. H. Cox, Sequential Minimax Estimation, *IEEE Trans. Automatic Control* 11, 323–324 (1966).

15. D. M. Detchmendy and R. Sridhar, Sequential Estimation of States and Parameters in Noisy Non-Linear Dynamical Systems, *Proc. 1965 Joint Automatic Control Conf.*, pp. 56–63, Troy, New York, June 1965.

16. R. Deutsch, "Estimation Theory." Prentice-Hall, Englewood Cliffs, New Jersey, 1965.

17. B. Friedland and I. Bernstein, Estimation of the State of a Nonlinear Process in the Presence of Nongaussian Noise and Disturbances, *J. Franklin Inst.* 281, 455–480 (1966).

18. Y. C. Ho, The Method of Least Squares and Optimal Filtering Theory, The RAND Corporation, Santa Monica, California, RM-3329-PR, 1962.

19. R. E. Kalman, A New Approach to Linear Filtering and Prediction Problems, *Trans. ASME, Ser. D: J. Basic Eng.* 82, 35–45 (1960).

20. R. E. Kalman and R. S. Bucy, New Results in Linear Filtering and Prediction Theory, *Trans. ASME, Ser. D: J. Basic Eng.* 83, 95–108 (1961).

21. R. E. Kalman, New Methods in Wiener Filtering Theory, in *Proc. Symp. Appl. Random Function Theory and Probability, 1st.* (J. L. Bogdanoff and F. Kozin, eds.). Wiley, New York, 1963.

22. H. J. Kushner, On the Dynamical Equations of Conditional Probability Density Functions, with Applications to Optimal Stochastic Control Theory, *J. Math. Anal. Appl.* 8, 332–344 (1964).

23. H. J. Kushner, On the Differential Equations Satisfied by Conditional Probability Densities of Markov Processes, *SIAM J. Control* 2, 106–119 (1964).

24. J. H. Laning and R. H. Battin, "Random Processes in Automatic Control." McGraw-Hill, New York, 1956.

25. R. E. Larson and J. Peschon, A Dynamic Programming Approach to Trajectory Estimation, *IEEE Trans. Automatic Control* 11, 537–540 (1966).

26. J. S. Meditch. Orthogonal Projection and Discrete Optimal Linear Smoothing, *SIAM J. Control* 5, 74–89 (1967).

27. R. K. Mehra and A. E. Bryson, Smoothing for Time-Varying Systems Using Measurements Containing Colored Noise, Harvard Univ., Division of Eng. and Applied Phys., Cambridge, Massachusetts, Technical Report No. 1, June 1967.

28. V. O. Mowery, Least Squares Recursive Differential-Correction Estimation in Nonlinear Problems, *IEEE Trans. Automatic Control* 10, 399–407 (1965).

29. E. Parzen, An Approach to Time Series Analysis, *Ann. Math. Statist.* 32, 951–989 (1961).

30. E. Parzen, Extraction and Detection Problems and Reproducing Kernel Hilbert Spaces, *SIAM J. Control* 1, 35–62 (1962).

31. H. E. Rauch, Solutions to the Linear Smoothing Problem, *IEEE Trans. Automatic Control* 8, 371–372 (1963).

32. H. E. Rauch, F. Tung, and C. T. Striebel, Maximum Likelihood Estimates of Linear Dynamic Systems, *AIAA J.* 3, 1445–1450 (1965).

33. S. F. Schmidt, Application of State-Space Methods to Navigation Problems, *Advances in Control Systems* 3. Academic Press, New York, 1966.

34. S. Sherman, A Theorem on Convex Sets with Applications, *Ann. Math. Statist.* 26, 763–767 (1955).

35. S. Sherman, Non-Mean-Square Error Criteria, *IRE Trans. Inform. Theory* **4**, 125–126 (1958).

36. A. N. Shiryaev, On Stochastic Equations in the Theory of Conditional Markov Processes, *Theor. Probability Appl.* **11** (1966).

37. R. L. Stratonovich, On the Theory of Optimal Nonlinear Filtration of Random Functions, *Theor. Probability Appl.* **4**, 223–225 (1959).

38. R. L. Stratonovich, Conditional Markov Processes, *Theor. Probability Appl.* **5**, 156–178 (1960).

39. P. Swerling, First Order Error Propagation in a Stagewise Smoothing Procedure for Satellite Observations, *J. Astronautical Sci.* **6**, 46–52 (1959).

40. P. Swerling, Parameter Estimation Accuracy Formulas, *IEEE Trans. Inform. Theory* **10**, 302–314 (1964).

41. P. Swerling, Topics in Generalized Least Squares Signal Estimation, *SIAM J. Appl. Math.* **14**, 998–1031 (1966).

42. N. Wiener, "The Extrapolation, Interpolation and Smoothing of Stationary Time Series." Wiley, New York, 1949.

43. N. Wiener, "Nonlinear Problems in Random Theory." Wiley, New York, 1958.

44. W. M. Wonham, Some Applications of Stochastic Differential Equations to Optimal Nonlinear Filtering, *SIAM J. Control* **2**, 347–369 (1965).

6

Nonlinear Filtering Theory

1. INTRODUCTION

In this chapter, we pursue the probabilistic approach to nonlinear filtering and prediction problems. As we concluded in Section 5.2, our approach begins with the determination of the equation of evolution for the conditional probability density function. We then develop equations of evolution for the conditional moments. Our interest lies especially in the conditional mean, which is the minimum variance estimate (Theorem 5.3). In addition, we obtain the equation of evolution for the conditional mode, which is the maximum likelihood (Bayesian) estimate (Section 5.3). Finally, the concept of limiting the memory (conditioning) of the filter is introduced, and a limited memory filter is developed.

The probabilistic approach to nonlinear filtering was pioneered by Stratonovich [13], and independently by Kushner [8, 9]. Much of the subsequent work in continuous filtering[1] is due to Kushner [10, 11]. Bucy [1] rederives Kushner's [9] result using stochastic calculus. This approach to continuous filtering is also taken by Wonham [14], Kashyap

---

[1] A new approach to continuous estimation has recently been developed by P. A. Frost [Nonlinear Estimation in Continuous Time Systems, Stanford Univ. Rept. SU-SEL-68-032, May 1968]. This approach, utilizing innovation processes, facilitates the treatment of nonlinear smoothing as well as filtering and prediction problems.

[7]², and Fisher [2]. The results of Stratonovich [13] and Wonham [14] should be interpreted in the sense of Stratonovich.³ Fisher treats a more general filtering problem than the one we have formulated (Section 5.1), but his analysis is completely formal. Mortensen [12] contains related theoretical work. The probabilistic approach to discrete filtering is developed in Ho and Lee [3]. Jazwinski [4, 5] treats the continuous-discrete problem.

2. CONTINUOUS-DISCRETE FILTERING

Our interest in filtering and prediction problems, in which the dynamics are continuous and observations discrete, stems from our interest in orbit determination. The dynamics of space vehicle trajectories are naturally continuous, and observations are usually taken at discrete time instants. This is especially true in the case of on-board manual navigation. Even in the case of radar tracking, where the observation rates can be quite high, limitations on computing speed and storage often dictate the discrete observation mode. Orbital dynamics can be effectively discretized, at least to first order. In highly nonlinear problems, however, the continuous dynamical model is more convenient.

Referring to Section 5.1, the dynamics will be described by the vector (Itô) stochastic differential equation (5.4):

$$dx_t = f(x_t, t) \, dt + G(x_t, t) \, d\beta_t, \qquad t \geqslant t_0 \,. \tag{6.1}$$

Observations (5.2) y_k are taken at discrete time instants t_k :

$$y_k = h(x_k, t_k) + v_k \,; \qquad k = 1, 2, ...; \qquad t_{k+1} > t_k \geqslant t_0 \,. \tag{6.2}$$

The dimensions of the vectors and matrices appearing in (6.1) and (6.2), and the noise statistics, are defined in Section 5.1. The a priori data, $p(x_{t_0})$, is assumed given. We suppose that (6.1) satisfies the hypotheses

---

² Kashyap's report is apparently out of print and was not personally examined by this author.

³ There is a problem in interpreting the results of Stratonovich [13] and Wonham [14] in the nonlinear case even if their stochastic integral is taken as a Stratonovich integral. This is because the Stratonovich integral (Section 4.6) is defined for explicit functions of η_t (Brownian motion), whereas [see (6.79)] in filtering problems the integrand is a functional on $[\eta_\tau : t_0 \leqslant \tau \leqslant t]$ through $\hat{h}_t$. In essence, the Stratonovich integral is not defined for the class of functions appearing in nonlinear filtering. Nevertheless, the pioneering work of Stratonovich stands as an inspiration to later investigators and should be recognized as such.

of Theorem 4.5, and that the function h is continuous in both arguments and bounded for each t_k wp 1.

Let

$$Y_\tau = \{y_l : t_l \leqslant \tau\}. \tag{6.3}$$

To solve the filtering problem, we must determine the evolution in time t of the conditional probability density function

$$p(x, t \mid Y_t). \tag{6.4}$$

3. EVOLUTION OF THE CONDITIONAL DENSITY

(CONTINUOUS-DISCRETE)

Now in Section 4.9 we have already determined the evolution of the prior density $p(x, t)$, which is clearly the same as the conditional density (6.4) in the absence of observations. Therefore, *between* observations, $p(x, t \mid Y_t)$ satisfies Kolmogorov's forward equation (4.121)

$$dp(x, t \mid Y_t) = \mathscr{L}(p) \, dt, \qquad t_k \leqslant t < t_{k+1}, \tag{6.5}$$

where

$$\mathscr{L}(\cdot) = -\sum_{i=1}^{n} \frac{\partial(\cdot f_i)}{\partial x_i} + \frac{1}{2} \sum_{i,j=1}^{n} \frac{\partial^2(\cdot (GQG^{\mathrm{T}})_{ij})}{\partial x_i \, \partial x_j} \tag{6.6}$$

is the forward diffusion operator. The initial condition (at t_k) for (6.5) is

$$p(x, t_k \mid Y_{t_k}). \tag{6.7}$$

It remains to determine how p changes *at an observation* (at t_k); that is, to determine the relationship between $p(x, t_k \mid Y_{t_k})$ in (6.7) and

$$p(x, t_k \mid Y_{t_k^-}) \equiv p(x, t_k \mid Y_{t_{k-1}}). \tag{6.8}$$

Since

$$p(x, t_k \mid Y_{t_k}) = p(x, t_k \mid y_k, Y_{t_{k-1}}),$$

we have by Bayes' rule

$$p(x, t_k \mid Y_{t_k}) = \frac{p(y_k \mid x_k, Y_{t_{k-1}}) \, p(x, t_k \mid Y_{t_{k-1}})}{p(y_k \mid Y_{t_{k-1}})}. \tag{6.9}$$

Now, since the noise $\{v_k\}$ in (6.2) is white,

$$p(y_k \mid x_k, Y_{t_{k-1}}) = p(y_k \mid x_k).$$

Similarly, it is easy to compute

$$p(y_k \mid Y_{t_{k-1}}) = \int p(y_k \mid x)\, p(x, t_k \mid Y_{t_{k-1}})\, dx.$$

Therefore, Eq. (6.9) becomes

$$p(x, t_k \mid Y_{t_k}) = \frac{p(y_k \mid x)\, p(x, t_k \mid Y_{t_k^-})}{\int p(y_k \mid \xi)\, p(\xi, t_k \mid Y_{t_k^-})\, d\xi}. \tag{6.10}$$

Once $p(y_k \mid x)$ is determined, (6.10) will provide the desired difference equation for the conditional density at an observation. By Theorem 2.7 and (6.2),

$$p(y_k \mid x) = p_{v_k}(y_k - h(x, t_k)), \tag{6.11}$$

and since $v_k \sim N(0, R_k)$,

$$p(y_k \mid x) = (1/(2\pi)^{m/2} \mid R_k \mid^{1/2})$$
$$\times \exp\{-\tfrac{1}{2}[y_k - h(x, t_k)]^{\mathrm{T}} R_k^{-1}[y_k - h(x, t_k)]\}. \tag{6.12}$$

We summarize these results in the following theorem.

Theorem 6.1 (Conditional Density for Continuous-Discrete Problems). *Let system* (6.1) *satisfy the hypotheses of Theorem 4.5. Suppose the prior density $p(x, t)$ for* (6.1) *exists and is once continuously differentiable with respect to t and twice with respect to x. Let h be continuous in both arguments and bounded for each t_k wp 1. Then, between observations, the conditional density $p(x, t \mid Y_t)$* (6.4) *satisfies Kolmogorov's forward equation*

$$dp(x, t \mid Y_t) = \mathscr{L}(p)\, dt, \qquad t_k \leqslant t < t_{k+1}, \qquad p(x, t_0 \mid Y_{t_0}) = p(x_{t_0}), \tag{6.13}$$

where $\mathscr{L}$ (6.6) *is the forward diffusion operator. At an observation (at t_k), the conditional density satisfies the difference equation*

$$p(x, t_k \mid Y_{t_k}) = \frac{p(y_k \mid x)\, p(x, t_k \mid Y_{t_k^-})}{\int p(y_k \mid \xi)\, p(\xi, t_k \mid Y_{t_k^-})\, d\xi}, \tag{6.14}$$

where $p(y_k \mid x)$ is given in (6.12). *Prediction for $t > \tau$, based on Y_τ* (6.3), *is accomplished via Kolmogorov's forward equation*

$$dp(x, t \mid Y_\tau) = \mathscr{L}(p)\, dt, \qquad t > \tau, \tag{6.15}$$

where the initial condition $p(x, \tau \mid Y_\tau)$ is the filtering solution.

Note that the normalizing constant

$$1/(2\pi)^{m/2} \mid R_k \mid^{1/2}$$

in $p(y_k \mid x)$ is independent of x and therefore cancels between the numerator and denominator in (6.14). $(\exp\{-\frac{1}{2}y_k{}^{\mathrm{T}}R_k^{-1}y_k\}$ also cancels.) As a result of this, and since

$$\int p(\xi, t_k \mid Y_{t_k^-}) \, d\xi = 1,$$

if $R_k^{-1} \equiv 0$, then

$$p(x, t_k \mid Y_{t_k}) = p(x, t_k \mid Y_{t_k^-}).$$

That is, if the observations are valueless (or there are no observations), then the conditional density does not change at the observation times. In that case the conditional density is identical to the prior density $p(x, t)$, whose evolution is then governed by Kolmogorov's forward equation (6.13).

One might in fact inquire as to the *value* of observations. Having processed the set of observations

$$\{y_1, ..., y_N\},$$

what have we learned, if anything, about the state x_{t_N} that is not contained in the prior data $p(x_{t_0})$? Such questions are clearly related to the properties of the system (6.1) and to the properties of the observation or measurement function h (6.2). The hypotheses we have imposed on the system and the function h merely guarantee the existence of the conditional probability density function. They say nothing about the actual value of the observations in improving our knowledge of the state.

Such questions as these are quite difficult, and no satisfactory theory is presently available in the general nonlinear problem. We shall return to this subject in Chapter 7, where we define and discuss the concepts of *observability* and *information* for linear filtering problems.

4. EVOLUTION OF MOMENTS (CONTINUOUS-DISCRETE)

Let $\varphi(x)$ be a twice continuously differentiable scalar function of the n-vector x. Define

$$\hat{\varphi}^\tau(x_t) = \mathscr{E}^\tau\{\varphi(x_t)\} \triangleq \mathscr{E}\{\varphi(x_t) \mid Y_\tau\} = \int \varphi(x) \, p(x, t \mid Y_\tau) \, dx, \qquad (6.16)$$

the expectation of φ using the conditional density $p(x, t \mid Y_\tau)$. We determine the evolution of $\hat{\varphi}^t(x_t)$ and $\hat{\varphi}^\tau(x_t)$ $(t > \tau)$ for the continuous-discrete filtering and prediction problems.

Now we have already determined the evolution of $\mathscr{E}\{\varphi(x_t)\}$ for scalar x_t in Examples 4.19 and 4.20, and the generalization to a vector x_t is straightforward. Since between observations both $p(x, t)$ and $p(x, t \mid Y_t)$ satisfy Kolmogorov's forward equation (6.13), the laws of evolution of $\mathscr{E}\{\varphi(x_t)\}$ and $\hat{\varphi}^t(x_t)$ are the same; specifically,

$$d\hat{\varphi}^t(x_t) = \mathscr{E}^t\{\varphi_x{}^T f\}\, dt + \tfrac{1}{2}\, \mathrm{tr}\, \mathscr{E}^t\{GQG^T\varphi_{xx}\}\, dt, \qquad t_k \leqslant t < t_{k+1}. \quad (6.17)$$

φ_x is the gradient, and φ_{xx} is the matrix of second partials. See Lemma 4.2.

We next compute the change in $\hat{\varphi}^t(x_t)$ at an observation (at t_k) using the difference equation (6.14) for the conditional density. Let

$$\hat{\varphi}^{t_k^+}(x_{t_k}) = \mathscr{E}^{t_k^+}\{\varphi(x_{t_k})\} = \int \varphi(x)\, p(x, t_k \mid Y_{t_k})\, dx, \qquad (6.18)$$

$$\hat{\varphi}^{t_k^-}(x_{t_k}) = \mathscr{E}^{t_k^-}\{\varphi(x_{t_k})\} = \int \varphi(x)\, p(x, t_k) \mid Y_{t_k^-})\, dx. \qquad (6.19)$$

Then, multiplying (6.14) by $\varphi(x)$ and integrating over x, we get

$$\hat{\varphi}^{t_k^+}(x_{t_k}) = \frac{\mathscr{E}^{t_k^-}\{\varphi(x_{t_k})\, p(y_k \mid x_{t_k})\}}{\mathscr{E}^{t_k^-}\{p(y_k \mid x_{t_k})\}}. \qquad (6.20)$$

Note that y_k is fixed with respect to the expectation $\mathscr{E}^{t_k^-}$ in (6.20). Thus, the right-hand side of (6.20) is a function of y_k.

These results are summarized in the following lemma.

Lemma 6.1. *Hypotheses of Theorem 6.1. Let $\varphi(x)$ be a twice continuously differentiable scalar function of the n-vector x. Then, between observations $\hat{\varphi}^t(x_t)$ satisfies*

$$d\hat{\varphi}^t(x_t) = \mathscr{E}^t\{\varphi_x{}^T f\}\, dt + \tfrac{1}{2}\, \mathrm{tr}\, \mathscr{E}^t\{GQG^T\varphi_{xx}\}\, dt, \qquad t_k \leqslant t < t_{k+1}. \quad (6.21)$$

At an observation at t_k,

$$\hat{\varphi}^{t_k^+}(x_{t_k}) = \frac{\mathscr{E}^{t_k^-}\{\varphi(x_{t_k})\, p(y_k \mid x_{t_k})\}}{\mathscr{E}^{t_k^-}\{p(y_k \mid x_{t_k})\}}. \qquad (6.22)$$

Fix τ. For $t > \tau$, $\hat{\varphi}^\tau(x_t)$ satisfies (6.21) with $\mathscr{E}^t$ replaced by $\mathscr{E}^\tau$.

From Lemma 6.1, we can determine the evolution of all the moments of the conditional density (6.4) for the continuous-discrete problem. We will be principally interested in the conditional mean

$$\hat{x}_t{}^\tau = \mathscr{E}^\tau\{x_t\} = \mathscr{E}\{x_t \mid Y_\tau\} \tag{6.23}$$

and the conditional covariance matrix

$$P_t{}^\tau = \mathscr{E}\{(x_t - \hat{x}_t{}^\tau)(x_t - \hat{x}_t{}^\tau)^{\mathrm{T}} \mid Y_\tau\} = \mathscr{E}^\tau\{x_t x_t{}^{\mathrm{T}}\} - \hat{x}_t{}^\tau \hat{x}_t{}^{\tau\mathrm{T}}. \tag{6.24}$$

The desired results clearly follow by setting $\varphi(x) = x_i$ and $\varphi(x) = x_i x_j$, respectively, in Lemma 6.1, where x_i is the ith component of the vector x. This was already done in Example 4.19 for (6.21) and x scalar.

Theorem 6.2 (Conditional Mean and Covariance Matrix for Continuous-Discrete Problems). *Hypotheses of Theorem 6.1. Between observations, the conditional mean and covariance matrix satisfy*

$$d\hat{x}_t{}^t/dt = f^t(x_t, t),$$

$$dP_t{}^t/dt = (\mathscr{E}^t\{x_t f^{\mathrm{T}}\} - \hat{x}_t{}^t f^{t\mathrm{T}}) + (\mathscr{E}^t\{f x_t{}^{\mathrm{T}}\} - f^t \hat{x}_t{}^{t\mathrm{T}}) + \mathscr{E}^t\{GQG^{\mathrm{T}}\}, \tag{6.25}$$

$$t_k \leqslant t < t_{k+1},$$

and, at an observation at t_k,

$$\hat{x}_{t_k}^{t_k^+} = \frac{\mathscr{E}^{t_k^-}\{x_{t_k} p(y_k \mid x_{t_k})\}}{\mathscr{E}^{t_k^-}\{p(y_k \mid x_{t_k})\}},$$

$$P_{t_k}^{t_k^+} = \frac{\mathscr{E}^{t_k^-}\{x_{t_k} x_{t_k}^{\mathrm{T}} p(y_k \mid x_{t_k})\}}{\mathscr{E}^{t_k^-}\{p(y_k \mid x_{t_k})\}} - \hat{x}_{t_k}^{t_k^+} \hat{x}_{t_k}^{t_k^{+\mathrm{T}}}. \tag{6.26}$$

$\hat{x}_t{}^\tau$ *and* $P_t{}^\tau$, *fixed* $\tau < t$ *(prediction based on Y_τ), satisfy (6.25) with $\mathscr{E}^t$ replaced by $\mathscr{E}^\tau$.*

As was already noted in Example 4.19, Eqs. (6.25) are not ordinary differential equations. Neither are (6.26) ordinary difference equations. The right-hand sides of (6.25) and (6.26) involve expectations that require the whole conditional density for their evaluation. Thus the first two moments of the conditional density depend on all the other moments. Apparently, in order to obtain a computationally realizable and practical filter and predictor in the general nonlinear case, some approximation must be made. We shall consider such approximations in Chapter 9.

Equations (6.25) were specialized to the linear case in Example 4.19. See also Examples 4.15–4.18. In that case, the right-hand sides of (6.25) depend only on $\hat{x}_t{}^t$ and $P_t{}^t$. In fact, Eqs. (6.25) are uncoupled. Equations (6.26) will be specialized to the linear case in Chapter 7, where linear filters and predictors are developed.

5. EVOLUTION OF THE MODE (CONTINUOUS-DISCRETE)

The mode of the conditional density $p(x, t \mid Y_t)$ is an intuitively appealing estimate of the state x_t because of its maximum likelihood interpretation. According to the terminology introduced in Section 5.3, it is the maximum likelihood (Bayesian) estimate. In this section, we study the evolution of the conditional mode in continuous-discrete problems.

Consider first the evolution of the conditional mode *between* observations. The conditional density satisfies Kolmogorov's forward equation (6.13) of Theorem 6.1. That is,

$$\partial p / \partial t = \mathscr{L}(p). \tag{6.27}$$

If m_t is the mode of $p(x, t \mid Y_t)$, then we must have, by definition,

$$p_x(m, t \mid Y_t) \equiv 0, \tag{6.28}$$

identically in m and t, where p_x is the gradient (column vector) of p with respect to x. Consequently,

$$dp_x = (\partial p_x / \partial t)\, dt + p_{xx}\, dm = 0, \tag{6.29}$$

where p_{xx} is the matrix of second partials and where, henceforth, p and all its derivatives are evaluated at (m, t). We assume that

$$p_{xx}(m, t \mid Y_t) > 0$$

(is positive definite) at m, for all t, so that the mode is well defined, and that all the required partial derivatives of p exist.

Now, using (6.27),

$$\frac{\partial p_x}{\partial t} = \frac{\partial}{\partial x} \frac{\partial p}{\partial t} = (\mathscr{L}(p))_x \, ,$$

and (6.29) produces

$$dm = -p_{xx}^{-1}(\mathscr{L}(p))_x \, dt, \tag{6.30}$$

which is the desired differential equation for the mode.

An explicit evaluation of the right-hand side of (6.30) reveals that it depends on pp_{xx}^{-1}, $p_{x_ix_j}$, and $p_{x_ix_jx_k}$ ($p_x \equiv 0$). If these derivatives were known, then (6.30) could be solved for m. Differential equations for these derivatives can be derived, but these depend on still higher-order derivatives. Apparently, as in the case of the conditional mean, approximations are required in order to obtain a realizable maximum likelihood filter (predictor).

Because in the linear-Gaussian case $-pp_{xx}^{-1}$ is the conditional covariance matrix, we shall derive a differential equation for it. Let

$$\Sigma_t = -pp_{xx}^{-1}. \tag{6.31}$$

Then

$$-d\Sigma_t = -pp_{xx}^{-1}\,dp_{xx}p_{xx}^{-1} + p_{xx}^{-1}\,dp. \tag{6.32}$$

Now

$$dp = p_x{}^{\mathrm{T}}\,dm + p_t\,dt = \mathscr{L}(p)\,dt, \tag{6.33}$$

since $p_x \equiv 0$. Proceeding,

$$(dp_{xx})_{ij} = \sum_k p_{x_ix_jx_k}\,dm_k + (\mathscr{L}(p))_{x_ix_j}\,dt,$$

and, substituting for dm from (6.30),

$$(dp_{xx})_{ij} = -\sum_k p_{x_ix_jx_k}[p_{xx}^{-1}(\mathscr{L}(p))_x]_k\,dt + (\mathscr{L}(p))_{x_ix_j}\,dt. \tag{6.34}$$

Substitution of (6.33) and (6.34) in (6.32) gives the desired equation.

There appears to be no simple means of computing the change in the mode at an observation. According to (6.14) of Theorem 6.1, the mode $m_{t_k^+}$ after the observation is the solution of the maximization problem

$$\max_x p(x, t_k \mid Y_{t_k}) = \max_x (1/c)\,p(y_k \mid x)\,p(x, t_k \mid Y_{t_k^-}), \tag{6.35}$$

where

$$c = \int p(y_k \mid \xi)\,p(\xi, t_k \mid Y_{t_k^-})\,d\xi$$

is independent of x. Therefore,

$$p_x(m_{t_k^+}, t_k \mid Y_{t_k})$$
$$= (1/c)[p_x(y_k \mid m_{t_k^+})\,p(m_{t_k^+}, t_k \mid Y_{t_k^-}) + p(y_k \mid m_{t_k^+})\,p_x(m_{t_k^+}, t_k \mid Y_{t_k^-})] = 0, \tag{6.36}$$

and, carrying out the differentiation of $p(y_k \mid x)$ and introducing the matrix notation

$$M = \left[\frac{\partial h_i}{\partial x_j}\right] \quad (m \times n),$$

we have

$$M^T(m_{t_k^+}) R_k^{-1}(y_k - h(m_{t_k^+}, t_k)) p(m_{t_k^+}, t_k \mid Y_{t_k^-}) + p_x(m_{t_k^+}, t_k \mid Y_{t_k^-}) = 0. \quad (6.37)$$

Equation (6.37) is an n-vector nonlinear equation for $m_{t_k^+}$. It depends on the whole conditional density and its gradient. By differentiating (6.36) again and introducing the abbreviations

$$p(y_k \mid x) = p^v, \qquad p(x, t_k \mid Y_{t_k^-}) = p^x, \quad (6.38)$$

we get

$$\Sigma_{t_k^+}^{-1} = -(1/p^v p^x)(p_{xx}^v p^x + p_x^{\,v} p_x^{xT} + p_x^{\,x} p_x^{vT} + p^v p_{xx}^x). \quad (6.39)$$

Summarizing, we have

Theorem 6.3 (Conditional Mode for Continuous-Discrete Problems). *Hypotheses of Theorem 6.1. Let the conditional density be sufficiently differentiable, and suppose $p_{xx} > 0$ at the mode for all t. Then, between observations, the conditional mode satisfies*

$$dm_t = -p_{xx}^{-1}(\mathscr{L}(p))_x \, dt, \quad (6.40)$$

and $\Sigma_t \, (= -pp_{xx}^{-1})$ satisfies

$$-d\Sigma_t = -pp_{xx}^{-1} \, dp_{xx} p_{xx}^{-1} + p_{xx}^{-1} \, dp, \quad (6.41)$$

where dp and dp_{xx} are given in (6.33) and (6.34). At an observation at t_k, $m_{t_k^+}$ satisfies

$$M^T(m_{t_k^+}) R_k^{-1}(y_k - h(m_{t_k^+}, t_k)) p(m_{t_k^+}, t_k \mid Y_{t_k^-}) + p_x(m_{t_k^+}, t_k \mid Y_{t_k^-}) = 0, \quad (6.42)$$

and

$$\Sigma_{t_k^+}^{-1} = -(1/p^v p^x)(p_{xx}^v p^x + p_x^{\,v} p_x^{xT} + p_x^{\,x} p_x^{vT} + p^v p_{xx}^x), \quad (6.43)$$

where p^v and p^x are defined in (6.38).

Example 6.1. Compute the right-hand sides of (6.40) and (6.41) in the scalar case ($n = m = 1$).

The results are

$$dm_t/dt = f - \Sigma_t f_{xx} - \frac{\sigma^2}{2} \frac{p_{xxx}}{p_{xx}} - \frac{3\sigma_x{}^2}{2} + \frac{1}{2} \Sigma_t \sigma_{xxx}^2 , \tag{6.44}$$

$$d\Sigma_t/dt = 2\Sigma_t f_x - \frac{\sigma^2}{2} \left(1 + \Sigma_t \frac{p_{xxxx}}{p_{xx}} \right)$$

$$+ \Sigma_t \frac{p_{xxx}}{p_{xx}} \left(\Sigma_t f_{xx} + \frac{\sigma^2}{2} \frac{p_{xxx}}{p_{xx}} + \frac{\sigma_x{}^2}{2} - \frac{1}{2} \Sigma_t \sigma_{xxx}^2 \right)$$

$$+ \frac{1}{2} \Sigma_t^2 \left(\sigma_{xxxx}^2 - 2f_{xxx} \right) - \frac{5}{2} \Sigma_t \sigma_{xx}^2 , \tag{6.45}$$

where $\sigma^2 = G Q G^{\mathrm{T}}$ and all the functions are evaluated at (m_t , t). Note that these equations depend on p_{xxx}/p_{xx} and p_{xxxx}/p_{xx}, and that equations for these quantities can be derived, or they can be approximated in some way.

Consider a special case in which p is symmetric $[p_{xxx}(m_t , t \mid Y_t) \equiv 0]$ and there is no system noise $(\sigma^2 = \sigma_x{}^2 = \sigma_{xx}^2 = \sigma_{xxx}^2 = \sigma_{xxxx}^2 = 0)$. Then (6.44) and (6.45) reduce to

$$dm_t/dt = f - \Sigma_t f_{xx} , \tag{6.46}$$

$$d\Sigma_t/dt = 2\Sigma_t f_x - \Sigma_t^2 f_{xxx} . \tag{6.47}$$

Suppose f is quadratic and $f_{xx} > 0$. Then the equation for the conditional mean (4.164) is

$$d\hat{x}_t/dt = f(\hat{x}_t , t) + \tfrac{1}{2} P_t f_{xx}(\hat{x}_t , t), \tag{6.48}$$

where $P_t > 0$ is the conditional variance. (Recall that $\Sigma_t > 0$.) Comparing (6.46) with (6.48), we see that the evolution of the mode and the mean are radically different. For example, with $m_t = \hat{x}_t$ and $f(m_t , t) = 0$, $dm_t < 0$, whereas $d\hat{x}_t > 0$.

Now specialize (6.44) and (6.45) to the linear-Gaussian case;

$$f = F(t) x_t , \quad \sigma_x{}^2 = \sigma_{xx}^2 = \sigma_{xxx}^2 = \sigma_{xxxx}^2 = p_{xxx} = 0, \quad p_{xxxx}/p_{xx} = -3\Sigma_t^{-1}.$$

We have

$$dm_t/dt = F(t) m_t , \tag{6.49}$$

$$d\Sigma_t/dt = 2\Sigma_t F(t) + \sigma^2, \tag{6.50}$$

which are in agreement with the equations for the mean and variance (4.138) previously obtained in Example 4.15. ▲

Example 6.2. In case observations are scalar[4] ($m = 1$),

$$p_x{}^v = p^v M^T(m_{t_k^+}) R_k^{-1}(y_k - h(m_{t_k^+}, t_k))$$

and

$$p_{xx}^v = p^v [M^T M R_k^{-2}(y_k - h)^2 + h_{xx} R_k^{-1}(y_k - h) - M^T M R_k^{-1}],$$

where

$$h_{xx} = \left[\frac{\partial h}{\partial x_i \, \partial x_j} \right] \quad (n \times n),$$

and (6.43) becomes

$$\Sigma_{t_k^+}^{-1} = M^T R_k^{-1} M - M^T M R_k^{-2}(y_k - h)^2 - h_{xx} R_k^{-1}(y_k - h)$$
$$- (1/p^x)[M^T p_x^{x^T} R_k^{-1}(y_k - h) + p_x{}^x M R_k^{-1}(y_k - h) + p_{xx}^x]. \quad (6.51)$$

If, in addition, p^x is assumed Gaussian with mean $m_{t_k^-}$ and covariance matrix $\Sigma_{t_k^-}$, then (6.42) and (6.51) become

$$-M^T(m_{t_k^+}) R_k^{-1}(y_k - h(m_{t_k^+}, t_k)) + \Sigma_{t_k^+}^{-1}(m_{t_k^+} - m_{t_k^-}) = 0, \quad (6.52)$$

$$\Sigma_{t_k^+}^{-1} = \Sigma_{t_k^-}^{-1} + M^T(m_{t_k^+}) R_k^{-1} M(m_{t_k^+}) - h_{xx}(m_{t_k^+}) R_k^{-1}(y_k - h(m_{t_k^+}, t_k)). \quad (6.53)$$

In the linear-Gaussian (in general vector observation) case, $h = M(t_k) x_{t_k}$, these equations reduce to

$$m_{t_k^+} = (\Sigma_{t_k^-}^{-1} + M^T R_k^{-1} M)^{-1} (M^T R_k^{-1} y_k + \Sigma_{t_k^-}^{-1} m_{t_k^-}), \quad (6.54)$$

$$\Sigma_{t_k^+}^{-1} = \Sigma_{t_k^-}^{-1} + M^T R_k^{-1} M. \quad (6.55)$$

Equations (6.49) and (6.50), together with (6.54) and (6.55), constitute the maximum likelihood (Bayesian) filter and predictor in the linear case for continuous-discrete problems. This is clearly the same as the minimum variance filter. In the scalar case ($n = m = 1$), (6.54) and (6.55) can be easily manipulated into the following form:

$$m_{t_k^+} = m_{t_k^-} + \frac{\Sigma_{t_k^-} M}{M^2 \Sigma_{t_k^-} + R}(y_k - M m_{t_k^-}), \quad (6.56)$$

$$\Sigma_{t_k^+} = \Sigma_{t_k^-} - \frac{\Sigma_{t_k^-}^2 M^2}{M^2 \Sigma_{t_k^-} + R}. \quad (6.57)$$

The results of this example will be discussed in some detail in Chapters 7 and 9. ▲

[4] This special case is considered only for notational convenience.

6. DISCRETE FILTERING

Our discrete dynamical system is described by the vector difference equation (5.1),

$$x_{k+1} = \varphi(x_k, t_{k+1}, t_k) + \Gamma(x_k, t_k) w_{k+1}, \qquad k = 0, 1, ..., \qquad (6.58)$$

and the discrete observations are given by (5.2),

$$y_k = h(x_k, t_k) + v_k, \qquad k = 1, ...; \qquad t_{k+1} > t_k. \qquad (6.59)$$

The statistics of the noise inputs w_k and v_k, and the dimensions of the vectors and matrices appearing in the foregoing, are given in Section 5.1. $p(x_0)$ is, of course, given.

Let

$$Y_l = \{y_1, ..., y_l\}. \qquad (6.60)$$

We determine the evolution in k of the conditional densities

$$p(x_k, t_k \mid Y_k)$$

and

$$p(x_k, t_k \mid Y_l),$$

fixed $l \leqslant k$. These densities provide, respectively, solutions to the discrete filtering and prediction problems.

Theorem 6.4 (Conditional Density for Discrete Problems). *Assume that the required densities exist. $p(x_k, t_k \mid Y_k)$ satisfies the integral difference equation*

$$p(x_{k+1}, t_{k+1} \mid Y_{k+1}) = \frac{p(y_{k+1} \mid x_{k+1}) \int p(x_{k+1} \mid x_k) p(x_k, t_k \mid Y_k) \, dx_k}{\iint p(y_{k+1} \mid x_{k+1}) p(x_{k+1} \mid x_k) p(x_k, t_k \mid Y_k) \, dx_k \, dx_{k+1}}, \qquad (6.61)$$

where $p(y_k \mid x_k)$ is given in (6.12), and $p(x_{k+1} \mid x_k)$ is the transition probability density of (6.58). Prediction for $k \geqslant l$ (l fixed) is accomplished via

$$p(x_{k+1}, t_{k+1} \mid Y_l) = \int p(x_{k+1} \mid x_k) p(x_k, t_k \mid Y_l) \, dx_k. \qquad (6.62)$$

PROOF: Actually, these results have already been derived. Equation (6.62) is just the Chapman–Kolmogorov equation (3.100). Equation (6.14) of Theorem 6.1 can be rewritten as

$$p(x_{k+1}, t_{k+1} \mid Y_{k+1}) = \frac{p(y_{k+1} \mid x_{k+1}) p(x_{k+1}, t_{k+1} \mid Y_k)}{\int p(y_{k+1} \mid x_{k+1}) p(x_{k+1}, t_{k+1} \mid Y_k) \, dx_{k+1}}. \qquad (6.63)$$

Setting $l = k$ in (6.62) and substituting into (6.63) gives (6.61). ∎

It remains to specify the transition probability densities $p(x_{k+1} \mid x_k)$. Unfortunately, these cannot be written conveniently in general; the various special cases are discussed in Section 3.9. As a result, we shall not pursue the nonlinear discrete filtering and prediction problem any further, but we shall return to Theorem 6.4 in Chapter 7, where we develop the linear case.

7. CONTINUOUS FILTERING

We now turn to continuous filtering and prediction problems in which the state evolves according to the vector (Itô) stochastic differential equation (5.4)

$$dx_t = f(x_t, t) \, dt + G(x_t, t) \, d\beta_t, \qquad t \geqslant t_0, \qquad (6.64)$$

and the observed process is the solution of the vector Itô equation (5.5)

$$dz_t = h(x_t, t) \, dt + d\eta_t, \qquad t \geqslant t_0. \qquad (6.65)$$

The dimensions of (6.64) and (6.65) are given in Section 5.1. $\{\beta_t, t \geqslant t_0\}$ and $\{\eta_t, t \geqslant t_0\}$ are independent Brownian motion processes with $\mathscr{E}\{d\beta_t \, d\beta_t{}^{\mathrm{T}}\} = Q(t) \, dt$, $\mathscr{E}\{d\eta_t \, d\eta_t{}^{\mathrm{T}}\} = R(t) \, dt$, $R(t) > 0$. $p(x_{t_0})$ is given, and x_{t_0} is independent of $\{\beta_t\}$ and $\{\eta_t\}$.

Let

$$Y_\tau = \{z_s, t_0 \leqslant s \leqslant \tau\}. \qquad (6.66)$$

In this section, we characterize the conditional density

$$p(x, t \mid Y_t). \qquad (6.67)$$

Subsequent sections will be devoted to the evolution of the conditional density, its moments, and its mode.

Partition the interval $[t_0, t]$,

$$t_0 = \tau_0 < \tau_1 < \cdots < \tau_n = t,$$

and let $\rho = \max_i (\tau_{i+1} - \tau_i)$. We shall construct the conditional density

$$p(x, t \mid z_{\tau_0}, ..., z_{\tau_n})$$

and formally *define*

$$p(x, t \mid Y_t) = \underset{\substack{\rho \to 0 \\ n \to \infty}}{\mathrm{l.i.m.}} \, p(x, t \mid z_{\tau_0}, ..., z_{\tau_n}). \qquad (6.68)$$

See Section 3.5, particularly (3.82).

Lemma 6.2. *Let* (6.64) *satisfy the hypotheses of Theorem 4.5, and let the prior density* $p(x, t)$ *be sufficiently differentiable. Let* h *be continuous in both arguments, and*

$$\int_{t_0}^{t} h^{\mathrm{T}}(x_\tau, \tau) R^{-1}(\tau) h(x_\tau, \tau) \, d\tau < \infty \tag{6.69}$$

for every $t > t_0$ *wp 1 and in mean square, so that*

$$\int_{t_0}^{t} \mathscr{E}\{h_\tau^{\mathrm{T}} R^{-1}(\tau) h_\tau\} \, d\tau < \infty \,^5 \tag{6.70}$$

for every $t > t_0$. *Suppose the expectation below exists. Then*

$p(x, t \mid Y_t)$

$$= \frac{\mathscr{E}^t \left\{ \exp\left[-\frac{1}{2} \int_{t_0}^{t} h_\tau^{\mathrm{T}} R^{-1}(\tau) h_\tau \, d\tau + \int_{t_0}^{t} h_\tau^{\mathrm{T}} R^{-1}(\tau) \, dz_\tau \right] \middle| x_t \right\} p(x, t)}{\mathscr{E}^t \left\{ \exp\left[-\frac{1}{2} \int_{t_0}^{t} h_\tau^{\mathrm{T}} R^{-1}(\tau) h_\tau \, d\tau + \int_{t_0}^{t} h_\tau^{\mathrm{T}} R^{-1}(\tau) \, dz_\tau \right] \right\}}, \tag{6.71}$$

where the expectation $\mathscr{E}^t$ *is over the* $\{x_\tau, \tau \in [t_0, t]\}$ *process with* Y_t *fixed.*

The result (6.71), a characterization of the conditional density, is stated by Bucy [1] and proved in Mortensen [12]. We will not prove it here, since the mathematics involved is beyond the scope of this book. For example, the expectation in (6.71) is over functions and requires the tools of function space integration. Rather, using the mean square definition of conditional density (6.68), and considering the expectation in (6.71) as a formal average, we give a formal demonstration of the lemma.

A rigorous proof of (6.71) apparently requires that h be uniformly bounded [12]. This, for example, excludes the linear case which is studied in Chapter 7. However, in Chapter 7, we essentially need only Lemma 6.3 which has been proved by Kushner [10] without this boundedness condition. Furthermore, the continuous linear filter is derived independently in the examples of Chapter 7.

Demonstration. Note that the hypotheses in (6.69) and (6.70) ensure that the integrals in (6.71) exist. The second integral is

$$\int_{t_0}^{t} h_\tau^{\mathrm{T}} R^{-1}(\tau) \, dz_\tau = \int_{t_0}^{t} h_\tau^{\mathrm{T}} R^{-1}(\tau) h_\tau \, d\tau + \int_{t_0}^{t} h_\tau^{\mathrm{T}} R^{-1}(\tau) \, d\eta_\tau,$$

where the last integral is an Itô integral.

[5] (6.69) in mean square implies (6.70) by Theorem 3.7 and its Corollary 1.

Now, by Bayes' rule,

$$p(x, t \mid z_{\tau_0}, ..., z_{\tau_n}) = \frac{p(z_{\tau_0}, ..., z_{\tau_n} \mid x_t) \, p(x, t)}{p(z_{\tau_0}, ..., z_{\tau_n})} . \tag{6.72}$$

Also,

$$p(z_{\tau_0}, ..., z_{\tau_n} \mid x_t) = \mathscr{E}^t\{p(z_{\tau_0}, ..., z_{\tau_n} \mid x_{\tau_0}, ..., x_{\tau_{n-1}}, x_t) \mid x_t\} \tag{6.73}$$

and

$$p(z_{\tau_0}, ..., z_{\tau_n}) = \mathscr{E}^t\{p(z_{\tau_0}, ..., z_{\tau_n} \mid x_t)\}, \tag{6.74}$$

and so we need only compute

$$p(z_{\tau_0}, ..., z_{\tau_n} \mid x_{\tau_0}, ..., x_{\tau_n}).$$

Integrating (6.65), and assuming a fine partition, we have

$$z_{\tau_{k+1}} - z_{\tau_k} = h(x_{\tau_k}, \tau_k)(\tau_{k+1} - \tau_k) + (\eta_{\tau_{k+1}} - \eta_{\tau_k}) + o(\rho). \tag{6.75}$$

We shall henceforth ignore the $o(\rho)$ terms. In view of (6.75),

$$\begin{aligned} p(z_{\tau_0}, ..., z_{\tau_n} \mid x_{\tau_0}, ..., x_{\tau_n}) &= p(z_{\tau_n} \mid z_{\tau_{n-1}}, ..., z_{\tau_0}, x_{\tau_0}, ..., x_{\tau_n}) \\ &\quad \cdot p(z_{\tau_{n-1}}, ..., z_{\tau_0} \mid x_{\tau_0}, ..., x_{\tau_n}) \\ &= p(z_{\tau_n} \mid z_{\tau_{n-1}}, x_{\tau_0}, ..., x_{\tau_n}) \, p(z_{\tau_{n-1}}, ..., z_{\tau_0} \mid x_{\tau_0}, ..., x_{\tau_n}) \\ &\ \vdots \\ &= p(z_{\tau_n} \mid z_{\tau_{n-1}}, x_{\tau_0}, ..., x_{\tau_n}) \, p(z_{\tau_{n-1}} \mid z_{\tau_{n-2}}, x_{\tau_0}, ..., x_{\tau_n}) \\ &\quad \cdots p(z_{\tau_0}), \end{aligned} \tag{6.76}$$

since $\{\eta_\tau\}$ has independent increments. Therefore, using (6.75) and Theorem 2.7, we get

$$p(z_{\tau_0}, ..., z_{\tau_n} \mid x_{\tau_0}, ..., x_{\tau_n})$$

$$= p(z_{\tau_0}) \, c(\rho) \exp\left[-\frac{1}{2} \sum_{k=0}^{n-1} (\tau_{k+1} - \tau_k)^{-1} \{(z_{\tau_{k+1}} - z_{\tau_k}) - h_{\tau_k}(\tau_{k+1} - \tau_k)\}^{\mathrm{T}} \right.$$

$$\left. \times R^{-1}(\tau_k)\{(z_{\tau_{k+1}} - z_{\tau_k}) - h_{\tau_k}(\tau_{k+1} - \tau_k)\} \right]. \tag{6.77}$$

When we substitute (6.77) into (6.73) and (6.74), and subsequently into (6.72), $p(z_{\tau_0})$, $c(\rho)$, and the

$$(z_{\tau_{k+1}} - z_{\tau_k})^{\mathrm{T}} R^{-1}(\tau_k)(z_{\tau_{k+1}} - z_{\tau_k})$$

terms cancel between numerator and denominator, and we get

$$p(x, t \mid z_{\tau_0}, ..., z_{\tau_n})$$

$$= K\mathscr{E}^t \left\{ \exp\left[-\frac{1}{2} \sum_{k=0}^{n-1} h_{\tau_k}^{\mathrm{T}} R^{-1}(\tau_k)\, h_{\tau_k}(\tau_{k+1} - \tau_k) \right.\right.$$

$$\left.\left. + \sum_{k=0}^{n-1} h_{\tau_k}^{\mathrm{T}} R^{-1}(\tau_k)(z_{\tau_{k+1}} - z_{\tau_k}) \right] \middle| x_t \right\} p(x, t), \qquad (6.78)$$

where K is a normalization constant that we shall determine later. Now

$$\underset{\substack{\rho \to 0 \\ n \to \infty}}{\mathrm{l.i.m.}} \sum_{k=0}^{n-1} h_{\tau_k}^{\mathrm{T}} R^{-1}(\tau_k)\, h_{\tau_k}(\tau_{k+1} - \tau_k) = \int_{t_0}^{t} h_\tau^{\mathrm{T}} R^{-1}(\tau)\, h_\tau\, d\tau,$$

and

$$\underset{\substack{\rho \to 0 \\ n \to \infty}}{\mathrm{l.i.m.}} \sum_{k=0}^{n-1} h_{\tau_k}^{\mathrm{T}} R^{-1}(\tau_k)(z_{\tau_{k+1}} - z_{\tau_k}) = \int_{t_0}^{t} h_\tau^{\mathrm{T}} R^{-1}(\tau)\, dz_\tau.$$

Therefore, by Example 3.11,

$$\underset{\substack{\rho \to 0 \\ n \to \infty}}{\mathrm{l.i.m.}}\, p(x, t \mid z_{\tau_0}, ..., z_{\tau_n})$$

$$= K\mathscr{E}^t \left\{ \exp\left[-\frac{1}{2} \int_{t_0}^{t} h_\tau^{\mathrm{T}} R^{-1}(\tau)\, h_\tau\, d\tau + \int_{t_0}^{t} h_\tau^{\mathrm{T}} R^{-1}(\tau)\, dz_\tau \right] \middle| x_t \right\} p(x, t).$$

$1/K$ is clearly the denominator in (6.71), and this completes the demonstration.

8. EVOLUTION OF THE CONDITIONAL DENSITY (CONTINUOUS)

Using Lemma 6.2 and the Itô stochastic calculus (Section 4.5), we now develop the equation of evolution for $p(x, t \mid Y_t)$ (6.67).

Theorem 6.5 (Conditional Density for Continuous Problems). *Hypotheses of Lemma 6.2. Then the conditional density $p(x, t \mid Y_t)$ satisfies Kushner's equation*

$$dp = \mathscr{L}(p)\, dt + (h_t - \hat{h}_t)^{\mathrm{T}} R^{-1}(t)(dz_t - \hat{h}_t\, dt)\, p, \qquad t \geqslant t_0, \quad (6.79)$$

where

$$\hat{h}_t = \mathscr{E}^t\{h(x_t, t)\} = \int h(x, t)\, p(x, t \mid Y_t)\, dx.$$

Equation (6.79) was first obtained by Kushner [8, 9] and later by Bucy [1]. Our derivation follows that of Bucy; Kushner's derivation is developed in Example 6.3. The Stratonovich analog of (6.79) is derived in [13] (see footnote 3 in this chapter).

Kushner's equation is a stochastic partial differential equation for the conditional density. No theory for equations of this type is currently available, and (6.79) can only be interpreted formally. Using (6.79) we shall, in the next two sections, develop equations of evolution for the moments and the mode of p. Fortunately, the equations for the moments have been rigorously established by Kushner [10].

With valueless or no observations [$R^{-1}(t) \equiv 0$], Kushner's equation reduces to Kolmogorov's forward equation (6.13). Thus, as in the continuous-discrete problem, prediction is accomplished via Kolmogorov's forward equation.

PROOF: Defining

$$d\zeta_t = -\tfrac{1}{2}h_t^{\mathrm{T}}R^{-1}(t)\,h_t\,dt + h_t^{\mathrm{T}}R^{-1}(t)\,dz_t$$
$$= \tfrac{1}{2}h_t^{\mathrm{T}}R^{-1}(t)\,h_t\,dt + h_t^{\mathrm{T}}R^{-1}(t)\,d\eta_t\,, \qquad (6.80)$$

Eq. (6.71) of Lemma 6.2 becomes

$$p(x, t \mid Y_t) = \frac{\mathscr{E}^t\{e^{\zeta_t} \mid x_t\}\,p(x, t)}{\mathscr{E}^t\{e^{\zeta_t}\}} \triangleq \frac{Q}{P}\,, \qquad (6.81)$$

where Q and P are the numerator and denominator, respectively. We see that Q is a function of ζ_t and, through $p(x, t)$, a function of t. Applying Itô's Lemma 4.2, we get

$$dQ = Q_t\,dt + Q_\zeta\,d\zeta_t + \tfrac{1}{2}h_t^{\mathrm{T}}R^{-1}(t)\,h_tQ_{\zeta\zeta}\,dt.$$

But

$$Q_t = \mathscr{E}^t\{e^{\zeta_t} \mid x_t\}\,\frac{\partial p(x, t)}{\partial t} = \mathscr{L}(Q),$$

and

$$Q_\zeta = Q_{\zeta\zeta} = Q.$$

Therefore,

$$dQ = \mathscr{L}(Q)\,dt + Qh_t^{\mathrm{T}}R^{-1}(t)\,h_t\,dt + Qh_t^{\mathrm{T}}R^{-1}(t)\,d\eta_t$$
$$= \mathscr{L}(Q)\,dt + Qh_t^{\mathrm{T}}R^{-1}(t)\,dz_t\,. \qquad (6.82)$$

Now

$$P = \int Q\,dx,$$

so that

$$dP = \int (dQ)\, dx = P \int (P^{-1}\, dQ)\, dx$$

$$= P \left[\int \mathscr{L}(QP^{-1})\, dt\, dx + \int QP^{-1}h_t^{\mathrm{T}}R^{-1}(t)\, dz_t\, dx \right]$$

$$= P \left[\left\{ \int \mathscr{L}(QP^{-1})\, dx \right\} dt + \left\{ \int QP^{-1}h_t^{\mathrm{T}}R^{-1}(t)\, dx \right\} dz_t \right]$$

$$= P \left[\mathscr{L} \left\{ \int QP^{-1}\, dx \right\} dt + \left\{ \int QP^{-1}h_t^{\mathrm{T}}R^{-1}(t)\, dx \right\} dz_t \right]$$

$$= P\mathscr{L}(1)\, dt + P\hat{h}_t^{\mathrm{T}}R^{-1}(t)\, dz_t = 0 + P\hat{h}_t^{\mathrm{T}}R^{-1}(t)\, dz_t\,,$$

that is,

$$dP = P\hat{h}_t^{\mathrm{T}}R^{-1}(t)\, dz_t = P\hat{h}_t^{\mathrm{T}}R^{-1}(t)\, h_t\, dt + P\hat{h}_t^{\mathrm{T}}R^{-1}(t)\, d\eta_t\,. \qquad (6.83)$$

Now, in Corollary 3 of Lemma 4.2, let $\varphi = x = Q$ and $\psi = y = P$. Then

$$d(QP^{-1}) = P^{-1}\, dQ - QP^{-2}\, dP - P^{-2}Qh_t^{\mathrm{T}}R^{-1}(t)\, R(t)\, R^{-1}(t)\, \hat{h}_t P\, dt$$

$$+ QP^{-3}P\hat{h}_t^{\mathrm{T}}R^{-1}(t)\, R(t)\, R^{-1}(t)\, \hat{h}_t P\, dt$$

$$= \mathscr{L}(QP^{-1})\, dt + (QP^{-1})\, h_t^{\mathrm{T}}R^{-1}(t)\, dz_t - (QP^{-1})\, \hat{h}_t^{\mathrm{T}}R^{-1}(t)\, dz_t$$

$$- (QP^{-1})\, h_t^{\mathrm{T}}R^{-1}(t)\, \hat{h}_t\, dt + (QP^{-1})\, \hat{h}_t^{\mathrm{T}}R^{-1}(t)\, \hat{h}_t\, dt$$

$$= \mathscr{L}(QP^{-1})\, dt + (QP^{-1})(h_t - \hat{h}_t)^{\mathrm{T}}\, R^{-1}(t)(dz_t - \hat{h}_t\, dt),$$

which is (6.79). ∎

Example 6.3. Instead of using the characterization of the conditional density in Lemma 6.2 and Itô's calculus, we compute dp via an expansion procedure similar to that employed in Section 4.2. Terms involving $d\eta_t\, d\eta_t^{\mathrm{T}}$ are replaced by their expectation (see the demonstration of Lemma 4.2).

We wish to compute the change in $p(x, t \mid Y_t)$ due to the dynamics and the differential observation dz_t. Now

$$\delta p = p(x, t + \delta t \mid Y_t, \delta z_t) - p(x, t \mid Y_t)$$

$$= p(x, t + \delta t \mid Y_t, \delta z_t) - p(x, t \mid Y_t, \delta z_t) + p(x, t \mid Y_t, \delta z_t) - p(x, t \mid Y_t)$$

$$\triangleq \delta p_d + \delta p_0\,, \qquad (6.84)$$

where δp_d is the change due to the dynamics and δp_0 is the change due to the differential observation δz_t. δp_d and δp_0 are independent, since $\{\beta_t\}$ and $\{\eta_t\}$ are assumed independent. Now dp_d has already been computed, and is given by Kolmogorov's forward equation:

$$dp_d = \mathscr{L}(p)\, dt.$$

It remains to compute dp_0.

By Theorem 6.1,

$$p(x, t \mid Y_t, \delta z_t) = \frac{p(\delta z_t \mid x_t)\, p(x, t \mid Y_t)}{\int p(\delta z_t \mid x_t)\, p(x, t \mid Y_t)\, dx}, \tag{6.85}$$

where

$$p(\delta z_t \mid x_t)$$
$$= C \exp\{(-\tfrac{1}{2}\, \delta t)[\delta z_t - h(x, t)\, \delta t]^{\mathrm{T}}\, R^{-1}(t)[\delta z_t - h(x, t)\, \delta t]\}. \tag{6.86}$$

Substituting (6.86) into (6.85),

$$p(x, t \mid Y_t, \delta z_t)$$
$$= \frac{p(x, t \mid Y_t) \exp[\delta z_t{}^{\mathrm{T}} R^{-1}(t)\, h(x, t) - \tfrac{1}{2} h^{\mathrm{T}}(x, t)\, R^{-1}(t)\, h(x, t)\, \delta t]}{\int [\text{numerator}]\, dx}. \tag{6.87}$$

Define

$$E(\delta t, \delta z_t) = \frac{p(x, t \mid Y_t, \delta z_t)}{p(x, t \mid Y_t)}, \tag{6.88}$$

and expand $E(\delta t, \delta z_t)$ in Taylor series about $(0, 0)$, retaining terms of order δt. Since

$$\mathscr{E}\{\delta z_t\, \delta z_t{}^{\mathrm{T}}\} = \mathscr{E}\{h_t h_t{}^{\mathrm{T}}\}\, \delta t^2 + \mathscr{E}\{\delta\eta_t\, \delta\eta_t{}^{\mathrm{T}}\}$$
$$= R(t)\, \delta t + o(\delta t), \tag{6.89}$$

the expansion must be carried out to second degree in δz_t and to first degree in δt. $\delta z_t\, \delta t = o(\delta t)$ in root mean square, so that the mixed terms need not be retained.

It is easy to compute

$$E(0, 0) = 1,$$
$$E_{\delta t}(0, 0) = -\tfrac{1}{2}[h^{\mathrm{T}} R^{-1} h - (h^{\mathrm{T}} R^{-1} h)\widehat{}], \qquad [\text{see } (6.91)],$$
$$E_{\delta z_t}(0, 0) = R^{-1}(h - \hat{h}),$$
$$E_{\delta z_t, \delta z_t}(0, 0) = R^{-1} h h^{\mathrm{T}} R^{-1} - 2R^{-1} h \hat{h}^{\mathrm{T}} R^{-1} + 2R^{-1} \hat{h} \hat{h}^{\mathrm{T}} R^{-1} - R^{-1}\widehat{h h^{\mathrm{T}}} R^{-1},$$

so that

$$E(\delta t, \delta z_t) = 1 - \tfrac{1}{2}[h^T R^{-1} h - (h^T R^{-1} h)^\wedge]\,\delta t + (h - \hat{h})^T R^{-1}\,\delta z_t$$
$$+ \tfrac{1}{2}\delta z_t^T[R^{-1} h h^T R^{-1} - 2R^{-1} h \hat{h}^T R^{-1} + 2R^{-1} \hat{h} h^T R^{-1}$$
$$- R^{-1}\widehat{h h^T} R^{-1}]\,\delta z_t + r,$$

where

$$\mathscr{E}\{r\} = o(\delta t), \qquad \mathscr{E}\{r^2\} = o(\delta t^2).$$

Replacing terms containing $\delta z_t\,\delta z_t^T$ by their expectations, we get

$$E(\delta t, \delta z_t) = 1 + (h - \hat{h})^T R^{-1}(\delta z_t - \hat{h}\,\delta t) + r,$$

and, in the mean square sense,

$$dp_0 = (h - \hat{h})^T R^{-1}(dz_t - \hat{h}\,dt)\,p. \quad \blacktriangle \qquad (6.90)$$

9. EVOLUTION OF MOMENTS (CONTINUOUS)

Let $\varphi(x)$ be a twice continuously differentiable scalar function of the n-vector x. Using Kushner's equation (6.79), we develop a stochastic differential equation for

$$\hat{\varphi}^t(x_t) = [\varphi(x_t)]^{\wedge t} = \mathscr{E}^t\{\varphi(x_t)\} = \mathscr{E}\{\varphi(x_t)|\ Y_t\}. \qquad (6.91)$$

For convenience of notation, we omit the superscript t and write simply $\hat{\varphi}(x_t)$.

By definition,

$$\hat{\varphi}(x_t) = \int \varphi(x)\,p(x, t \mid Y_t)\,dx,$$

so that

$$d\hat{\varphi} = \int \varphi(x)\,(dp)\,dx$$

$$= \left[\int \varphi(x)\,\mathscr{L}(p)\,dx\right] dt + \int \varphi(x)(h_t - \hat{h}_t)^T R^{-1}(t)(dz_t - \hat{h}_t\,dt)\,p\,dx$$

$$\triangleq I_1 + I_2\,, \qquad (6.92)$$

where we have used (6.79). From Lemma 6.1,

$$I_1 = \widehat{\varphi_x^T f}\,dt + \tfrac{1}{2}\,\mathrm{tr}\,(GQG^T\varphi_{xx})^\wedge\,dt, \qquad (6.93)$$

and we easily evaluate

$$I_2 = (\widehat{\varphi h} - \hat{\varphi}\hat{h})^{\mathrm{T}} R^{-1}(dz_t - \hat{h}\, dt).$$ (6.94)

To summarize, we have

Lemma 6.3. *Hypotheses of Lemma 6.2. Let* $\varphi(x)$ *be a twice continuously differentiable scalar function of the n-vector x. Then* $\hat{\varphi}(x_t) = \mathscr{E}\{\varphi(x_t)|\ Y_t\}$ *satisfies*

$$d\hat{\varphi}(x_t) = [\widehat{\varphi_x^{\mathrm{T}}f} + \tfrac{1}{2}\,\mathrm{tr}\,(GQG^{\mathrm{T}}\varphi_{xx})^{\wedge}]\, dt$$

$$+ (\widehat{\varphi h} - \hat{\varphi}\hat{h})^{\mathrm{T}} R^{-1}(t)(dz_t - \hat{h}\, dt), \qquad t \geqslant t_0.$$ (6.95)

Equation (6.95) has been established only formally, inasmuch as it was implicitly assumed that the conditional density is sufficiently differentiable and well behaved at infinity. Under a multitude of technical conditions, Kushner [10] proves (6.95) rigorously, but we cannot pursue that proof here. Note that, with $R^{-1}(t) \equiv 0$ (valueless or no observations), (6.95) reduces to (6.21) of Lemma 6.1.

Lemma 6.3 can be used to determine (stochastic) differential equations for all the moments of $p(x, t\mid Y_t)$. We now develop the equations for the conditional mean and the conditional covariance matrix. The development is similar to Theorem 6.2, except that here we must use Itô's stochastic calculus.

Setting $\varphi(x) = x_i$, the *i*th component of the vector x, we immediately get the equations for $\hat{x}_t$:

$$d\hat{x}_i = \hat{f}_i\, dt + (\widehat{x_i h} - \hat{x}_i\hat{h})^{\mathrm{T}} R^{-1}(t)\, (dz_t - \hat{h}\, dt), \qquad i = 1,...,n.$$ (6.96)

Now

$$P_t = \mathscr{E}\{(x_t - \hat{x}_t)(x_t - \hat{x}_t)^{\mathrm{T}} \mid Y_t\} = \widehat{x_t x_t^{\mathrm{T}}} - \hat{x}_t\hat{x}_t^{\mathrm{T}},$$

so that

$$dP_t = d(\widehat{x_t x_t^{\mathrm{T}}}) - d(\hat{x}_t\hat{x}_t^{\mathrm{T}}).$$ (6.97)

Setting $\varphi(x) = x_i x_j$ in (6.95), we have

$$d(\widehat{x_i x_j}) = (\widehat{x_i f_j} + \widehat{f_i x_j} + (GQG^{\mathrm{T}})_{ij}^{\wedge})\, dt$$

$$+ (\widehat{x_i x_j h} - \widehat{x_i x_j}\hat{h})^{\mathrm{T}} R^{-1}(t)\, (dz_t - \hat{h}\, dt).$$ (6.98)

Now $\hat{x}_t$ is the solution of the stochastic differential equation (6.96). Therefore, by Itô's Lemma 4.2,

$$d(\hat{x}_i\hat{x}_j) = \hat{x}_i\, d\hat{x}_j + (d\hat{x}_i)\, \hat{x}_j + (\widehat{x_i h} - \hat{x}_i\hat{h})^{\mathrm{T}} R^{-1}(t)(\widehat{hx_j} - \hat{h}\hat{x}_j)\, dt.$$ (6.99)

Substituting from (6.96) for $d\hat{x}_i$, combining (6.99), (6.98), and (6.97), we get the desired equation for dP_t.

Theorem 6.6 (Conditional Mean and Covariance Matrix for Continuous Problems). *Hypotheses of Lemma 6.2. Then the conditional mean $\hat{x}_t$ and conditional covariance matrix P_t satisfy the stochastic differential equations*

$$d\hat{x}_t = \widehat{f(x_t, t)}\, dt + (\widehat{x_t h^T} - \hat{x}_t \hat{h}^T)\, R^{-1}(t)\, (dz_t - \hat{h}\, dt), \tag{6.100}$$

$$(dP_t)_{ij} = [(\widehat{x_i f_j} - \hat{x}_i \hat{f}_j) + (\widehat{f_i x_j} - \hat{f}_i \hat{x}_j) + (GQG^T)_{\widehat{ij}}$$

$$- (\widehat{x_i h} - \hat{x}_i \hat{h})^T\, R^{-1}(t)(\widehat{hx_j} - \hat{h}\hat{x}_j)]\, dt$$

$$+ (\widehat{x_i x_j h} - \widehat{x_i x_j}\hat{h} - \hat{x}_i \widehat{x_j h} - \hat{x}_j \widehat{x_i h} + 2\hat{x}_i \hat{x}_j \hat{h})^T\, R^{-1}(t)\, (dz_t - \hat{h}\, dt). \tag{6.101}$$

The remarks made in Section 6.4 concerning Theorem 6.2 apply to Eqs. (6.100) and (6.101) as well. Approximations to these equations are considered in Chapter 9. They are specialized to the linear problem in Chapter 7.

10. EVOLUTION OF THE MODE (CONTINUOUS)

We develop here the stochastic differential equation for the mode of $p(x, t \mid Y_t)$ (6.67). The derivation is similar in principle to that of Section 5 except that we use Kushner's equation (6.79) instead of Kolmogorov's forward equation. We *assume* that the mode satisfies a stochastic differential equation and use the formal rules of Itô's stochastic calculus. The derivation is based in part on the work of Kushner [11]. We will need the following formal result from Itô's stochastic calculus.

Lemma 6.4. *Suppose that, for a fixed vector m, x_t is the unique solution of the Itô equation*

$$dx_i = f_i(x, m, t)\, dt + \sum_j G_{ij}(x, m, t)\, d\beta_j, \tag{6.102}$$

where $\mathscr{E}\{d\beta_t\, d\beta_t^T\} = Q(t)\, dt$, and where all the functions are assumed to be sufficiently differentiable with respect to all arguments. Now, if m_t is the unique solution of the Itô equation

$$dm_i = \varphi_i(x, m, t)\, dt + \sum_j \Gamma_{ij}(x, m, t)\, d\beta_j, \tag{6.103}$$

then dx_t (6.102) is the partial (time) differential. The total differential of x_t is given by

$$dx_i = \left[f_i + \sum_j (x_i)_{m_j} \varphi_j + \sum_{j,k,l} (G_{ik})_{m_j} \Gamma_{jl} Q_{kl} + \frac{1}{2} \sum_{j,k} (\Gamma Q \Gamma^T)_{jk} (x_i)_{m_j m_k} \right] dt$$

$$+ \sum_j G_{ij} d\beta_j + \sum_{j,k} (x_i)_{m_j} \Gamma_{jk} d\beta_k . \tag{6.104}$$

In this notation, $(x_i)_{m_j}$ is the partial derivative of x_i with respect to m_j, and so on. $(x_i)_m$ will be the gradient (column) vector of x_i with respect to m, $(x_i)_{mm}$ will be the matrix of second partials, and x_m is the matrix whose rows are the gradients with respect to m. x_i and its partials are evaluated at (m_t, t).

Lemma 6.4 is the stochastic generalization of the analogous deterministic rule that was already used in (6.29) and (6.33) of Section 5. We give now a formal demonstration of the lemma using the expansion procedure used in the demonstration of Lemma 4.2. The expansion must be carried to second order in differentials involving $d\beta_t$, and $d\beta_t\, d\beta_t^T$ is replaced by its expectation $Q(t)\, dt$.

Demonstration. The total differential of x_t is

$$dx_i = x_i(m + dm, t + dt) - x_i(m, t)$$

$$= x_i(m + dm, t + dt) - x_i(m, t + dt) + x_i(m, t + dt) - x_i(m, t).$$

Now

$$x_i(m, t + dt) - x_i(m, t) = dx_i$$

and

$$x(m + dm, t + dt) - x_i(m, t + dt)$$

$$= \sum_j (x_i(m, t + dt))_{m_j}\, dm_j + \frac{1}{2} \sum_{j,k} (x_i(m, t + dt))_{m_j m_k}\, dm_j\, dm_k + o(dt).$$

But

$$\sum_j (x_i(m, t + dt))_{m_j}\, dm_j = \sum_j [(x_i(m, t))_{m_j} + d(x_i)_{m_j}]\, dm_j$$

and

$$\frac{1}{2} \sum_{j,k} (x_i(m, t + dt))_{m_j m_k}\, dm_j\, dm_k = \frac{1}{2} \sum_{j,k} (x_i(m, t))_{m_j m_k}\, dm_j\, dm_k + o(dt).$$

Therefore,

$$dx_i = dx_i + \sum_j [(x_i)_{m_j} + d(x_i)_{m_j}]\, dm_j + \frac{1}{2} \sum_{j,k} (x_i)_{m_j m_k}\, dm_j\, dm_k + o(dt). \tag{6.105}$$

From (6.102),

$$d(x_i)_{m_j} = (f_i)_{m_j} \, dt + \sum_k (G_{ik})_{m_j} \, d\beta_k \,.$$ (6.106)

Substituting this, (6.102), and (6.103) into (6.105), we get

$$dx_i = \left[f_i + \sum_j (x_i)_{m_j}\varphi_j \right] dt + \sum_j G_{ij} \, d\beta_j + \sum_{j,k} (x_i)_{m_j}\Gamma_{jk} \, d\beta_k$$

$$+ \sum_{j,k,l} (G_{ik})_{m_j}\Gamma_{jl} \, d\beta_k \, d\beta_l + \frac{1}{2} \sum_{j,k,l,q} (x_i)_{m_j m_k}\Gamma_{jl}\Gamma_{kq} \, d\beta_l \, d\beta_q + o(dt).$$ (6.107)

Replacing $d\beta_i \, d\beta_j$ by its expectation $Q_{ij} \, dt$ produces (6.104).

Now we proceed to the derivation of the equation for the mode. If m_t is the mode of $p(x, t \mid Y_t)$, then, by definition,

$$p_m(m, t \mid Y_t) \equiv 0,$$ (6.108)

and consequently

$$dp_m(m, t \mid Y_t) = 0.$$ (6.109)

Now, from Kushner's equation (6.79),

$$dp_m = (\mathscr{L}(p))_m \, dt + h_m{}^{\mathrm{T}}R^{-1}(dz_t - \hat{h} \, dt) \, p + p_m(h - \hat{h})^{\mathrm{T}} R^{-1} (dz_t - \hat{h} \, dt),$$ (6.110)

or, equivalently,

$$dp_m = [(\mathscr{L}(p))_m + ph_m{}^{\mathrm{T}}R^{-1}(h - \hat{h}) + p_m(h - \hat{h})^{\mathrm{T}} R^{-1}(h - \hat{h})] \, dt$$

$$+ [ph_m{}^{\mathrm{T}}R^{-1} + p_m(h - \hat{h})^{\mathrm{T}} R^{-1}] \, d\eta.$$ (6.110a)

Suppose m_t satisfies (6.103) with β_t replaced by η_t. Then, by Lemma 6.4 (with $x = p_m$),

$$0 = dp_{m_i} = \left\{ [(\mathscr{L}(p))_m + ph_m{}^{\mathrm{T}}R^{-1}(h - \hat{h})]_i + \sum_j p_{m_i m_j}\varphi_j + p \sum_{j,k} (h_k)_{m_i m_j}\Gamma_{jk} \right.$$

$$\left. + \sum_{j,k} p_{m_i m_j}(h - \hat{h})_k \, \Gamma_{jk} + \frac{1}{2} \sum_{j,k} (\Gamma R \Gamma^{\mathrm{T}})_{jk} \, p_{m_i m_j m_k} \right\} dt$$

$$+ p \sum_j (h_m{}^{\mathrm{T}}R^{-1})_{ij} \, d\eta_j + \sum_{j,k} p_{m_i m_j}\Gamma_{jk} \, d\eta_k \,,$$ (6.111)

where we have used (6.108).

Now the coefficients of dt and $d\eta$ in (6.111) must be zero. Setting the coefficient of $d\eta$ equal to zero produces

$$\Gamma = -p p_{mm}^{-1} h_m{}^{\mathrm{T}}R^{-1}.$$ (6.112)

Putting this into the zero coefficient of dt gives

$$\varphi = -p_{mm}^{-1}(\mathscr{L}(p))_m + p^2 p_{mm}^{-1} u - \tfrac{1}{2} p^2 p_{mm}^{-1} s, \tag{6.113}$$

where u and s are vectors with components

$$u_i = \sum_{j,k} (h_k)_{m_i m_j} (p_{mm}^{-1} h_m^{\mathrm{T}} R^{-1})_{jk}, \tag{6.114}$$

$$s_i = \operatorname{tr} p_{mm}^{-1} h_m^{\mathrm{T}} R^{-1} h_m p_{mm}^{-1} (p_{m_i})_{mm}. \tag{6.115}$$

Therefore, the mode m_t satisfies (6.103):

$$dm_t = [-p_{mm}^{-1}(\mathscr{L}(p))_m + p^2 p_{mm}^{-1} u - \tfrac{1}{2} p^2 p_{mm}^{-1} s] \, dt$$
$$- p p_{mm}^{-1} h_m^{\mathrm{T}} R^{-1}(t) \, (dz_t - h \, dt), \tag{6.116}$$

where p and all its partials are evaluated at (m_t, t). We have, of course, assumed that $p_{mm} > 0$. This is the desired stochastic differential equation for the conditional mode.

The right-hand side of (6.116) depends on p, $p_{m_i m_j}$ and $p_{m_i m_j m_k}$. Equations for these derivatives can be derived, and these will depend on still higher-order derivatives. Thus, in order to obtain a practical, realizable maximum likelihood filter, these equations have to be truncated or the partials have to be approximated in some way.

Since in the linear-Gaussian case

$$\Sigma_t = -p p_{mm}^{-1} \tag{6.117}$$

is the conditional covariance matrix, we will derive a stochastic differential equation for it. Because of the tedious algebra, we limit ourselves to the scalar case $(n = m = 1)$.

By Lemma 6.4 $(x = p)$,

$$d\bar{p} = dp - \tfrac{1}{2} \Sigma_t^2 p_{mm} h_m^2 R^{-1} \, dt$$
$$= [\mathscr{L}(p) + p(h - \hat{h})^2 \, R^{-1} - \tfrac{1}{2} \Sigma_t^2 p_{mm} h_m^2 R^{-1}] \, dt + p(h - \hat{h}) R^{-1} \, d\eta, \tag{6.118}$$

where we have used (6.108). From (6.110),

$$dp_{mm} = (\mathscr{L}(p))_{mm} \, dt + (p h_{mm} R^{-1} + p_{mm}(h - \hat{h}) \, R^{-1} + 2 p_m h_m R^{-1}) \, (dz_t - \hat{h} \, dt). \tag{6.119}$$

and, using Lemma 6.4 (with $x = p_{mm}$),

$$\bar{d}p_{mm} = dp_{mm} + p_{mmm}\, dm_t$$
$$+ (ph_{mmm} + p_{mmn}(h - \hat{h}) + 3p_{mm}h_m + \tfrac{1}{2}\Sigma_t h_m p_{mmmm}) \, h_m R^{-1}\Sigma_t \, dt$$
$$= [(\mathscr{L}(p))_{mm} + (ph_{mm} + p_{mm}(h - \hat{h})) \, R^{-1}(h - \hat{h})$$
$$+ p_{mmm}(-p_{mm}^{-1}(\mathscr{L}(p))_m + \Sigma_t^2 h_m h_{mm} R^{-1} - \tfrac{1}{2}\Sigma_t^2 R^{-1} h_m^2 p_{mm}^{-1} p_{mmm})$$
$$+ (ph_{mmm} + p_{mmm}(h - \hat{h}) + 3p_{mm}h_m + \tfrac{1}{2}\Sigma_t h_m p_{mmmm}) \, h_m R^{-1}\Sigma_t] \, dt$$
$$+ [ph_{mm} + p_{mm}(h - \hat{h}) + \Sigma_t h_m p_{mmm}] \, R^{-1} \, d\eta, \qquad (6.120)$$

where again we have used (6.108). Now applying Corollary 3 of Lemma 4.2 with $\varphi = x = p$, $\psi = y = p_{mm}$, we finally get

$$-\bar{d}\Sigma_t = \bar{d}(pp_{mm}^{-1})$$
$$= p_{mm}^{-1}\, \bar{d}p + \Sigma_t p_{mm}^{-1}\, \bar{d}p_{mm}$$
$$+ [-\Sigma_t h_{mm}^2 + h_{mm}(h - \hat{h}) + 2\Sigma_t p_{mm}^{-1} h_m h_{mm} p_{mmm}$$
$$- p_{mm}^{-1}(h - \hat{h}) \, h_m p_{mmm} - \Sigma_t p_{mm}^{-2} h_m^2 p_{mmm}^2] \, \Sigma_t^2 R^{-1} \, dt. \qquad (6.121)$$

Substituting (6.118) and (6.120) into (6.121) yields the desired result. We summarize these results in the following theorem.

Theorem 6.7 (Conditional Mode for Continuous Problems). *Hypotheses of Lemma 6.2. Let $p(x, t \mid Y_t)$ be sufficiently differentiable and $p_{mm}(m_t, t \mid Y_t) > 0$. Suppose the conditional mode m_t satisfies an (Itô) stochastic differential equation. Then the conditional mode satisfies*

$$dm_t = [-p_{mm}^{-1}(\mathscr{L}(p))_m - p\Sigma_t(u - \tfrac{1}{2}s)] \, dt + \Sigma_t h_m^{\mathrm{T}} R^{-1}(t) \, (dz_t - h\, dt), \qquad (6.122)$$

where the n-vectors u and s are defined in (6.114) and (6.115), and where p, its partials, h, and all other functions are evaluated at (m_t, t). In the scalar case $(n = m = 1)$, $\Sigma_t \, (= -pp_{mm}^{-1})$ satisfies

$$\bar{d}\Sigma_t = -p_{mm}^{-1}[\mathscr{L}(p) + \Sigma_t(\mathscr{L}(p))_{mm} - p_{mmm}\Sigma_t p_{mm}^{-1}(\mathscr{L}(p))_m] \, dt$$
$$+ \Sigma_t^2(h_{mm} - p_{mm}^{-1} h_m p_{mmm}) \, R^{-1}(t) \, (dz_t - \hat{h}\, dt)$$
$$- \Sigma_t p_{mm}^{-1}[3p_{mmm}\Sigma_t^2 h_m(h_{mm} - \tfrac{1}{2}h_m p_{mm}^{-1} p_{mmm}) \, R^{-1}(t) \, dt$$
$$+ (ph_{mmm} + \tfrac{5}{2}p_{mm}h_m + \tfrac{1}{2}\Sigma_t h_m p_{mmmm}) \, h_m R^{-1}(t) \, \Sigma_t \, dt]$$
$$+ \Sigma_t^3 h_{mm}^2 R^{-1}(t) \, dt + \Sigma_t^2(h - \hat{h})(p_{mm}^{-1} h_m p_{mmm} - h_{mm}) \, R^{-1}(t) \, dt. \qquad (6.123)$$

Note that (6.122) and (6.123) reduce to (6.40) and (6.41), respectively, of Theorem 6.3 in case $R^{-1}(t) \equiv 0$. Comparing the equation for the conditional mode with that for the conditional mean (6.100), we see an important qualitative difference. The equation for the mean depends on average properties of the functions f, h, and so on, whereas the equation for the mode depends on local properties of these functions. Only indirectly through Σ_t does the mode depend on average properties. See also Example 6.1.

Example 6.4. In the scalar case ($n = m = 1$),

$$\mathcal{L}(p)(m, t) = -pf_m + \tfrac{1}{2}(p_{mm}\sigma^2 + p\sigma^2_{mm}),$$

$$(\mathcal{L}(p))_m(m, t) = -p_{mm}f - pf_{mm} + \tfrac{1}{2}(p_{mmm}\sigma^2 + 3p_{mm}\sigma_m{}^2 + p\sigma^2_{mmm}),$$

$$(\mathcal{L}(p))_{mm}(m, t) = -p_{mmm}f - 3p_{mm}f_m - pf_{mmm}$$
$$+ \tfrac{1}{2}(p_{mmmm}\sigma^2 + 4p_{mmm}\sigma_m{}^2 + 6p_{mm}\sigma^2_{mm} + p\sigma^2_{mmmm}), \quad (6.124)$$

where $\sigma^2 = GQG^\mathrm{T}$ and all the functions are evaluated at (m, t).

We specialize (6.122) and (6.123) to the linear-Gaussian, scalar case ($f = F(t)\, x$, $h = M(t)\, x$, $\sigma_x{}^2 \equiv 0$, $p_{mmm} = 0$, $p_{mmmm}/p_{mm} = -3\Sigma_t^{-1}$):

$$dm_t = F(t)\, m_t\, dt + \Sigma_t M(t)\, R^{-1}(t)\, (dz_t - M(t)\, m_t\, dt),$$
$$d\Sigma_t/dt = 2\Sigma_t F(t) + \sigma^2 - \Sigma_t M(t)\, R^{-1}(t)\, M(t)\, \Sigma_t. \quad (6.125)$$

Equations (6.125) constitute the maximum likelihood (Bayesian) filter in the continuous, linear-Gaussian, scalar case. This is clearly also the minimum variance filter; the mode m_t is the conditional mean, and Σ_t is the conditional variance. Compare this with Examples 6.1 and 6.2. Equations (6.125) will be discussed in some detail in the next chapter. ▲

11. LIMITED MEMORY FILTER

The filters (predictors) developed in the preceding sections utilize all the available data, including all observations and prior data. It is clearly optimal to base an estimate on all the available information. This is predicated, however, on the knowledge of the dynamics, measurement function, and the statistical parameters associated with the dynamical system. If, for example, the dynamics are imprecisely known, then the filter might learn the state "too well." Because of imprecise dynamics used in the filter, the "good estimate" may be wrong.

This problem is treated in detail in Chapter 8. It is referred to as the problem of filter *divergence*. A reasonable approach to this problem is to limit the filter memory so that the estimate does not become "too good." By limiting the filter memory, we mean computing an estimate based (conditioned) on data from only the recent past. Alternatively, this means discarding the conditioning (of the estimate) on the distant past. The memory length of the filter should coincide with the (time) length over which the approximate dynamical system represents a satisfactory approximation to reality. We develop the theory of the limited memory filter in the present section. The development is based on the work of Jazwinski [6].

Let the dynamical system be discrete (6.58) or continuous (6.64), but without the random forcing term. That is, the state evolves either according to

$$x_{k+1} = \varphi(x_k, t_{k+1}, t_k), \qquad k = 0, 1, \dots, \tag{6.126D}$$

or according to

$$dx_t/dt = f(x_t, t), \qquad t \geqslant t_0, \tag{6.126C}$$

where x_k is x_{t_k}. Discrete observations (6.59) are taken as

$$y_k = h(x_k, t_k) + v_k, \qquad k = 1, 2, \dots, \tag{6.127}$$

where $\{v_k\}$ is white with the usual statistics.

Consider the sequence of data

$$y_1, \dots, y_m, y_{m+1}, \dots, y_n, \dots; \qquad n - m = N > 0, \tag{6.128}$$

and define

$$Y_m = \{y_1, \dots, y_m\},$$
$$Y_n = \{y_1, \dots, y_n\}, \tag{6.129}$$
$$Y_N = \{y_{m+1}, \dots, y_n\}.$$

Lemma 6.5. *Assume dynamical system* (6.126) *with observations* (6.127). *Suppose that the required density functions exist. Then*

$$p(Y_N \mid x_n) = c \, \frac{p(x_n \mid Y_n)}{p(x_n \mid Y_m)}, \tag{6.130}$$

where

$$c = \frac{p(Y_n)}{p(Y_m)}$$

is a constant, independent of x_n. *Also,*

$$p(x_n \mid Y_N) = c_1 \, \frac{p(x_n \mid Y_n)}{p(x_n \mid Y_m)}, \tag{6.131}$$

where

$$c_1 = \frac{p(Y_n)}{p(Y_m)\,p(Y_N)}\,p(x_n).$$

PROOF: By Bayes' rule,

$$p(x_n \mid Y_n) = \frac{p(Y_n \mid x_n)\,p(x_n)}{p(Y_n)} = \frac{p(Y_m,\,Y_N \mid x_n)\,p(x_n)}{p(Y_n)}.$$

But

$$p(Y_m,\,Y_N \mid x_n) = p(Y_m \mid Y_N,\,x_n)\,p(Y_N \mid x_n),$$

and, since the dynamical system (6.126) is noise-free,

$$p(Y_m \mid Y_N,\,x_n) = p(Y_m \mid x_n).$$

Therefore,

$$p(x_n \mid Y_n) = \frac{p(Y_m \mid x_n)\,p(Y_N \mid x_n)\,p(x_n)}{p(Y_n)}.$$

Now, applying Bayes' rule to $p(Y_m \mid x_n)$,

$$p(Y_m \mid x_n) = \frac{p(x_n \mid Y_m)\,p(Y_m)}{p(x_n)},$$

and rearranging terms,

$$p(x_n \mid Y_n) = \frac{p(Y_m)}{p(Y_n)}\,p(x_n \mid Y_m)\,p(Y_N \mid x_n),$$

which is (6.130). Equation (6.131) follows after application of Bayes' rule to $p(Y_N \mid x_n)$:

$$p(Y_N \mid x_n) = \frac{p(x_n \mid Y_N)\,p(Y_N)}{p(x_n)}. \qquad \blacksquare \qquad (6.132)$$

Consider the case in which the dynamical system is continuous (6.126C). $p(x_n \mid Y_n)$ is the output of the continuous-discrete filter, and $p(x_n \mid Y_m)$ is the output of the predictor (Theorem 6.1). Then Eq. (6.130) gives the appropriate density for computing the maximum likelihood estimate of x_n based on Y_N.

The conditional density $p(x_n \mid Y_N)$ in (6.131) is the density of x_n given Y_N *and* the density of x_0. It is defined in terms of the joint density $p(x_n,\,Y_N)$ and the marginal density $p(Y_N)$ (see Section 2.5). What we wish to do is to define the conditional density of x_n given *only* Y_N, excluding any prior information about x_0. To do this, we suppose formally that, for each n, $p(x_n)$ is Gaussian with zero-mean and covariance matrix

$$(1/\epsilon) I,$$

and *define* $p(x_n \mid Y_N \mid)$, the density of x_n given only Y_N, by

$$p(x_n \mid Y_N \mid) = \lim_{\epsilon \to 0} p(x_n \mid Y_N), \tag{6.133}$$

assuming the limit exists.

Theorem 6.8. *Hypotheses of Lemma 6.5. Then*

$$p(x_n \mid Y_N \mid) = c_2 \frac{p(x_n \mid Y_n)}{p(x_n \mid Y_m)}, \tag{6.134}$$

where c_2 is a constant, independent of x_n. [*The normalizing constant can obviously be determined by*

$$\frac{1}{c_2} = \int \frac{p(x_n \mid Y_n)}{p(x_n \mid Y_m)} dx_n \cdot \bigg]$$

PROOF: From (6.132),

$$\frac{p(Y_N \mid x_n)}{p(x_n \mid Y_N)} = \frac{p(Y_N)}{p(x_n)}. \tag{6.135}$$

Now $p(Y_N \mid x_n)$ does not depend on the statistics of x_n. Assuming the limit in (6.133) exists,[6] and taking the limit in (6.135), we get

$$\frac{p(Y_N \mid x_n)}{p(x_n \mid Y_N \mid)} = \lim_{\epsilon \to 0} \frac{p(Y_N)}{p(x_n)} = c_3, \tag{6.136}$$

which proves that c_3 exists. c_3 is clearly independent of x_n. (6.136), together with (6.130) of Lemma 6.5, proves the theorem. ∎

As was already noted, $p(x_n \mid Y_n)$ and $p(x_n \mid Y_m)$ may be computed via Theorem 6.1. Equation (6.134) then produces the density of x_n given only Y_N. From it, we can compute the moments. In particular, the mean of

$$p(x_n \mid Y_N \mid)$$

is the minimum variance estimate of x_n based on the most recent N observations. In view of (6.136), $p(Y_N \mid x_n)$ and $p(x_n \mid Y_N \mid)$ are equal up to a multiplicative constant. Thus, any estimate of x_n obtained from

---

[6] This is proved for linear-Gaussian problems in Chapter 7.

$p(Y_N \mid x_n)$ can also be obtained from $p(x_n \mid Y_N \mid)$ by the same operation, and vice versa. We shall return to the limited memory filter in the next chapter, where we will consider its linear version and its mechanization.

REFERENCES

1. R. S. Bucy, Nonlinear Filtering Theory, *IEEE Trans. Automatic Control* **10**, 198 (1965).
2. J. R. Fisher, Optimal Nonlinear Filtering, *Advan. Control Systems* **5**, (1967). [Also see J. R. Fisher and E. B. Stear, Optimal Nonlinear Filtering for Independent Increment Processes, Parts I and II, *IEEE Trans. Inform. Theory* **3**, 558–578 (1967).]
3. Y. C. Ho and R. C. K. Lee, A Bayesian Approach to Problems in Stochastic Estimation and Control, *IEEE Trans. Automatic Control* **9**, 333–339 (1964).
4. A. H. Jazwinski, Nonlinear Filtering with Discrete Observations, Paper No. 66-38, *AIAA 3rd Aerospace Sciences Meeting, New York*, 1966.
5. A. H. Jazwinski, Filtering for Nonlinear Dynamical Systems, *IEEE Trans. Automatic Control* **11**, 765–766 (1966).
6. A. H. Jazwinski, Limited Memory Optimal Filtering, *1968 Joint Automatic Control Conf.*, Ann Arbor, Michigan, 1968. Also in *IEEE Trans. Automatic Control* **13**, 558–563 (1968).
7. R. L. Kashyap, On the Partial Differential Equation for the Conditional Probability Distribution for Nonlinear Dynamic Systems with Noisy Observations. Tech. Rept. 432, Cruft Lab., Harvard Univ., Cambridge, Massachusetts, 1963.
8. H. J. Kushner, On the Dynamical Equations of Conditional Probability Density Functions, with Applications to Optimal Stochastic Control Theory, *J. Math. Anal. Appl.* **8**, 332–344 (1964).
9. H. J. Kushner, On the Differential Equations Satisfied by Conditional Probability Densities of Markov Processes, *SIAM J. Control* **2**, 106–119 (1964).
10. H. J. Kushner, Dynamical Equations for Optimal Nonlinear Filtering, *J. Differential Equations* **3**, 179–190 (1967).
11. H. J. Kushner, Nonlinear Filtering: The Exact Dynamical Equations Satisfied by the Conditional Mode, *IEEE Trans. Automatic Control* **12**, 262–267 (1967).
12. R. E. Mortensen, Optimal Control of Continuous-Time Stochastic Systems, Ph.D. thesis (engineering), Univ. of California, Berkeley, California, 1966.
13. R. L. Stratonovich, Conditional Markov Processes, *Theor. Probability Appl.* **5**, 156–178 (1960).
14. W. M. Wonham, Stochastic Problems in Optimal Control, *IEEE Conv. Record, Part 2* **11**, 114–124 (1963).

7

Linear Filtering Theory

1. INTRODUCTION

In this chapter, we specialize the nonlinear theory developed in Chapter 6 to linear filtering problems. As was already noted in Section 5.2, in linear problems the conditional density function is Gaussian and is, therefore, completely characterized by its mean vector and covariance matrix. This fact makes linear filtering problems particularly tractable. As we have seen in some special cases (Examples 6.1, 6.2, and 6.4), the equation for the conditional covariance matrix is independent of the observations. Thus, it is also the unconditional covariance matrix and may be precomputed in advance. We shall see that this is true, in general, in linear problems. As a result, the filtering algorithm essentially consists of the equation for the conditional mean. Appropriately, the conditional mean is sometimes referred to as a *sufficient statistic*. We recall that, according to the Corollary of Theorem 5.2, the conditional mean minimizes the expected loss for a large class of loss functions. In any case, it is, of course, the minimum variance estimate (Theorem 5.3).

By specializing the nonlinear results to the linear case, we are following the probabilistic approach outlined in Section 5.2. We will use some of the statistical methods (Section 5.3) to derive these same results in various examples. In addition, some examples will be devoted to the

194

generalizations described in Section 5.4 and to the derivation of a linear smoother.

The bulk of the remainder of this chapter is devoted to the study of some important properties of linear filters. The concepts of observability, information, and controllability are defined and exploited in the study of the asymptotic behavior of linear filters. Useful bounds and stability properties are developed. Of great importance in applications is the sensitivity of the filtering algorithm to dynamical model and statistical model errors. Considerable attention is devoted to this subject. Finally, linear limited memory filter algorithms are developed. These, as we shall see in Chapter 8, have important applications in problems of optimal filter divergence.

2. CONTINUOUS-DISCRETE FILTER

Our linear dynamical system is described by the linear vector (Itô) stochastic differential equation

$$dx_t = F(t) x_t \, dt + G(t) \, d\beta_t, \qquad t \geqslant t_0, \qquad (7.1)$$

where x_t is the n-vector state, F and G are, respectively, $n \times n$ and $n \times r$ nonrandom, continuous matrix time-functions, and $\{\beta_t, t \geqslant t_0\}$ is an r-vector Brownian motion process with $\mathscr{E}\{d\beta_t \, d\beta_t^\mathrm{T}\} = Q(t) \, dt$. Discrete, linear observations are taken at time instants t_k :

$$y_k = M(t_k) x_{t_k} + v_k ; \qquad k = 1, 2,\dots ; \qquad t_{k+1} > t_k \geqslant t_0, \qquad (7.2)$$

where y_k is the m-vector observation, M is an $m \times n$ nonrandom, bounded matrix function, and $\{v_k, k = 1, 2,\dots\}$ is an m-vector, white Gaussian sequence, $v_k \sim N(0, R_k)$, $R_k > 0$. The distribution of x_{t_0} is Gaussian, $x_{t_0} \sim N(\hat{x}_{t_0}, P_{t_0})$, and x_{t_0}, $\{\beta_t\}$, and $\{v_k\}$ are assumed independent.

For the linear system (7.1), Kolmogorov's forward equation (6.5) becomes

$$\partial p/\partial t = -p \, \mathrm{tr}(F) - p_x{}^\mathrm{T}Fx + \tfrac{1}{2} \, \mathrm{tr}(GQG^\mathrm{T}p_{xx}). \qquad (7.3)$$

The conditional density in Eq. (6.12) is

$$p(y_k \mid x) = (1/(2\pi)^{m/2} \mid R_k \mid^{1/2}) \exp\{-\tfrac{1}{2}(y_k - M(t_k) x)^\mathrm{T} R_k^{-1}(y_k - M(t_k) x)\}, \qquad (7.4)$$

and Theorem 6.1 is specialized to the linear case.

We obtain the equations for the conditional mean and covariance matrix, between observations, directly from Eq. (6.25) of Theorem 6.2. Since

$$\dot{f}^t = F(t)\,\hat{x}_t{}^t,$$

$$\mathscr{E}^t\{x_t f^{\mathrm{T}}\} - \hat{x}_t{}^t \dot{f}^{t^{\mathrm{T}}} = \widehat{(x_t x_t^{\mathrm{T}}} - \hat{x}_t{}^t \hat{x}_t^{t^{\mathrm{T}}})F^{\mathrm{T}}(t) = P_t{}^t F^{\mathrm{T}}(t),$$

$$\mathscr{E}^t\{GQG^{\mathrm{T}}\} = GQG^{\mathrm{T}},$$

we have

$$
\begin{aligned}
d\hat{x}_t{}^t/dt &= F(t)\,\hat{x}_t{}^t, \\
dP_t{}^t/dt &= F(t)\,P_t{}^t + P_t{}^t F^{\mathrm{T}}(t) + G(t)Q(t)\,G^{\mathrm{T}}(t).
\end{aligned}
\tag{7.5}
$$

To compute the change in $\hat{x}$ and P at an observation at t_k, we use Eq. (6.14) of Theorem 6.1.[1] Carrying out the integration in the denominator of (6.14), we have

$$p(x, t_k \mid Y_{t_k}) = \frac{p(y_k \mid x)\,p(x, t_k \mid Y_{t_k^-})}{p(y_k \mid Y_{t_k^-})}.^2 \tag{7.6}$$

All the densities appearing in (7.6) are Gaussian. By definition,[3]

$$p(x, t_k \mid Y_{t_k^-}) \sim N(\hat{x}_{t_k}^{t_k^-}, P_{t_k}^{t_k^-}).$$

$p(y_k \mid x)$ has already been computed in (7.4). Now

$$\mathscr{E}\{y_k \mid Y_{t_k^-}\} = M(t_k)\,\hat{x}_{t_k}^{t_k^-},$$

in view of (7.2), and

$$\mathscr{E}\{(y_k - \mathscr{E}\{y_k \mid Y_{t_k^-}\})(y_k - \mathscr{E}\{y_k \mid Y_{t_k^-}\})^{\mathrm{T}} \mid Y_{t_k^-}\}$$

$$= M(t_k)\,P_{t_k}^{t_k^-}\,M^{\mathrm{T}}(t_k) + R_k,$$

so that

$$p(y_k \mid Y_{t_k^-}) \sim N(M(t_k)\,\hat{x}_{t_k}^{t_k^-},\ M(t_k)\,P_{t_k}^{t_k^-}M^{\mathrm{T}}(t_k) + R_k).$$

Combining these results, and dropping the t_k subscripts and superscripts for the moment, we have

$$p(x, t_k \mid Y_{t_k}) = \frac{\mid MP^- M^{\mathrm{T}} + R \mid^{1/2}}{(2\pi)^{n/2} \mid R \mid^{1/2} \mid P^- \mid^{1/2}}\exp[-\tfrac{1}{2}\{\cdot\}], \tag{7.7}$$

[1] The following derivation was given by Ho and Lee [27] and Lee [37].

[2] Recall that $Y_\tau = \{y_l : t_l \leqslant \tau\}$ (6.3); $Y_{t_k} = \{y_1, ..., y_k\}$; $Y_{t_k^-} = Y_{t_{k-1}}$.

[3] Recall that $\hat{\varphi}^\tau(x_t) = [\varphi(x_t)]^{\wedge \tau} = \mathscr{E}^\tau\{\varphi(x_t)\} = \mathscr{E}\{\varphi(x_t) \mid Y_\tau\}$ (6.16).

where

$$\{\cdot\} = (y - Mx)^{\mathrm{T}} R^{-1}(y - Mx) + (x - \hat{x}^-)^{\mathrm{T}} P^{-1}(x - \hat{x}^-)$$
$$- (y - M\hat{x}^-)^{\mathrm{T}} (MP^-M^{\mathrm{T}} + R)^{-1} (y - M\hat{x}^-). \tag{7.8}$$

Now, by definition,

$$p(x, t_k \mid Y_{t_k}) \sim N(\hat{x}_{t_k}^{t_k^+}, P_{t_k}^{t_k^+}),$$

so that the term in braces in (7.8) must be

$$\{\cdot\} = (x - \hat{x}^+)^{\mathrm{T}} P^{+^{-1}}(x - \hat{x}^+). \tag{7.9}$$

To put (7.8) in the form of (7.9), we rewrite it as

$$\{\cdot\} = x^{\mathrm{T}}(M^{\mathrm{T}}R^{-1}M + P^{-1})\,x - 2(y^{\mathrm{T}}R^{-1}M + \hat{x}^{-\mathrm{T}}P^{-1})(M^{\mathrm{T}}R^{-1}M + P^{-1})^{-1}$$
$$\times (M^{\mathrm{T}}R^{-1}M + P^{-1})\,x + y^{\mathrm{T}}R^{-1}y + \hat{x}^{-\mathrm{T}}P^{-1}\hat{x}^-$$
$$- (y - M\hat{x}^-)^{\mathrm{T}} (MP^-M^{\mathrm{T}} + R)^{-1} (y - M\hat{x}^-)$$

and complete the squares by adding and subtracting

$$(y^{\mathrm{T}}R^{-1}M + \hat{x}^{-\mathrm{T}}P^{-1})(M^{\mathrm{T}}R^{-1}M + P^{-1})^{-1} (M^{\mathrm{T}}R^{-1}y + P^{-1}\hat{x}^-).$$

We thus obtain

$$\{\cdot\} = [x - (M^{\mathrm{T}}R^{-1}M + P^{-1})^{-1} (M^{\mathrm{T}}R^{-1}y + P^{-1}\hat{x}^-)]^{\mathrm{T}} [M^{\mathrm{T}}R^{-1}M + P^{-1}]$$
$$\times [x - (M^{\mathrm{T}}R^{-1}M + P^{-1})^{-1} (M^{\mathrm{T}}R^{-1}y + P^{-1}\hat{x}^-)] + r, \tag{7.10}$$

where, with the aid of the matrix equalities developed in Appendix 7B, r can be shown to be zero. Comparing (7.10) with (7.9), we conclude that

$$\hat{x}_{t_k}^{t_k^+} = (M^{\mathrm{T}}(t_k)\, R_k^{-1}M(t_k) + P_{t_k}^{t_k^{-1}})^{-1} (M^{\mathrm{T}}(t_k)\, R_k^{-1}y_k + P_{t_k}^{t_k^{-1}}\hat{x}_{t_k}^{t_k}),$$
$$P_{t_k}^{t_k^{+^{-1}}} = P_{t_k}^{t_k^{-1}} + M^{\mathrm{T}}(t_k)\, R_k^{-1}M(t_k). \tag{7.11}$$

Equations (7.11) are the desired relations. We see [(7.5), (7.11)] that the conditional covariance matrix is independent of observations and is, therefore, the unconditional covariance matrix:

$$P_t^{\,t} \triangleq \mathscr{E}\{(x_t - \hat{x}_t^{\,t})(x_t - \hat{x}_t^{\,t})^{\mathrm{T}} \mid Y_t\} = \mathscr{E}\{(x_t - \hat{x}_t^{\,t})(x_t - \hat{x}_t^{\,t})^{\mathrm{T}}\}. \tag{7.12}$$

Consequently, we shall sometimes write $P_t^{\,t}$ as $P(t \mid t)$ and even as $P(t)$ when no confusion can arise.

The computation of Eqs. (7.11) involves the inversion of $n \times n$ matrices. We next reduce (7.11) to a form that requires the inversion of an $m \times m$ matrix (m is the dimension of the observation vector). Often, in applications, $m < n$, and, in case R is diagonal, we may always take $m = 1$. This is done by processing the components of y_k one by one, and supposing that there is no change in $\hat{x}$ or P due to the dynamics (7.5) in the meantime. That this is equivalent to processing the whole observation vector at once is conceptually clear and can be proved algebraically (very tedious!).

Using (7B.5) of Appendix 7B, we rewrite the second of Eqs. (7.11) as

$$P_{t_k}^{t_k^+} = P_{t_k}^{t_k^-} - P_{t_k}^{t_k^-} M^{\mathrm{T}}(t_k)[M(t_k) P_{t_k}^{t_k^-} M^{\mathrm{T}}(t_k) + R_k]^{-1} M(t_k) P_{t_k}^{t_k^-}. \quad (7.13)$$

With the aid of (7B.5) and (7B.6) of Appendix 7B, the first of Eqs. (7.11) becomes

$$\hat{x}_{t_k}^{t_k^+} = \hat{x}_{t_k}^{t_k^-} + P_{t_k}^{t_k^-} M^{\mathrm{T}}(t_k)[M(t_k) P_{t_k}^{t_k^-} M^{\mathrm{T}}(t_k) + R_k]^{-1} (y_k - M(t_k) \hat{x}_{t_k}^{t_k^-}). \quad (7.14)$$

If we define the *Kalman gain* [32, 36]

$$K(t_k) \triangleq P_{t_k}^{t_k^-} M^{\mathrm{T}}(t_k)[M(t_k) P_{t_k}^{t_k^-} M^{\mathrm{T}}(t_k) + R_k]^{-1}, \quad (7.15)$$

we have

$$\hat{x}_{t_k}^{t_k^+} = \hat{x}_{t_k}^{t_k^-} + K(t_k)(y_k - M(t_k) \hat{x}_{t_k}^{t_k^-}),$$

$$P_{t_k}^{t_k^+} = P_{t_k}^{t_k^-} - K(t_k) M(t_k) P_{t_k}^{t_k^-} \quad (7.16)$$

$$= [I - K(t_k) M(t_k)] P_{t_k}^{t_k^-}[I - K(t_k) M(t_k)]^{\mathrm{T}} + K(t_k) R_k K^{\mathrm{T}}(t_k).$$

We summarize these results in the following theorem.

Theorem 7.1. *The optimal (minimum variance) filter for the continuous-discrete system* (7.1), (7.2) *consists of equations of evolution for the conditional mean $\hat{x}_t^t$ and covariance matrix P_t^t. Between observations, these satisfy the differential equations*

$$d\hat{x}_t^t/dt = F(t) \hat{x}_t^t,$$

$$dP_t^t/dt = F(t) P_t^t + P_t^t F^{\mathrm{T}}(t) + G(t) Q(t) G^{\mathrm{T}}(t), \qquad t_k \leqslant t < t_{k+1}. \quad (7.17)$$

At an observation at t_k, they satisfy the difference equations

$$\hat{x}_{t_k}^{t_k^+} = \hat{x}_{t_k}^{t_k^-} + K(t_k)(y_k - M(t_k) \hat{x}_{t_k}^{t_k^-}),$$

$$P_{t_k}^{t_k^+} = P_{t_k}^{t_k^-} - K(t_k) M(t_k) P_{t_k}^{t_k^-}, \quad (7.18)$$

where the Kalman gain K is given in (7.15). *Prediction for* $t > \tau$ $(\hat{x}_t{}^\tau, P_t{}^\tau)$, *based on* Y_τ (6.3), *is clearly accomplished via* (7.17), *with initial condition* $(\hat{x}_\tau{}^\tau, P_\tau{}^\tau)$.

Theorem 7.1 gives the celebrated Kalman–Bucy filter [32, 33, 36] for the continuous-discrete problem. Compare this with Examples 6.1 and 6.2. The equations for the estimate are linear, with a forcing term that is linear in the observation. The forcing term is proportional to the *residual*

$$y_k - M(t_k)\, \hat{x}_{t_k}^{t_k^-} = y_k - \mathscr{E}\{y_k \mid Y_{t_k^-}\}.$$

The initial conditions $(\hat{x}_{t_0}, P_{t_0})$ are part of the problem statement and reflect the *prior* knowledge of x_t before the filtering begins. As will be seen from other derivations of these same results, we need only that $P_{t_0} \geqslant 0$. In connection with the question of prior information, see Appendix 7A and also Sections 5 and 10. The parameters $(\hat{x}_t{}^t, P_t{}^t)$ comprise the *state* of the linear filter. They determine the density $p(x, t \mid Y_t)$. However, as has already been noted, $P_t{}^t$ and the filter gain $K(t_k)$ may be *precomputed*, since they do not depend on the noise samples, but only on the noise statistics which are part of the problem statement. Thus, once the observation schedule has been settled, the real-time implementation of the Kalman–Bucy filter requires only the computation of the first of Eqs. (7.17) and the first of Eqs. (7.18). This happy situation is, of course, unique to the linear case.

Now Eq. (7.1) can be integrated over intervals $[t_k, t_{k+1}]$:

$$x_{t_{k+1}} = \Phi(t_{k+1}, t_k)\, x_{t_k} + \int_{t_k}^{t_{k+1}} \Phi(t_{k+1}, \tau)\, G(\tau)\, d\beta_\tau, \qquad k = 0, 1, \ldots, \quad (7.19)$$

where the $n \times n$ matrix Φ is the *fundamental matrix* of the homogeneous part of (7.1), or the *state transition matrix* of (7.1). That is,

$$d\Phi(t, \tau)/dt = F(t)\, \Phi(t, \tau), \qquad \Phi(\tau, \tau) = I \quad \text{all} \quad \tau, \qquad (7.20)$$

$$\Phi(t, \tau)\, \Phi(\tau, \xi) = \Phi(t, \xi) \qquad \text{all} \quad t, \tau, \xi. \qquad (7.21)$$

From (7.21), it follows that

$$\Phi^{-1}(t, \tau) = \Phi(\tau, t) \qquad \text{all} \quad t, \tau. \qquad (7.22)$$

We write (7.19) as

$$x_{t_{k+1}} = \Phi(t_{k+1}, t_k)\, x_{t_k} + w_{k+1}, \qquad (7.23)$$

where

$$w_{k+1} = \int_{t_k}^{t_{k+1}} \Phi(t_{k+1}, \tau)\, G(\tau)\, d\beta_\tau. \qquad (7.24)$$

It is easy to see that $\{w_k\}$ is a zero-mean, white Gaussian sequence with (see Theorem 4.1)

$$\mathscr{E}\{w_{k+1}w_{k+1}^{\mathrm{T}}\} = \int_{t_k}^{t_{k+1}} \Phi(t_{k+1}, \tau) G(\tau) Q(\tau) G^{\mathrm{T}}(\tau) \Phi^{\mathrm{T}}(t_{k+1}, \tau) \, d\tau. \quad (7.25)$$

Thus the continuous-discrete filter can be imbedded in a discrete filter, and we need not study the properties of the continuous-discrete filter separately.

It is easy to verify that (7.17) of Theorem 7.1 are equivalent to

$$\hat{x}_t^t = \Phi(t, t_k) \hat{x}_{t_k}^t,$$

$$P_t^t = \Phi(t, t_k) P_{t_k}^t \Phi^{\mathrm{T}}(t, t_k) + \mathscr{E}\{w_{k+1}w_{k+1}^{\mathrm{T}}\}, \qquad t_k \leqslant t < t_{k+1}. \quad (7.26)$$

3. DISCRETE FILTER

The discrete linear system is described by the vector difference equation

$$x_{k+1} = \Phi(t_{k+1}, t_k) x_k + \Gamma(t_k) w_{k+1}, \qquad k = 0, 1, \dots, \quad (7.27)$$

where x_k is the n-vector state at t_k, Φ is the $n \times n$, nonsingular state transition matrix, Γ is $n \times r$, and $\{w_k, k = 1, \dots\}$ is an r-vector, white Gaussian sequence, $w_k \sim N(0, Q_k)$. The discrete, linear observations are given in (7.2):

$$y_k = M(t_k) x_k + v_k. \quad (7.2)$$

x_0, $\{w_k\}$, and $\{v_k\}$ are assumed independent.

Difference equations for $\hat{x}$ and P at an observation are already given in Theorem 7.1. We compute difference equations between observations directly from (7.27). This procedure avoids the computation of the transition probability density $p(x_{k+1} \mid x_k)$ and the Chapman–Kolmogorov equation (6.62) of Theorem 6.4. We know that all the densities are Gaussian. Now

$$\hat{x}_{k+1}^k = \mathscr{E}\{x_{k+1} \mid Y_k\} = \Phi(t_{k+1}, t_k) \hat{x}_k^k,$$

in view of (7.27). Also,

$$P_{k+1}^k = \mathscr{E}\{(x_{k+1} - \hat{x}_{k+1}^k)(x_{k+1} - \hat{x}_{k+1}^k)^{\mathrm{T}} \mid Y_k\}$$

$$= \mathscr{E}\{[\Phi(x_k - \hat{x}_k^k) + \Gamma w_{k+1}][\Phi(x_k - \hat{x}_k^k) + \Gamma w_{k+1}]^{\mathrm{T}} \mid Y_k\}$$

$$= \Phi P_k^k \Phi^{\mathrm{T}} + \Gamma Q_{k+1} \Gamma^{\mathrm{T}}.$$

Compare this with (7.26).

The discrete filter is summarized in the following theorem.

Theorem 7.2. *The optimal (minimum variance) filter for the discrete system (7.27), (7.2) consists of difference equations for the conditional mean and covariance matrix. Between observations,*

$$\hat{x}_{k+1}^k = \Phi(t_{k+1}, t_k)\, \hat{x}_k^{\,k},$$

$$P_{k+1}^k = \Phi(t_{k+1}, t_k)\, P_k^{\,k}\Phi^{\mathrm{T}}(t_{k+1}, t_k) + \Gamma(t_k)\, Q_{k+1}\Gamma^{\mathrm{T}}(t_k). \tag{7.28}$$

At observations,

$$\hat{x}_k^{\,k} = \hat{x}_k^{k-1} + K(t_k)(y_k - M(t_k)\, \hat{x}_k^{k-1}),$$

$$P_k^{\,k} = P_k^{k-1} - K(t_k)\, M(t_k)\, P_k^{k-1}, \tag{7.29}$$

where

$$K(t_k) = P_k^{k-1}M^{\mathrm{T}}(t_k)[M(t_k)\, P_k^{k-1}M^{\mathrm{T}}(t_k) + R_k]^{-1} \tag{7.15}$$

is the Kalman gain. Prediction for $t_l > t_k$ $(\hat{x}_l^{\,k}, P_l^{\,k})$ is accomplished via (7.28) with initial condition $(\hat{x}_k^{\,k}, P_k^{\,k})$.

This is the Kalman–Bucy filter [32, 36] for the discrete problem. The discussion of Theorem 7.1 applies in this case as well. We may note that, if $P_0 \geqslant 0$, then $P_k^{\,k} \geqslant 0$ for all k, since P is a covariance matrix.

Example 7.1. (Orthogonal Projections). In his original derivation of the discrete filter, Kalman [32] used the concept of orthogonal projection. We develop this concept here and, using it, rederive the discrete filter.

As was pointed out in footnote 7 of Chapter 3, the space of random variables with finite second moments, with scalar product

$$\langle x, y \rangle = \mathscr{E}\{xy\}$$

and norm

$$\| x \| = \mathscr{E}^{1/2}\{x^2\},$$

is a (complete) normed space (an infinite-dimensional Hilbert space). Now, by definition, the minimum variance estimate minimizes the error norm

$$\mathscr{E}\{(x - \hat{x})^2\}.$$

The following lemma gives a condition that is equivalent to the minimization of the norm.

Orthogonal Projection Lemma. *Let X be a normed space, $x \in X$, and let Y be a subspace of X. Then*

$$\min_{\alpha \in Y} \| x - \alpha \|^2 = \| x - \hat{x} \|^2$$

if, and only if,

$$\langle x - \hat{x}, \alpha \rangle = 0 \quad \text{for all} \quad \alpha \in Y.$$

The space Y is the "approximation space." That is, we want to find that $\alpha \in Y$ which minimizes the error norm for every $x \in X$. The orthogonal projection lemma states that the error is orthogonal to the approximation space Y. That is, if we write

$$x = \tilde{x} + \hat{x},$$

where $\tilde{x} = x - \hat{x}$ is the error and $\hat{x}$ is the best approximation (estimate), then $\hat{x}$ is the orthogonal projection of x into Y.

PROOF: We first prove the "if" part. Suppose $\langle x - \hat{x}, Y \rangle = 0$. Take any $\alpha \in Y$, $\alpha \neq 0$. Then

$$\langle x - \hat{x} + \alpha, x - \hat{x} + \alpha \rangle = \langle x - \hat{x}, x - \hat{x} \rangle + 2\langle x - \hat{x}, \alpha \rangle + \langle \alpha, \alpha \rangle$$

$$= \langle x - \hat{x}, x - \hat{x} \rangle + \langle \alpha, \alpha \rangle$$

$$> \langle x - \hat{x}, x - \hat{x} \rangle.$$

Now we shall prove the "only if" part. Suppose there exists an α such that

$$\langle x - \hat{x}, \alpha \rangle = \beta \neq 0.$$

Then, for any scalar λ,

$$\langle x - \hat{x} + \lambda \alpha, x - \hat{x} + \lambda \alpha \rangle = \| x - \hat{x} \|^2 + 2\lambda\beta + \lambda^2 \| \alpha \|^2.$$

Then, for $\lambda = -\beta / \| \alpha \|^2$,

$$\| x - \hat{x} + \lambda \alpha \|^2 = \| x - \hat{x} \|^2 - \frac{2\beta^2}{\| \alpha \|^2} + \frac{\beta^2}{\| \alpha \|^2} < \| x - \hat{x} \|^2. \quad \blacksquare$$

In our filtering problem, the approximation space is $Y_k = \{$space spanned by $y_1, ..., y_k\}$. This is a km-dimensional space. The space X is infinite-dimensional. Let the set of m-vectors $\{u_1, ..., u_k\}$ be an orthonormal basis for Y_k, that is,

$$\mathscr{E}\{u_i u_i^T\} = I, \quad \mathscr{E}\{u_i u_j^T\} = 0, \quad i \neq j.$$

It is interesting that the minimization of

$$\mathscr{E}\{(x_k - \hat{x}_k{}^k)^{\mathsf{T}} (x_k - \hat{x}_k{}^k)\} \tag{7.30}$$

is accomplished componentwise. That is, in minimizing (7.30), we shall minimize each component of the expected error.

Now, from the orthogonal projection lemma, we have

$$\mathscr{E}\{(x_k - \hat{x}_k{}^k) u_i{}^{\mathsf{T}}\} = 0 \qquad \text{for all} \quad i,$$

or

$$\mathscr{E}\{x_k u_i{}^{\mathsf{T}}\} = \mathscr{E}\{\hat{x}_k{}^k u_i{}^{\mathsf{T}}\}.$$

Multiplying by u_i and summing,

$$\sum_{i=1}^{k} \mathscr{E}\{x_k u_i{}^{\mathsf{T}}\} u_i = \sum_{i=1}^{k} \mathscr{E}\{\hat{x}_k{}^k u_i{}^{\mathsf{T}}\} u_i . \tag{7.31}$$

According to the lemma, $\hat{x}_k{}^k$ is the orthogonal projection of x_k into Y. Therefore,

$$\hat{x}_k{}^k = \sum_{i=1}^{k} A_i u_i \tag{7.32}$$

for some set of constant matrices A_i. Using (7.32), it is easy to show that the right-hand side of (7.31) is simply $\hat{x}_k{}^k$; thus,

$$\hat{x}_k{}^k = \sum_{i=1}^{k} \mathscr{E}\{x_k u_i{}^{\mathsf{T}}\} u_i$$

or

$$\hat{x}_k{}^k = \sum_{i=1}^{k-1} \mathscr{E}\{x_k u_i{}^{\mathsf{T}}\} u_i + \mathscr{E}\{x_k u_k{}^{\mathsf{T}}\} u_k ,$$

and, using (7.27),

$$\hat{x}_k{}^k = \sum_{i=1}^{k-1} \mathscr{E}\{[\Phi(k, k - 1) x_{k-1} + \Gamma w_k] u_i{}^{\mathsf{T}}\} u_i + \mathscr{E}\{x_k u_k{}^{\mathsf{T}}\} u_k .$$

Now w_k is independent of Y_{k-1}, and so we get

$$\hat{x}_k{}^k = \Phi(k, k - 1) \hat{x}_{k-1}^{k-1} + \mathscr{E}\{x_k u_k{}^{\mathsf{T}}\} u_k . \tag{7.33}$$

Now u_k is orthogonal to Y_{k-1}. We next show that

$$[y_k - M(t_k) \Phi(k, k - 1) \hat{x}_{k-1}^{k-1}]$$

is also orthogonal to Y_{k-1} , and, since it clearly lies in Y_k , we must have

$$\mathscr{E}\{x_k u_k{}^{\mathrm{T}}\} u_k = K_k[y_k - M(t_k)\, \Phi(k, k-1)\, \hat{x}_{k-1}^{k-1}]$$

for some constant matrix K_k . Clearly,

$$\mathscr{E}\{(x_{k-1} - \hat{x}_{k-1}^{k-1})\, y_i{}^{\mathrm{T}}\} = 0, \qquad i = 1,..., k-1,$$

and, multiplying this by Φ and using (7.27),

$$\mathscr{E}\{(x_k - \Phi \hat{x}_{k-1}^{k-1})\, y_i{}^{\mathrm{T}}\} = 0, \qquad i = 1,..., k-1.$$

Multiplying this by M and using (7.2),

$$\mathscr{E}\{(y_k - M\Phi \hat{x}_{k-1}^{k-1})\, y_i{}^{\mathrm{T}}\} = 0, \qquad i = 1,..., k-1.$$

Therefore,

$$\hat{x}_k{}^k = \Phi(k, k-1)\, \hat{x}_{k-1}^{k-1} + K_k(y_k - M(t_k)\, \Phi(k, k-1)\, \hat{x}_{k-1}^{k-1}). \qquad (7.34)$$

It remains to determine the gain K_k .

The estimation error is given by

$$\tilde{x}_k{}^k = x_k - \hat{x}_k{}^k = \Phi(k, k-1)\, \tilde{x}_{k-1}^{k-1} + \Gamma(k-1)\, w_k - K_k M(k)\, \Phi(k, k-1)\, \tilde{x}_{k-1}^{k-1}$$

$$- K_k[M(k)\, \Gamma(k-1)\, w_k + v_k], \qquad (7.35)$$

where we have used (7.27) and (7.2); and

$$y_k = M(k)\, \Phi(k, k-1)(\tilde{x}_{k-1}^{k-1} + \hat{x}_{k-1}^{k-1}) + M(k)\, \Gamma(k-1)\, w_k + v_k .$$

Therefore, by the orthogonal projection lemma,

$$0 = \mathscr{E}\{\tilde{x}_k{}^k y_k{}^{\mathrm{T}}\} = \Phi(k, k-1)\, P_{k-1}^{k-1} \Phi^{\mathrm{T}}(k, k-1)\, M^{\mathrm{T}}(k)$$

$$+ \Gamma(k-1)\, Q_k \Gamma^{\mathrm{T}}(k-1)\, M^{\mathrm{T}}(k)$$

$$- K_k[M(k)\, \Phi(k, k-1)\, P_{k-1}^{k-1} \Phi^{\mathrm{T}}(k, k-1)\, M^{\mathrm{T}}(k)$$

$$+ M(k)\, \Gamma(k-1)\, Q_k \Gamma^{\mathrm{T}}(k-1)\, M^{\mathrm{T}}(k) + R_k], \qquad (7.36)$$

where we have defined

$$P_{k-1}^{k-1} \triangleq \mathscr{E}\{\tilde{x}_{k-1}^{k-1} \tilde{x}_{k-1}^{k-1}{}^{\mathrm{T}}\}.$$

Defining in addition

$$P_k^{k-1} \triangleq \Phi(k, k-1)\, P_{k-1}^{k-1} \Phi^{\mathrm{T}}(k, k-1) + \Gamma(k-1)\, Q_k \Gamma^{\mathrm{T}}(k-1),$$

we get from (7.36)

$$K_k = P_k^{k-1} M^{\mathrm{T}}(k)[M(k)\, P_k^{k-1} M^{\mathrm{T}}(k) + R_k]^{-1}.$$

Using (7.35), it is easy to compute the difference equation for

$$P_k{}^k = \mathcal{E}\{\tilde{x}_k{}^k \tilde{x}_k^{k\mathrm{T}}\}$$

and thus to see that we have derived the Kalman–Bucy filter in Theorem 7.2. ▲

In the next several examples, we shall rederive the Kalman–Bucy filter using some of the statistical methods outlined in Chapter 5.

Example 7.2 (*Recursive Least Squares*). This method has been used by many authors (e.g., Swerling [53], Gainer [19], Fagin [16], Mowery [40], and Bryson and Ho [10]) to derive the discrete Kalman–Bucy filter. We give a version of the method here. For simplicity, we assume that the dynamical system is noise-free [$w_k \equiv 0$ in (7.27)].

In this case, the least squares problem (5.22), (5.23) at time t_k is to minimize, with respect to $\{x_0, ..., x_k\}$,

$$J_k' = \frac{1}{2}(x_0 - \bar{x}_0)^{\mathrm{T}} P_0^{-1}(x_0 - \bar{x}_0) + \frac{1}{2}\sum_{i=1}^{k}(y_i - M(i)x_i)^{\mathrm{T}} R_i^{-1}(y_i - M(i)x_i),$$

subject to the constraints

$$x_{i+1} = \Phi(i+1, i) x_i, \qquad i = 0, ..., k-1.$$

In view of the constraints, we may alternatively minimize

$$J_k = \frac{1}{2}(x_k - \bar{x}_k)^{\mathrm{T}} \Phi^{\mathrm{T}}(0, k) P_0^{-1}\Phi(0, k)(x_k - \bar{x}_k)$$

$$+ \frac{1}{2}\sum_{i=1}^{k}(y_i - M(i)\Phi(i, k) x_k)^{\mathrm{T}} R_i^{-1}(y_i - M(i)\Phi(i, k) x_k) \quad (7.37)$$

with respect to x_k. Note that $\Phi(i, k) = \Phi^{-1}(k, i)$. Since the *prior* estimate of x_0 is $\bar{x}_0$, that of x_k is $\bar{x}_k = \Phi(k, 0)\bar{x}_0$.

To minimize J_k in (7.37), we set its gradient with respect to x_k equal to zero:

$$\Phi^{\mathrm{T}}(0, k) P_0^{-1}\Phi(0, k)(x_k - \bar{x}_k) - \sum_{i=1}^{k}\Phi^{\mathrm{T}}(i, k) M^{\mathrm{T}}(i) R_i^{-1}(y_i - M(i)\Phi(i, k) x_k) = 0.$$

$$(7.38)$$

Now defining

$$\mathscr{I}_{k,1} \triangleq \sum_{i=1}^{k} \Phi^{\mathrm{T}}(i, k) M^{\mathrm{T}}(i) R_i^{-1}M(i) \Phi(i, k) \quad (7.39)$$

(this is the *information matrix* defined in Section 5; see also Appendix 7A) and

$$H_k \triangleq \sum_{i=1}^{k} \Phi^{\mathrm{T}}(i, k) M^{\mathrm{T}}(i) R_i^{-1} y_i , \tag{7.40}$$

the solution $\bar{x}_k{}^k$ of (7.38) is written as

$$\bar{x}_k{}^k = [\Phi^{\mathrm{T}}(0, k) P_0^{-1} \Phi(0, k) + \mathscr{I}_{k,1}]^{-1} [H_k + \Phi^{\mathrm{T}}(0, k) P_0^{-1} \Phi(0, k) \bar{x}_k]. \tag{7.41}$$

Note that the indicated inverse in (7.41) always exists since, by assumption, $P_0 > 0$, and $R_k > 0$ implies $\mathscr{I}_{k,1} \geq 0$ for all k.

Now clearly the minimization of J_{k+1} yields

$$\bar{x}_{k+1}^{k+1} = [\Phi^{\mathrm{T}}(0, k + 1) P_0^{-1} \Phi(0, k + 1) + \mathscr{I}_{k+1,1}]^{-1}$$
$$\times [H_{k+1} + \Phi^{\mathrm{T}}(0, k + 1) P_0^{-1} \Phi(0, k + 1) \bar{x}_{k+1}]. \tag{7.42}$$

It is easy to derive the following difference equations for $\mathscr{I}_{k,1}$ and H_k :

$$\mathscr{I}_{k+1,1} = \Phi^{\mathrm{T}}(k, k + 1) \mathscr{I}_{k,1} \Phi(k, k + 1) + M^{\mathrm{T}}(k + 1) R_{k+1}^{-1} M(k + 1),$$
$$H_{k+1} = \Phi^{\mathrm{T}}(k, k + 1) H_k + M^{\mathrm{T}}(k + 1) R_{k+1}^{-1} y_{k+1} , \tag{7.43}$$

and, using these and the definition

$$P_k^{k^{-1}} \triangleq \Phi^{\mathrm{T}}(0, k) P_0^{-1} \Phi(0, k) + \mathscr{I}_{k,1} , \tag{7.44}$$

to rewrite (7.42) as

$$\bar{x}_{k+1}^{k+1} = [\Phi^{\mathrm{T}}(k, k + 1) P_k^{k^{-1}} \Phi(k, k + 1) + M^{\mathrm{T}}(k + 1) R_{k+1}^{-1} M(k + 1)]^{-1}$$
$$\times [\Phi^{\mathrm{T}}(k, k + 1)(H_k + \Phi^{\mathrm{T}}(0, k) P_0^{-1} \Phi(0, k) \bar{x}_k) + M^{\mathrm{T}}(k + 1) R_{k+1}^{-1} y_{k+1}]. \tag{7.45}$$

Now, eliminating

$$(H_k + \Phi^{\mathrm{T}}(0, k) P_0^{-1} \Phi(0, k) \bar{x}_k)$$

via (7.41) and defining

$$P_{k+1}^{k^{-1}} \triangleq \Phi^{\mathrm{T}}(k, k + 1) P_k^{k^{-1}} \Phi(k, k + 1), \tag{7.46}$$

$$\bar{x}_{k+1}^{k} \triangleq \Phi(k + 1, k) \bar{x}_k{}^k, \tag{7.47}$$

Eq. (7.45) takes the form

$$\bar{x}_{k+1}^{k+1} = [P_{k+1}^{k^{-1}} + M^{\mathrm{T}}(k + 1) R_{k+1}^{-1} M(k + 1)]^{-1} [P_{k+1}^{k^{-1}} \bar{x}_{k+1}^{k} + M^{\mathrm{T}}(k + 1) R_{k+1}^{-1} y_{k+1}]. \tag{7.48}$$

It is easy to show by induction that the difference equations

$$P_{i+1}^{i-1} = \Phi^{\mathrm{T}}(i, i+1)\, P_i^{i-1}\Phi(i, i+1), \tag{7.49}$$

$$P_{i+1}^{i+1-1} = P_{i+1}^{i-1} + M^{\mathrm{T}}(i+1)\, R_{i+1}^{-1}M(i+1); \quad i = 0, 1, \dots \; ; \quad P_0^{0-1} = P_0^{-1}, \tag{7.50}$$

applied in the order listed, have P_k^{k-1} as defined in (7.44), as their solution.

Comparing (7.47), (7.49) with (7.28), and (7.48), (7.50) with (7.11), we see that we have derived the discrete Kalman–Bucy filter. At least, our recursive least squares filter has the same form as the Kalman–Bucy filter. We have not attached any statistical significance to the weighting matrices P_0 and R_k and need not do so. We simply have a recursive algorithm, which requires no data storage, for generating the least squares estimate. Once we make the identifications

$$x_0 \sim N(\bar{x}_0, P_0), \qquad v_k \sim N(0, R_k),$$

then we have the Kalman–Bucy filter. Then it is easy to compute from (7.41) directly that

$$\mathscr{E}\{(x_k - \bar{x}_k^{\,k})(x_k - \bar{x}_k^{\,k})^{\mathrm{T}}\} = P_k^{\,k},$$

where $P_k^{\,k}$ is given in (7.44). ▲

Example 7.3 (Maximum Likelihood). Ho [25], Schmidt [47], Rauch *et al.* [45], and others used a maximum likelihood approach to derive the discrete Kalman–Bucy filter (see also Smith *et al.* [50]). We derive the maximum likelihood (Bayesian) estimator (see Section 5.3) in this example.

Since all the densities in our linear problem are Gaussian, their means correspond to their respective modes. As a consequence, the mode (mean) and its error covariance matrix satisfy (7.28) between observations. The recursion for the mode (mean) at an observation is obtained by maximizing, with respect to x, the conditional density

$$p(x, t_k \mid Y_{t_k}),$$

which is given in Eq. (7.7). This is equivalent to minimizing the term in braces in (7.8). Setting the gradient of Eq. (7.8) with respect to x equal to zero, we get

$$-M^{\mathrm{T}}(k)\, R_k^{-1}(y_k - M(k)\, x_k) + P_k^{k-1-1}(x_k - \hat{x}_k^{k-1}) = 0. \tag{7.51}$$

The solution of (7.51), denote it by $\hat{x}_k{}^k$, is

$$\hat{x}_k{}^k = (M^{\mathrm{T}}(k)\, R_k^{-1} M(k) + P_k^{k-1}{}^{-1})^{-1}\, (M^{\mathrm{T}}(k)\, R_k^{-1} y_k + P_k^{k-1}{}^{-1} \hat{x}_k^{k-1}),$$

which is just the first of Eqs. (7.11). Using the matrix identities in Appendix 7B, we get, as before,

$$\hat{x}_k{}^k = \hat{x}_k^{k-1} + P_k^{k-1} M^{\mathrm{T}}(k)[M(k)\, P_k^{k-1} M^{\mathrm{T}}(k) + R_k]^{-1}\, (y_k - M(k)\, \hat{x}_k^{k-1}),$$

$$(7.14)$$

and the estimation error

$$\tilde{x}_k{}^k \triangleq x_k - \hat{x}_k{}^k = [I - K(k)\, M(k)]\, \tilde{x}_k^{k-1} - K(k)\, v_k\,, \qquad (7.52)$$

where we have used (7.14), (7.2), and (7.15). Therefore, the estimation error covariance matrix is computed to be

$$P_k{}^k = \mathscr{E}\{\tilde{x}_k{}^k \tilde{x}_k{}^{k\mathrm{T}}\} = [I - K(k)\, M(k)]\, P_k^{k-1}[I - K(k)\, M(k)]^{\mathrm{T}} + K(k)\, R_k K^{\mathrm{T}}(k)$$

$$= P_k^{k-1} - K(k)\, M(k)\, P_k^{k-1}. \qquad (7.53)$$

Comparing (7.14) and (7.53) with (7.29), we see that we have derived the Kalman–Bucy filter. ▲

Example 7.4 (Linear Minimum Variance Estimator). Suppose the estimate and its error covariance evolve according to (7.28) between observations. At an observation, assume a *linear* estimator of the form

$$\hat{x}_k{}^k = \hat{x}_k^{k-1} + K(k)[y_k - M(k)\, \hat{x}_k^{k-1}], \qquad (7.54)$$

where the gain $K(k)$ is to be determined so as to minimize the error variance

$$\mathscr{E}\{(x_k - \hat{x}_k{}^k)^{\mathrm{T}}\, (x_k - \hat{x}_k{}^k)\} = \operatorname{tr} \mathscr{E}\{(x_k - \hat{x}_k{}^k)(x_k - \hat{x}_k{}^k)^{\mathrm{T}}\} \triangleq \operatorname{tr} P_k{}^k. \qquad (7.55)$$

This approach to the linear filtering problem was taken by Battin [2, 3], Schmidt [47], and Sorenson [51].

With considerable hindsight, let

$$K(k) = P_k^{k-1} M^{\mathrm{T}}(k)[M(k)\, P_k^{k-1} M^{\mathrm{T}}(k) + R_k]^{-1} + \hat{K}(k), \qquad (7.56)$$

and find that $\hat{K}(k)$ which minimizes (7.55). Using (7.54)–(7.56), we get

$$\operatorname{tr} P_k{}^k = \operatorname{tr}[P_k^{k-1} - P_k^{k-1} M^{\mathrm{T}}(MP_k^{k-1} M^{\mathrm{T}} + R_k)^{-1}\, MP_k^{k-1}$$

$$+ \hat{K}(MP_k^{k-1} M^{\mathrm{T}} + R_k)\, \hat{K}^{\mathrm{T}}], \qquad (7.57)$$

where
$$P_k^{k-1} \triangleq \mathscr{E}\{(x_k - \hat{x}_k^{k-1})(x_k - \hat{x}_k^{k-1})^{\mathrm{T}}\}.$$
Now
$$MP_k^{k-1}M^{\mathrm{T}} + R_k > 0,$$
so that
$$\mathrm{tr}\,\hat{K}(MP_k^{k-1}M^{\mathrm{T}} + R_k)\,\hat{K}^{\mathrm{T}} \geqslant 0$$

for all $\hat{K}$. Therefore, the minimizing $\hat{K} = 0$, and $K(k)$ is the Kalman gain. From (7.57), we have the recursion for P. It is easy to rederive a known result. ▲

The several examples that follow consider some extensions to our filtering problem, as outlined in Section 5.4.

Example 7.5 (Correlated System and Measurement Noise: 1). Suppose the system noise (w_k) and measurement noise (v_k) are correlated, that is,

$$\mathscr{E}\{w_k v_l^{\mathrm{T}}\} = C_k\,\delta_{kl}\,. \tag{7.58}$$

The discrete Kalman–Bucy filter for this case was derived by Kalman [36]. [Note that his problem is slightly different from ours in that the noise in our system (7.27) is $\Gamma(k)w_{k+1}$, whereas Kalman has $\Gamma(k)w_k$.]

We shall use the orthogonal projection lemma of Example 7.1 in our derivation. The reader can verify that the analysis in Example 7.1 up to, but not including, Eq. (7.36) applies in the present case. Equation (7.36) now becomes

$$0 = P_k^{k-1}M^{\mathrm{T}}(k) + \Gamma(k-1)\,C_k - K_k^{\,c}[M(k)\,P_k^{k-1}M^{\mathrm{T}}(k) + M(k)\,\Gamma(k-1)\,C_k$$
$$+ C_k^{\mathrm{T}}\Gamma^{\mathrm{T}}(k-1)\,M^{\mathrm{T}}(k) + R_k], \tag{7.59}$$

where, as before,

$$P_k^{k-1} = \Phi(k, k-1)\,P_{k-1}^{k-1}\Phi^{\mathrm{T}}(k, k-1) + \Gamma(k-1)Q_k\Gamma^{\mathrm{T}}(k-1). \tag{7.60}$$

Therefore, the Kalman gain for this problem is

$$K_k^{\,c} = [P_k^{k-1}M^{\mathrm{T}}(k) + \Gamma(k-1)\,C_k][M(k)\,P_k^{k-1}M^{\mathrm{T}}(k) + M(k)\,\Gamma(k-1)\,C_k$$
$$+ C_k^{\mathrm{T}}\Gamma^{\mathrm{T}}(k-1)\,M^{\mathrm{T}}(k) + R_k]^{-1}. \tag{7.61}$$

Using (7.35) and (7.61), it is easy to compute the recursion for P:

$$P_k^{\,k} = P_k^{k-1} - K_k^{\,c}[M(k)\,P_k^{k-1} + C_k^{\mathrm{T}}\Gamma^{\mathrm{T}}(k-1)]. \tag{7.62}$$

Thus, between observations, the estimate and covariance matrix still evolve according to (7.28). At an observation, the estimate satisfies the first of Eqs. (7.29), but with K_k replaced by $K_k{}^c$. The covariance matrix satisfies (7.62). ▲

Example 7.6 (*Correlated System and Measurement Noise*: 2). Suppose that instead of (7.27), (7.2), our system is

$$x_{k+1} = \Phi(k+1, k)\, x_k + \Gamma(k)\, w_k,$$
$$y_k = M(k)\, x_k + v_k,$$
$$\tag{7.63}$$

with

$$\mathscr{E}\left\{\begin{bmatrix} w_k \\ v_k \end{bmatrix} [w_l{}^{\mathrm{T}}, v_l{}^{\mathrm{T}}]\right\} = \begin{bmatrix} Q_k & C_k \\ C_k{}^{\mathrm{T}} & R_k \end{bmatrix} \delta_{kl}. \tag{7.64}$$

This corresponds to the problem treated by Kalman [36] and others. We derive the optimal filter for this problem using the orthogonal projection lemma again, but in a slightly different form.

Instead of minimizing (7.30), we minimize

$$\mathscr{E}\{(x_{k+1} - \hat{x}_{k+1}^k)^{\mathrm{T}}\, (x_{k+1} - \hat{x}_{k+1}^k)\}.$$

This will enable us to compute a linear recursion for $\hat{x}_{k+1}^k$; that is, we will compute $\hat{x}_{k+1}^k$ as a linear function of $\hat{x}_k^{k-1}$. (In Example 7.1, we computed the recursion for $\hat{x}_k^k$.) The Kalman gain for this problem is more conveniently computed within this structure.

By the orthogonal projection lemma, we have

$$\mathscr{E}\{(x_{k+1} - \hat{x}_{k+1}^k)\, u_i^{\mathrm{T}}\} = 0 \qquad \text{for all } i \leqslant k.$$

Computations analogous to those in Example 7.1 lead to

$$\hat{x}_{k+1}^k = \Phi(k+1, k)\, \hat{x}_k^{k-1} + \mathscr{E}\{x_{k+1} u_k^{\mathrm{T}}\}\, u_k,$$

which is the analog of (7.33). It is easy to show that

$$y_k - M(k)\, \hat{x}_k^{k-1}$$

is orthogonal to Y_{k-1}, and, since it is contained in Y_k, that

$$\mathscr{E}\{x_{k+1} u_k^{\mathrm{T}}\}\, u_k = K_k{}^d[y_k - M(k)\, \hat{x}_k^{k-1}].$$

Thus, the filter for this problem has the structure

$$\hat{x}_{k+1}^k = \Phi(k+1, k)\, \hat{x}_k^{k-1} + K_k{}^d[y_k - M(k)\, \hat{x}_k^{k-1}]. \tag{7.65}$$

It remains to determine the Kalman gain $K_k{}^d$.

From (7.63) and (7.65),

$$\tilde{x}_{k+1}^k \triangleq x_{k+1} - \hat{x}_{k+1}^k = \Phi(k+1, k)\, \tilde{x}_k^{k-1} + \Gamma(k)\, w_k - K_k^d[M(k)\, \tilde{x}_k^{k-1} + v_k],$$
(7.66)

and

$$y_k = M(k)(\tilde{x}_k^{k-1} + \hat{x}_k^{k-1}) + v_k.$$

Therefore, by the orthogonal projection lemma,

$$0 = \mathscr{E}\{\tilde{x}_{k+1}^k y_k^{\mathrm{T}}\}$$
$$= \Phi(k+1, k)\, P_k^{k-1} M^{\mathrm{T}}(k) + \Gamma(k)\, C_k - K_k^d[M(k)\, P_k^{k-1} M^{\mathrm{T}}(k) + R_k],$$

so that the Kalman gain is

$$K_k^d = [\Phi(k+1, k)\, P_k^{k-1} M^{\mathrm{T}}(k) + \Gamma(k)\, C_k][M(k)\, P_k^{k-1} M^{\mathrm{T}}(k) + R_k]^{-1}.$$
(7.67)

Using (7.66) and (7.67), it is easy to compute the recursion for P:

$$P_{k+1}^k \triangleq \mathscr{E}\{\tilde{x}_{k+1}^k \tilde{x}_{k+1}^{k\mathrm{T}}\} = \Phi(k+1, k)\, P_k^{k-1} \Phi^{\mathrm{T}}(k+1, k) + \Gamma(k) Q_k \Gamma^{\mathrm{T}}(k)$$
$$- K_k^d[M(k)\, P_k^{k-1} \Phi^{\mathrm{T}}(k+1, k) + C_k^{\mathrm{T}} \Gamma^{\mathrm{T}}(k)].^4 \quad (7.68)$$

By setting $\Phi = I$, $\Gamma = 0$ in (7.65) and (7.68) (so that $x_{k+1} = x_k$), we have

$$\hat{x}_k^k = \hat{x}_k^{k-1} + K_k(y_k - M(k)\, \hat{x}_k^{k-1}),$$
(7.69)
$$P_k^k = P_k^{k-1} - K_k M(k)\, P_k^{k-1},$$

where

$$K_k = P_k^{k-1} M^{\mathrm{T}}(k)[M(k)\, P_k^{k-1} M^{\mathrm{T}}(k) + R_k]^{-1}.$$
(7.15)

Then, using (7.69) with (7.65) and (7.68),

$$\hat{x}_{k+1}^k = \Phi(k+1, k)\, \hat{x}_k^k + \Gamma(k)\, C_k[M(k)\, P_k^{k-1} M^{\mathrm{T}}(k) + R_k]^{-1}\, (y_k - M(k)\, \hat{x}_k^{k-1}),$$
$$P_{k+1}^k = \Phi(k+1, k)\, P_k^k \Phi^{\mathrm{T}}(k+1, k) + \Gamma(k) Q_k \Gamma^{\mathrm{T}}(k)$$
(7.70)
$$- \Gamma(k)\, C_k[M(k)\, P_k^{k-1} M^{\mathrm{T}}(k) + R_k]^{-1}\, C_k^{\mathrm{T}} \Gamma^{\mathrm{T}}(k)$$
$$- \Phi(k+1, k)\, K_k C_k^{\mathrm{T}} \Gamma^{\mathrm{T}}(k) - \Gamma(k)\, C_k K_k^{\mathrm{T}} \Phi^{\mathrm{T}}(k+1, k).$$

Prediction for $k > l + 1$, based on Y_l, is clearly accomplished via

$$\hat{x}_{k+1}^l = \Phi(k+1, k)\, \hat{x}_k^l,$$
$$P_{k+1}^l = \Phi(k+1, k)\, P_k^l \Phi^{\mathrm{T}}(k+1, k) + \Gamma(k) Q_k \Gamma^{\mathrm{T}}(k), \qquad k = l+1, l+2, \ldots,$$

[4] Equation (III$_d$), p. 303 in Kalman [36], which is analogous to this equation, is in error.

where $\hat{x}^l_{l+1}$, P^l_{l+1} is the filtering solution. This is easily verified directly from (7.63) (take the expectation with Y_l fixed). ▲

Example 7.7 (*Sequentially Correlated* (*Colored*) *Measurement Noise*). Consider the system (7.63), where $\{w_k\}$ is white, Gaussian, $w_k \sim N(0, Q_k)$, but $\{v_k\}$ is sequentially correlated, the output of the linear system (see Example 4.12)

$$v_{k+1} = \Psi(k+1, k)\, v_k + u_k, \qquad v_0 \sim N(0, V), \qquad \mathscr{E}\{x_0 v_0^{\mathrm{T}}\} = 0. \quad (7.71)$$

Here $\{u_k\}$ is also white, Gaussian, $u_k \sim N(0, S_k)$, independent of $\{w_k\}$. We could proceed as follows. Define the *augmented* state vector

$$\bar{x}_k^{\mathrm{T}} = [x_k^{\mathrm{T}}, \quad v_k^{\mathrm{T}}]$$

and the matrices

$$\Phi = \begin{bmatrix} \Phi, & 0 \\ 0, & \Psi \end{bmatrix}, \qquad \bar{\Gamma} = \begin{bmatrix} \Gamma, & 0 \\ 0, & I \end{bmatrix}, \qquad \bar{w}_k^{\mathrm{T}} = [w_k^{\mathrm{T}}, \quad u_k^{\mathrm{T}}],$$

$$\bar{M} = [M, \quad I], \qquad \bar{Q}_k = \begin{bmatrix} Q_k, & 0 \\ 0, & S_k \end{bmatrix}.$$

Then the augmented dynamical system is

$$\bar{x}_{k+1} = \bar{\Phi}(k+1, k)\, \bar{x}_k + \bar{\Gamma}(k)\, \bar{w}_k, \qquad y_k = \bar{M}(k)\, \bar{x}_k, \quad (7.72)$$

with *perfect* measurements uncontaminated by noise. Assuming $\bar{M}(k)\, \bar{P}_k^{k-1}\bar{M}^{\mathrm{T}}(k)$ is positive definite,[5] the Kalman–Bucy filter of Theorem 7.2 can be used in system (7.72). Recall that its derivation in Example 7.1 does not require a nonsingular measurement noise covariance matrix. In fact, the derivation stands in the absence of measurement noise ($R_k \equiv 0$). Thus the difference equations for the augmented state estimation error covariance matrix are

$$\bar{P}_k^k = \bar{P}_k^{k-1} - \bar{P}_k^{k-1}\bar{M}^{\mathrm{T}}(k)[\bar{M}(k)\, \bar{P}_k^{k-1}\bar{M}^{\mathrm{T}}(k)]^{-1}\, \bar{M}(k)\, \bar{P}_k^{k-1},$$

$$\bar{P}_{k+1}^k = \bar{\Phi}(k+1, k)\, \bar{P}_k^k \bar{\Phi}^{\mathrm{T}}(k+1, k) + \bar{\Gamma}(k)\, \bar{Q}_k \bar{\Gamma}^{\mathrm{T}}(k). \qquad (7.73)$$

Now since the linear combinations of state variables $\bar{M}(k)\bar{x}_k$ are known precisely at time t_k, $\bar{P}_k^k$ must be singular. In fact, from (7.73),

$$\bar{M}(k)\, \bar{P}_k^k \bar{M}^{\mathrm{T}}(k) = 0.$$

---

[5] In fact, this matrix need not be positive definite, if, following Kalman [36], we use the pseudo-inverse rather than the inverse in (7.73).

As a consequence, $\bar{P}_{k+1}^k$ can be ill-conditioned, leading to difficulties in the computation of $\bar{P}_{k+1}^{k+1}$. Furthermore, the dimension of the augmented state filter is $n + m$, which also compounds the computational problems.

Following Bryson and Hendrickson [9], we introduce a measurement differencing scheme that results in measurements with additive white noise and leads to an equivalent filter of dimension n, in which computational ill-conditioning is eliminated, or at least reduced. With the aid of (7.71), the measurement can be written as

$$y_k = M(k)\, x_k + \Psi(k, k - 1)\, v_{k-1} + u_{k-1} .$$

It is easily seen that v_{k-1} can be eliminated from y_k by subtracting

$$\Psi(k, k - 1)\, y_{k-1} .$$

Thus,

$$
\begin{aligned}
y_k - \Psi(k, k - 1)\, y_{k-1} &= M(k)\, x_k + u_{k-1} - \Psi(k, k - 1)\, M(k - 1)\, x_{k-1} \\
&= [M(k)\, \Phi(k, k - 1) - \Psi(k, k - 1)\, M(k - 1)]\, x_{k-1} \\
&\quad + M(k)\, \Gamma(k - 1)\, w_{k-1} + u_{k-1} .
\end{aligned}
$$

Let

$$H(k - 1) \triangleq M(k)\, \Phi(k, k - 1) - \Psi(k, k - 1)\, M(k - 1), \qquad (7.74)$$

and define the measurements

$$
\begin{aligned}
\xi_1 &\triangleq y_1 = M(1)\, x_1 + v_1 , \\
\xi_k &\triangleq y_k - \Psi(k, k - 1)\, y_{k-1} = H(k - 1)\, x_{k-1} + M(k)\, \Gamma(k - 1)\, w_{k-1} + u_{k-1} , \\
&\qquad\qquad\qquad\qquad\qquad\qquad\qquad\qquad\qquad\qquad\qquad\qquad k \geqslant 2.
\end{aligned}
\tag{7.75}
$$

The measurements ξ_k, $k \geqslant 2$, now contain additive white noise which, we note, is correlated with the system noise (w). The general approach is this. We process ξ_1 with the augmented-state filter described earlier. Subsequent observations ξ_k will be processed via the filter developed in Example 7.6, where the system is

$$x_k = \Phi(k, k - 1)\, x_{k-1} + \Gamma(k - 1)\, w_{k-1} .$$

Thus, aside from the starting procedure (processing ξ_1), the filter will be of dimension n. Our measurement noise covariance matrix will be

$$R_{k-1} = M(k)\, \Gamma(k - 1)\, Q_{k-1} \Gamma^{\mathrm{T}}(k - 1)\, M^{\mathrm{T}}(k) + S_{k-1} ,$$

which, we assume, is positive definite.

Assume, for simplicity, that $t_0 = t_1$. That is, the first observation is made at the initial time t_0. [If this is not the case, then a prediction via (7.28) propagates the initial conditions to t_1.] Then applying the augmented-state filter to ξ_1 results in the following estimate of x_1 (*not* $\bar{x}_1$) and the associated covariance matrix:

$$\hat{x}_1{}^1 = \hat{x}_0 + P_0 M^T(1)[M(1)\,P_0 M^T(1) + V]^{-1}\,(y_1 - M(1)\,\hat{x}_0),$$

$$P_1{}^1 = P_0 - P_0 M^T(1)[M(1)\,P_0 M^T(1) + V]^{-1}\,M(1)\,P_0. \tag{7.76}$$

Now we use the n-dimensional filter of Example 7.6 on measurements ξ_k, $k \geqslant 2$, with initial conditions $\hat{x}_1{}^1$ and $P_1{}^1$ from (7.76).
The system is

$$x_{k+1} = \Phi(k+1, k)\,x_k + \Gamma(k)\,w_k,$$

$$\xi_{k+1} = y_{k+1} - \Psi(k+1, k)\,y_k = H(k)\,x_k + M(k+1)\,\Gamma(k)\,w_k + u_k, \quad k \geqslant 1,$$

$$H(k) = M(k+1)\,\Phi(k+1, k) - \Psi(k+1, k)\,M(k). \tag{7.77}$$

Note that ξ_{k+1}, although it depends on x_k and is to be formally treated as a measurement at t_k, depends on y_{k+1}. As a consequence, in applying the filter equations (7.65) and (7.68) to (7.77), $\hat{x}_{k+1}$ and P_{k+1} in those filter equations are, in fact, $\hat{x}_{k+1}^{k+1}$ and P_{k+1}^{k+1}, respectively. That is to say, they are conditioned on Y_{k+1}, *not* on Y_k. Keeping this in mind, and noting that y_k in (7.65) is now ξ_{k+1}, and M is now H, we have

$$\hat{x}_{k+1}^{k+1} = \Phi(k+1, k)\,\hat{x}_k{}^k + [\Phi(k+1, k)\,P_k{}^k H^T(k) + \Gamma(k)\,C_k]$$

$$\times\,[H(k)\,P_k{}^k H^T(k) + R_k]^{-1}\,[\xi_{k+1} - H(k)\,\hat{x}_k{}^k],$$

$$P_{k+1}^{k+1} = \Phi(k+1, k)\,P_k{}^k \Phi^T(k+1, k) + \Gamma(k)\,Q_k \Gamma^T(k) \tag{7.78}$$

$$-\,[\Phi(k+1, k)\,P_k{}^k H^T(k) + \Gamma(k)\,C_k]$$

$$\times\,[H(k)\,P_k{}^k H^T(k) + R_k]^{-1}\,[H(k)\,P_k{}^k \Phi^T(k+1, k) + C_k{}^T \Gamma^T(k)],$$

where

$$C_k = Q_k \Gamma^T(k)\,M^T(k+1),$$

$$R_k = M(k+1)\,\Gamma(k)\,Q_k \Gamma^T(k)\,M^T(k+1) + S_k. \tag{7.79}$$

Note that the reduced filter (7.78) requires the storage of one measurement, since ξ_k is the difference of two consecutive measurements.
The reduced filter does not generate estimates of v_k, but these

estimates are, in general, not desired. It is easy to compute them, however. From the measurement equation (7.63), it readily follows that

$$\hat{v}_k{}^k = y_k - M(k)\,\hat{x}_k{}^k,$$

$$V_k{}^k \triangleq \mathscr{E}\{(v_k - \hat{v}_k{}^k)(v_k - \hat{v}_k{}^k)^{\mathrm{T}}\} = M(k)\,P_k{}^k M^{\mathrm{T}}(k). \tag{7.80}$$

From the system equations (7.77) and (7.71), it is easy to compute the prediction equations

$$\hat{x}_{k+1}^l = \Phi(k+1,k)\,\hat{x}_k{}^l,$$

$$v_{k+1}^l = \Psi(k+1,k)\,\hat{v}_k{}^l,$$

$$P_{k+1}^l = \Phi(k+1,k)\,P_k{}^l \Phi^{\mathrm{T}}(k+1,k) + \Gamma(k)Q_k\Gamma^{\mathrm{T}}(k), \tag{7.81}$$

$$V_{k+1}^l = \Psi(k+1,k)\,V_k{}^l \Psi^{\mathrm{T}}(k+1,k) + S_k, \qquad k = l, l+1,...,$$

where $\hat{x}_l{}^l$, $v_l{}^l$, $P_l{}^l$, $V_l{}^l$ are the filtering solutions.

Now the reduced filter (7.78) was derived formally, assuming that optimal filtering of observations ξ_k is equivalent to optimal filtering of observations y_k. This latter fact remains to be proved. Actually, it is easy to prove using the orthogonal projection lemma (see Rishel [46]). The fact that (7.78) is the (unique) minimum variance filter will be proved if we show that

$$\mathscr{E}\{(x_k - \hat{x}_k{}^k)\,y_l{}^{\mathrm{T}}\} = 0, \qquad \text{all} \quad l \leqslant k. \tag{7.82}$$

Now it was proved in Example 7.6 that

$$\mathscr{E}\{(x_k - \hat{x}_k{}^k)\,\xi_l{}^{\mathrm{T}}\} = 0, \qquad \text{all} \quad l \leqslant k. \tag{7.83}$$

But $\xi_1 = y_1$, so (7.82) is true for $l = 1$. Suppose (7.82) is true for l. Then

$$\mathscr{E}\{(x_k - \hat{x}_k{}^k)\,y_{l+1}^{\mathrm{T}}\} = \mathscr{E}\{(x_k - \hat{x}_k{}^k)\,\xi_{l+1}^{\mathrm{T}}\} + \mathscr{E}\{(x_k - \hat{x}_k{}^k)\,y_l{}^{\mathrm{T}}\Psi^{\mathrm{T}}\} = 0,$$

using (7.77), (7.83), and the inductive hypothesis. This proves (7.82). Since

$$v_k - \hat{v}_k{}^k = M(k)(x_k - \hat{x}_k{}^k),$$

(7.82) also implies that

$$\mathscr{E}\{(v_k - \hat{v}_k{}^k)\,y_l{}^{\mathrm{T}}\} = 0, \qquad \text{all} \quad l \leqslant k. \quad \blacktriangle$$

Example 7.8 (Linear Smoother). We develop here one form of the linear smoother using the Bayesian maximum likelihood approach of

Rauch *et al.* [45]. Other derivations and (equivalent) forms of the linear smoother may be found in the references cited in Section 5.1.

Our system is (7.27), (7.2) with $\{w_k\}$ and $\{v_k\}$ independent. We assume for simplicity (see Section 3.9) that $\Gamma(k)$ and $Q(k)$ are $n \times n$ and nonsingular. The solution in the general case, without this assumption, is the same, but the derivation requires the cumbersome machinery developed in Section 3.9 to get an expression for the transition density. We leave that case for the reader. Hint: the Dirac delta function is formally maximized when its argument is zero. Set its argument equal to zero, and use that relation as a constraint.

Assuming the filtering solution $(\hat{x}_k{}^k, P_k{}^k)$ is available, we seek a recursion for $\hat{x}_k{}^l$, $k < l$. Now the conditional means $\hat{x}_k{}^l$ and $\hat{x}_{k+1}^l$ are those values of x_k and x_{k+1}, respectively, which maximize the (Gaussian) density

$$p(x_k, x_{k+1} \mid Y_l).$$

Now

$$p(x_k, x_{k+1} \mid Y_l) = \frac{p(x_k, x_{k+1}, Y_l)}{p(Y_l)} = \frac{p(x_k, x_{k+1}, Y_k, y_{k+1}, \dots, y_l)}{p(Y_l)}$$

$$= \frac{p(Y_k)}{p(Y_l)} p(x_k, x_{k+1}, y_{k+1}, \dots, y_l \mid Y_k)$$

$$= \frac{p(Y_k)}{p(Y_l)} p(y_{k+1}, \dots, y_l \mid x_k, x_{k+1}, Y_k) p(x_k, x_{k+1} \mid Y_k).$$

But

$$p(y_{k+1}, \dots, y_l \mid x_k, x_{k+1}, Y_k) = p(y_{k+1}, \dots, y_l \mid x_{k+1}),$$

and

$$p(x_k, x_{k+1} \mid Y_k) = p(x_{k+1} \mid x_k, Y_k) p(x_k \mid Y_k) = p(x_{k+1} \mid x_k) p(x_k \mid Y_k),$$

in view of the properties of our system, so that

$$p(x_k, x_{k+1} \mid Y_l) = c(x_{k+1}) p(x_{k+1} \mid x_k) p(x_k \mid Y_k), \tag{7.84}$$

where c is independent of x_k.

The densities in (7.84) are Gaussian, specifically,

$$p(x_{k+1} \mid x_k) \sim N[\Phi(k+1, k) x_k, \Gamma(k) Q_{k+1} \Gamma^{\mathrm{T}}(k)],$$

$$p(x_k \mid Y_k) \sim N(\hat{x}_k{}^k, P_k{}^k).$$

Therefore, to maximize (7.84) with respect to x_k, x_{k+1}, we minimize

$$\tfrac{1}{2}[x_{k+1} - \Phi(k+1, k) x_k]^{\mathrm{T}} [\Gamma(k) Q_{k+1} \Gamma^{\mathrm{T}}(k)]^{-1} [x_{k+1} - \Phi(k+1, k) x_k]$$

$$+ \tfrac{1}{2}(x_k - \hat{x}_k{}^k)^{\mathrm{T}} P_k^{-1}(x_k - \hat{x}_k{}^k) + d(x_{k+1}),$$

where d is independent of x_k. Suppose $\hat{x}_{k+1}^l$, the minimizing x_{k+1}, is available. Then $\hat{x}_k{}^l$ minimizes

$$J = \tfrac{1}{2}[\hat{x}_{k+1}^l - \Phi x_k]^{\mathrm{T}} [\Gamma Q_{k+1} \Gamma^{\mathrm{T}}]^{-1} [\hat{x}_{k+1}^l - \Phi x_k]$$

$$+ \tfrac{1}{2}(x_k - \hat{x}_k{}^k)^{\mathrm{T}} P_k^{k-1}(x_k - \hat{x}_k{}^k)$$

with respect to x_k. Setting the gradient of J with respect to x_k equal to zero, we get

$$\hat{x}_k{}^l = [P_k^{k-1} + \Phi^{\mathrm{T}}(k+1, k)[\Gamma(k) Q_{k+1} \Gamma^{\mathrm{T}}(k)]^{-1} \Phi(k+1, k)]^{-1}$$

$$\times [P_k^{k-1}\hat{x}_k{}^k + \Phi^{\mathrm{T}}(k+1, k)[\Gamma(k) Q_{k+1} \Gamma^{\mathrm{T}}(k)]^{-1} \hat{x}_{k+1}^l]. \qquad (7.85)$$

Applying Eqs. (7B.5) and (7B.6) of Appendix 7B, we have

$$\hat{x}_k{}^l = \hat{x}_k{}^k + S_k(\hat{x}_{k+1}^l - \Phi(k+1, k)\,\hat{x}_k{}^k), \qquad (7.86)$$

where

$$S_k = P_k{}^k \Phi^{\mathrm{T}}(k+1, k)[\Phi(k+1, k)\,P_k{}^k \Phi^{\mathrm{T}}(k+1, k) + \Gamma(k) Q_{k+1} \Gamma^{\mathrm{T}}(k)]^{-1} \qquad (7.87)$$

$$= P_k{}^k \Phi^{\mathrm{T}}(k+1, k)\,P_{k+1}^{k-1}.$$

Equation (7.86) is the smoothing algorithm. The computation goes backwards (in the index k) with $\hat{x}_l{}^l$, the filtering solution, as initial condition. The filter solutions $(\hat{x}_k{}^k, P_k{}^k)$ are required in the computations.

We now develop a recursion for the smoothing error covariance matrix. From (7.86),

$$x_k - \hat{x}_k{}^l = x_k - \hat{x}_k{}^k - S_k\,(\hat{x}_{k+1}^l - \Phi\hat{x}_k{}^k),$$

or

$$\tilde{x}_k{}^l + S_k\hat{x}_{k+1}^l = \tilde{x}_k{}^k + S_k\Phi\hat{x}_k{}^k.$$

Squaring both sides and taking the expectation, we get

$$P_k{}^l + S_k\mathscr{E}\{\hat{x}_{k+1}^l\hat{x}_{k+1}^{l\mathrm{T}}\}\,S_k{}^{\mathrm{T}} = P_k{}^k + S_k\Phi\mathscr{E}\{\hat{x}_k{}^k\hat{x}_k{}^{k\mathrm{T}}\}\,\Phi^{\mathrm{T}}S_k{}^{\mathrm{T}}, \qquad (7.88)$$

where we have used the fact that

$$\mathscr{E}\{\tilde{x}_k{}^l\hat{x}_{k+1}^{l\mathrm{T}}\} = \mathscr{E}\{\tilde{x}_k{}^k\hat{x}_k{}^{k\mathrm{T}}\} = 0. \qquad (7.89)$$

It is easy to see that (7.89) holds. For example,

$$\mathscr{E}\{\tilde{x}_k{}^l\hat{x}_{k+1}^{l\mathrm{T}}\} = \mathscr{E}\{\tilde{x}_k{}^l\hat{x}_k{}^{l\mathrm{T}}\}\,\Phi^{\mathrm{T}} = \mathscr{E}\{\mathscr{E}\{\tilde{x}_k{}^l\hat{x}_k{}^{l\mathrm{T}} \mid Y_l\}\}\,\Phi^{\mathrm{T}}$$

$$= \mathscr{E}\{\mathscr{E}\{\tilde{x}_k{}^l \mid Y_l\}\,\hat{x}_k{}^{l\mathrm{T}}\}\,\Phi^{\mathrm{T}} = \mathscr{E}\{(\hat{x}_k{}^l - \hat{x}_k{}^l)\,\hat{x}_k{}^{l\mathrm{T}}\}\,\Phi^{\mathrm{T}} = 0.$$

Now

$$\mathscr{E}\{x_{k+1}x_{k+1}^{\mathrm{T}}\} = \mathscr{E}\{\hat{x}_{k+1}^{l}\hat{x}_{k+1}^{l\mathrm{T}}\} + P_{k+1}^{l} \tag{7.90}$$

in view of (7.89). Similarly,

$$\mathscr{E}\{x_{k}x_{k}^{\mathrm{T}}\} = \mathscr{E}\{\hat{x}_{k}^{k}\hat{x}_{k}^{k\mathrm{T}}\} + P_{k}^{k}. \tag{7.91}$$

Also,

$$\mathscr{E}\{x_{k+1}x_{k+1}^{\mathrm{T}}\} = \Phi\mathscr{E}\{x_{k}x_{k}^{\mathrm{T}}\}\Phi^{\mathrm{T}} + \Gamma Q_{k+1}\Gamma^{\mathrm{T}}. \tag{7.92}$$

Using (7.90)–(7.92) in (7.88), we get

$$P_{k}^{l} = P_{k}^{k} + S_{k}[P_{k+1}^{l} - \Phi(k+1, k) P_{k}^{k}\Phi^{\mathrm{T}}(k+1, k) - \Gamma(k)Q_{k+1}\Gamma^{\mathrm{T}}(k)] S_{k}^{\mathrm{T}} \tag{7.93}$$

or

$$P_{k}^{l} = P_{k}^{k} + S_{k}(P_{k+1}^{l} - P_{k+1}^{k}) S_{k}^{\mathrm{T}}, \tag{7.94}$$

which is the desired recursion. It is also computed backwards, with P_{l}^{l} as initial condition.

Suppose our system is noise-free ($\Gamma = 0$). Then, from (7.86) and (7.94), it is easy to see that smoothing is simply *backward* prediction:

$$\hat{x}_{k}^{l} = \Phi^{-1}(k+1, k) \hat{x}_{k+1}^{l},$$

$$P_{k}^{l} = \Phi^{-1}(k+1, k) P_{k+1}^{l}\Phi^{-\mathrm{T}}(k+1, k).$$

See Bryson and Henrikson [9] for the linear smoothing algorithm in the case of sequentially correlated measurement noise. ▲

4. CONTINUOUS FILTER

Our continuous linear system is described by the vector (Itô) stochastic differential equation (7.1), which we repeat here,

$$dx_t = F(t) x_t dt + G(t) d\beta_t, \qquad t \geqslant t_0, \tag{7.95}$$

and the vector (Itô) observation equation,

$$dz_t = M(t) x_t dt + d\eta_t, \qquad t \geqslant t_0. \tag{7.96}$$

Equation (7.95) is described in Section 2. The observed process $\{z_t, t \geqslant t_0\}$ is an m-vector process, $M(t)$ is an $m \times n$, nonrandom, continuous matrix time-function, and $\{\eta_t, t \geqslant t_0\}$ is an m-vector Brownian motion process with $\mathscr{E}\{d\eta_t d\eta_t^{\mathrm{T}}\} = R(t) dt$, $R(t) > 0$. x_{t_0},

$\{\beta_t\}$, and $\{\eta_t\}$ are assumed independent. The formal analogy between (7.95), (7.96), and

$$dx_t/dt = F(t)\, x_t + G(t)\, w_t, \tag{7.95'}$$

$$y_t = M(t)\, x_t + v_t, \tag{7.96'}$$

where $\{w_t\}$ and $\{v_t\}$ are white Gaussian noise processes, $w_t \sim N(0, Q(t))$, $v_t \sim N(0, R(t))$, was noted in Section 5.1. (See also Chapter 4, especially Section 4.8.)

For this linear filtering problem, Kushner's equation (6.79) of Theorem 6.5 becomes

$$dp = -p\, \mathrm{tr}(F)\, dt - p_x{}^{\mathrm{T}} Fx\, dt + \tfrac{1}{2}\, \mathrm{tr}(GQG^{\mathrm{T}} p_{xx})\, dt$$

$$+ (x - \hat{x})^{\mathrm{T}}\, M^{\mathrm{T}} R^{-1}(dz_t - M\hat{x}\, dt)\, p. \tag{7.97}$$

We obtain the equations for the conditional mean and covariance matrix by specializing Theorem 6.6 to the linear case. Since $f = Fx$, $h = Mx$,

$$(\widehat{xh^{\mathrm{T}}} - \hat{x}\hat{h}^{\mathrm{T}}) = (\widehat{xx^{\mathrm{T}}} - \hat{x}\hat{x}^{\mathrm{T}})\, M^{\mathrm{T}},$$

and similarly for the terms involving xf^{T}. Since, in this case, the conditional density is Gaussian, its third central moments are zero,

$$[(x_i - \hat{x}_i)(x_j - \hat{x}_j)(x_k - \hat{x}_k)]^{\widehat{}} = 0,$$

and, as a consequence, the last term in (6.101) is zero. We therefore have

Theorem 7.3. *The optimal (minimum variance) filter for the continuous system* (7.95), (7.96) *consists of the following differential equations for the conditional mean and covariance matrix*:

$$d\hat{x}_t{}^t = F(t)\, \hat{x}_t{}^t\, dt + P_t{}^t M^{\mathrm{T}}(t)\, R^{-1}(t)(dz_t - M(t)\, \hat{x}_t{}^t\, dt), \tag{7.98}$$

$$dP_t{}^t/dt = F(t)\, P_t{}^t + P_t{}^t F^{\mathrm{T}}(t) + G(t) Q(t)\, G^{\mathrm{T}}(t) - P_t{}^t M^{\mathrm{T}}(t)\, R^{-1}(t)\, M(t)\, P_t{}^t,$$

$$t \geqslant t_0. \tag{7.99}$$

Prediction is accomplished as in the continuous-discrete problem (*Theorem* 7.1). [*Set* $R^{-1} \equiv 0$ *in* (7.98) *and* (7.99).]

This is the now well-known continuous Kalman–Bucy filter [33, 36]. The *variance equation* (7.99) is the well-known matrix *Riccati* equation. It is independent of the observed process and, as a result, the Kalman gain

$$K(t) = P_t{}^t M^{\mathrm{T}}(t)\, R^{-1}(t) \tag{7.100}$$

may be precomputed. The discussion of Theorem 7.1 applies to Theorem 7.3 as well. Since the Kalman gain is a nonrandom time-function, the coefficient of dz_t in the filter equation (7.98) is nonrandom. As a result, Eq. (7.98) may be manipulated by formal rules (see Section 4.6). We may write (7.98) as

$$d\hat{x}_t^{\,t}/dt = F(t)\,\hat{x}_t^{\,t} + P_t^{\,t}M^{\mathrm{T}}(t)\,R^{-1}(t)(y_t - M(t)\,\hat{x}_t^{\,t}).\tag{7.98}$$

Compare Theorem 7.3 with Example 6.4.

The mechanization of the continuous Kalman–Bucy filter requires the solution of the variance equation (7.99). We must verify that the variance equation has a unique solution existing for all $t \geqslant t_0$.

Now since the matrices F, G, M, Q, and R are assumed continuous, and, being quadratic in P, the right-hand side of (7.99) is locally Lipschitzian, standard existence and uniqueness proofs (see Coddington and Levinson [11]) show that the variance equation has a unique solution locally. We now show that the unique solution of (7.99) can be continued for all $t \geqslant t_0$, provided P_{t_0} is positive semidefinite (a covariance matrix). It suffices to show that $P_t^{\,t}$ is bounded for all $t \geqslant t_0$ [11]. By Theorem 2.13,

$$P_t^{\,t} = P_t^{\,t_0} - B,$$

where $B \geqslant 0$ and $P_t^{\,t_0}$ is the unconditioned covariance matrix of the solution of (7.95) given only the initial condition. That is,

$$P_t^{\,t} \leqslant P_t^{\,t_0}.\tag{7.101}$$

But it is easy to compute from (7.95) [see (7.19)–(7.25)] that

$$P_t^{\,t_0} = \Phi(t, t_0)\,P_{t_0}\Phi^{\mathrm{T}}(t, t_0) + \int_{t_0}^{t} \Phi(t, \tau)\,G(\tau)\,Q(\tau)\,G^{\mathrm{T}}(\tau)\,\Phi^{\mathrm{T}}(t, \tau)\,d\tau,\tag{7.102}$$

which is bounded for all $t \geqslant t_0$. This, together with (7.101), shows that $P_t^{\,t}$ is bounded for all $t \geqslant t_0$. As a result, the variance equation has a unique solution existing for all $t \geqslant t_0$.

It is useful to note, and can easily be verified, that the solution of the (nonlinear) variance equation can be obtained as a solution of the linear system:

$$\begin{aligned} dX/dt &= -F^{\mathrm{T}}(t)\,X + M^{\mathrm{T}}(t)\,R^{-1}(t)\,M(t)\,Y, & X(t_0) &= I, \\ dY/dt &= G(t)\,Q(t)\,G^{\mathrm{T}}(t)\,X + F(t)\,Y, & Y(t_0) &= P_{t_0}. \end{aligned}\tag{7.103}$$

For then

$$Y(t) = P_t^{\,t}X(t) \qquad \text{for all} \quad t \geqslant t_0,\tag{7.104}$$

so that

$$P_t{}^t = Y(t)\, X^{-1}(t). \tag{7.105}$$

This result was derived by Kalman [36].

The next few examples give alternate derivations of the continuous Kalman–Bucy filter.

Example 7.9 (Conditional Characteristic Function). We use here the method of Example 4.15 (characteristic function), together with Itô's stochastic calculus, to derive the Kalman–Bucy filter from Kushner's equation (7.97). The conditional characteristic function (2.122) is defined as

$$\varphi(u, t, z_t) = \int e^{iu^\mathrm{T}x} p(x, t \mid Y_t)\, dx.$$

Therefore,

$$d\varphi = \int e^{iu^\mathrm{T}x} \,(dp)\, dx.$$

Substituting from Kushner's equation and carrying out the integrations by parts, we get

$$d\varphi = u^\mathrm{T}F\varphi_u\, dt - \tfrac{1}{2}\varphi u^\mathrm{T}GQG^\mathrm{T}u\, dt - \left(i\varphi_u{}^\mathrm{T} + \varphi\hat{x}_t^{t\mathrm{T}}\right) M^\mathrm{T}(t)\, R^{-1}(t)\, (dz_t - M(t)\,\hat{x}_t{}^t\, dt). \tag{7.106}$$

This is a stochastic partial differential equation for the conditional characteristic function.

Now since p is Gaussian, we know that the conditional characteristic function is given by (2.130):

$$\varphi(u, t, z_t) = \exp(iu^\mathrm{T}\hat{x}_t{}^t - \tfrac{1}{2}u^\mathrm{T}P_t{}^tu), \tag{7.107}$$

so that

$$\varphi_u = \varphi(i\hat{x}_t{}^t - P_t{}^tu).$$

Substituting this in (7.106), we get

$$d\varphi = \varphi iu^\mathrm{T}[F\hat{x}_t{}^t\, dt + P_t{}^tM^\mathrm{T}R^{-1}\, (dz_t - M\hat{x}_t{}^t\, dt)]$$
$$- \tfrac{1}{2}\varphi u^\mathrm{T}(FP_t{}^t + P_t{}^tF^\mathrm{T} + GQG^\mathrm{T})\, u\, dt. \tag{7.108}$$

We next compute $d\varphi$ from (7.107). Since φ is a function of $\hat{x}_t{}^t$ and $P_t{}^t$, which may be solutions of stochastic differential equations, we must use Itô's Lemma 4.2. Applying Itô's lemma, we first compute the imaginary part of $d\varphi$ from (7.107). Note that this part will not involve partials with respect to the elements of $P_t{}^t$, nor will it involve $\varphi_{\hat{x}\hat{x}}$, since these are real. Thus we have

$$\text{Im}(d\varphi/\varphi) = iu^{\text{T}}\,d\hat{x}_t{}^t.$$

Comparing this with (7.108), we immediately have

$$d\hat{x}_t{}^t = F\hat{x}_t{}^t\,dt + P_t{}^t M^{\text{T}} R^{-1}\,(dz_t - M\hat{x}_t{}^t\,dt), \qquad (7.98)$$

which is (7.98).

Now we compute $\text{Re}(d\varphi)$ from (7.107). Using Itô's lemma,

$$\text{Re}(d\varphi/\varphi) = -\tfrac{1}{2}\,\text{tr}\,P_t{}^t M^{\text{T}} R^{-1} R R^{-1} M P_t{}^t u u^{\text{T}}\,dt - \tfrac{1}{2}\,\text{tr}\,u u^{\text{T}}\,dP_t{}^t$$

$$+ \text{(terms quartic in the elements of } u \text{ resulting from second}$$

$$\text{partials of } \varphi \text{ with respect to the elements of } P_t{}^t), \qquad (7.109)$$

where we have used the fact that $\hat{x}_t{}^t$ satisfies (7.98). Now there are no terms quartic in the elements of u in (7.108). Therefore, such terms must have zero coefficients in (7.109). Thus (7.109) becomes

$$\text{Re}(d\varphi/\varphi) = -\tfrac{1}{2}u^{\text{T}}P_t{}^t M^{\text{T}} R^{-1} M P_t{}^t u\,dt - \tfrac{1}{2}u^{\text{T}}\,dP_t{}^t u,$$

and, comparing this with (7.108), we get (7.99). ▲

Example 7.10 (*Limit of Discrete Filter*). We obtain the continuous filter (Theorem 7.3) from the discrete filter (Theorem 7.2) by the formal limiting process that was already employed in Example 4.18. For $\Delta t = t_{k+1} - t_k$ small, Eq. (7.95′) is integrated to produce the difference equation

$$x_{t_{k+1}} - x_{t_k} = F(t_k)\,x_{t_k}\,\Delta t + \Delta t\,G(t_k)\,w_{t_{k+1}} + o(\Delta t)$$

or

$$x_{t_{k+1}} = [I + \Delta t\,F(t_k)]\,x_{t_k} + \Delta t\,G(t_k)\,w_{t_{k+1}}, \qquad (7.110)$$

where

$$\mathscr{E}\{w_{t_k}w_{t_k}^{\text{T}}\} = Q(t_k)/\Delta t,$$

according to Example 3.20. The observations (7.96′) are

$$y_{t_k} = M(t_k)\,x_{t_k} + v_{t_k}, \qquad (7.111)$$

with

$$\mathscr{E}\{v_{t_k}v_{t_k}^{\text{T}}\} = R(t_k)/\Delta t.$$

Comparing (7.110), (7.111) with (7.27), (7.2),

$$\Phi(t_{k+1}, t_k) = I + \Delta t \, F(t_k), \qquad \Gamma(t_k) = \Delta t \, G(t_k). \tag{7.112}$$

Using these relations in Theorem 7.2, we get

$$(\hat{x}_{t_{k+1}}^{t_{k+1}} - \hat{x}_{t_k}^{t_k})/\Delta t = F(t_k) \, \hat{x}_{t_k}^{t_k} + (P_{t_k}^{t_k} + O(\Delta t)) \, M^{\mathrm{T}}(t_k)$$

$$\times [\Delta t \, M(P_{t_k}^{t_k} + O(\Delta t)) \, M^{\mathrm{T}} + R(t_k)]^{-1}$$

$$\times (y_{t_k} - M(t_k) \, \hat{x}_{t_k}^{t_k} - O(\Delta t)) + o(\Delta t)/\Delta t,$$

$$(P_{t_{k+1}}^{t_{k+1}} - P_{t_k}^{t_k})/\Delta t = F(t_k) \, P_{t_k}^{t_k} + P_{t_k}^{t_k} F^{\mathrm{T}}(t_k) + G(t_k) \, Q(t_{k+1}) \, G^{\mathrm{T}}(t_k)$$

$$- (P_{t_k}^{t_k} + O(\Delta t)) \, M^{\mathrm{T}}(t_k) \, [\Delta t \, M(P_{t_k}^{t_k} + O(\Delta t)) \, M^{\mathrm{T}} + R(t_k)]^{-1}$$

$$\times [M(t_k)(P_{t_k}^{t_k} + O(\Delta t))] + o(\Delta t)/\Delta t.$$

In the $\Delta t = 0$ limit, we get Theorem 7.3. ▲

Example 7.11 (Calculus of Variations). We solve the continuous least squares problem (5.24) using the calculus of variations [20]. Minimize

$$J_t' = \frac{1}{2}(x_{t_0} - \bar{x}_{t_0})^{\mathrm{T}} P_{t_0}^{-1}(x_{t_0} - \bar{x}_{t_0})$$

$$+ \frac{1}{2} \int_{t_0}^{t} (y_\tau - M(\tau) \, x_\tau)^{\mathrm{T}} R^{-1}(\tau)(y_\tau - M(\tau) \, x_\tau) \, d\tau$$

$$+ \frac{1}{2} \int_{t_0}^{t} w_\tau^{\mathrm{T}} Q^{-1}(\tau) \, w_\tau \, d\tau \tag{7.113}$$

with respect to x_τ and w_τ, $t_0 \leqslant \tau \leqslant t$, subject to the constraint

$$dx_\tau/d\tau = F(\tau) \, x_\tau + G(\tau) \, w_\tau, \qquad t_0 \leqslant \tau \leqslant t. \tag{7.114}$$

This variational approach to estimation was first taken by Bryson and Frazier [7]. The solution $\hat{x}_\tau^{\,t}$ is the smoothing solution.

In view of the constraint (7.114), x_{t_0} and w_τ, $t_0 \leqslant \tau \leqslant t$, determine x_τ, $t_0 \leqslant \tau \leqslant t$. Thus, we minimize J_t' with respect to x_{t_0} and w_τ, $t_0 \leqslant \tau \leqslant t$. $\hat{x}_\tau^{\,t}$ will then be determined by (7.114) once the minimizing x_{t_0} and w_τ are established. Using the standard calculus of variations technique, we adjoin the constraint (7.114) to J_t' via a vector Lagrange multiplier function $\lambda(\tau)$:

$$J_t = \frac{1}{2}(x_{t_0} - \bar{x}_{t_0})^T P_{t_0}^{-1}(x_{t_0} - \bar{x}_{t_0})$$

$$+ \frac{1}{2}\int_{t_0}^t [(y_\tau - Mx_\tau)^T R^{-1}(y_\tau - Mx_\tau) + w_\tau^T Q^{-1}w_\tau$$

$$+ 2\lambda^T(\dot{x}_\tau - Fx_\tau - Gw_\tau)]\, d\tau.$$

We compute the first variation of J_t :

$$\delta J_t = [(x_{t_0} - \bar{x}_{t_0})^T P_{t_0}^{-1} - \lambda^T(t_0)]\,\delta x_{t_0} + \lambda^T(t)\,\delta x_t$$

$$+ \int_{t_0}^t \{-[(y_\tau - Mx_\tau)^T R^{-1}M + \dot{\lambda}^T + \lambda^T F]\,\delta x_\tau + [w_\tau^T Q^{-1} - \lambda^T G]\,\delta w_\tau\}\, d\tau.$$

Necessary conditions for $\delta J_t = 0$ are

$$(x_{t_0} - \bar{x}_{t_0})^T P_{t_0}^{-1} - \lambda^T(t_0) = 0, \tag{7.115}$$

$$\lambda(t) = 0, \tag{7.116}$$

$$d\lambda/d\tau = -F^T\lambda - M^TR^{-1}(y_\tau - Mx_\tau), \tag{7.117}$$

$$w_\tau^T Q^{-1} - \lambda^T G = 0. \tag{7.118}$$

Eliminating w_τ from (7.117) and (7.114) via (7.118), we have the two-point boundary-value problem:

$$d\lambda(\tau)/d\tau = -F^T\lambda(\tau) + M^TR^{-1}Mx_\tau - M^TR^{-1}y_\tau, \tag{7.119}$$

$$dx_\tau/d\tau = GQG^T\lambda(\tau) + Fx_\tau, \tag{7.120}$$

$$x_{t_0} = \bar{x}_{t_0} + P_{t_0}\lambda(t_0), \tag{7.121}$$

$$\lambda(t) = 0. \tag{7.122}$$

Compare (7.119), (7.120) with (7.103).

This problem is linear and easy to solve [7]. Let $y(\tau)$ and $\xi(\tau)$ be particular solutions of (7.120) and (7.119), respectively, with initial conditions

$$y(t_0) = \bar{x}_{t_0}, \qquad \xi(t_0) = 0. \tag{7.123}$$

Let $Y(\tau)$ and $X(\tau)$ be as defined in (7.103). Then it is easy to show that the smoothing solution is

$$\hat{x}_\tau{}^t = y(\tau) - Y(\tau)\,X^{-1}(t)\,\xi(t) \tag{7.124}$$

and

$$\lambda(\tau) = \xi(\tau) - X(\tau) X^{-1}(t) \xi(t).$$

Now we are interested in the filtering solution

$$\hat{x}_t^t = y(t) - Y(t) X^{-1}(t) \xi(t) = y(t) - P_t^t \xi(t) \tag{7.125}$$

in view of (7.105). We already saw that P_t^t satisfies the variance equation. To obtain the differential equation for the estimate, we differentiate (7.125) with respect to t:

$$d\hat{x}_t^t/dt = dy(t)/dt - (dP_t^t/dt)\, \xi(t) - P_t^t\, d\xi(t)/dt.$$

Using the variance equation, (7.119), and (7.120), it is easy to show that

$$d\hat{x}_t^t/dt = F\hat{x}_t^t + P_t^t M^T R^{-1}(y_t - M\hat{x}_t^t). \quad \blacktriangle$$

Example 7.12 (Dynamic Programming). The solution of the continuous least squares problem of Example 7.11 via dynamic programming [4] was first given by Cox [13]. The problem of minimizing J_t' (7.113) subject to constraint (7.114) is *imbedded* in a family of problems parameterized by values of x_t by defining the *cost* function

$$V(x_t, t) = \min_{\{w_\tau, t_0 \leqslant \tau \leqslant t\}} J_t'. \tag{7.126}$$

Keeping in mind the probabilistic interpretation of the least squares problem (Sect. 5.3), $-V(x, t)$ can be interpreted as a measure of the likelihood of the most likely trajectory passing through $x_t = x$. The filtering solution $\hat{x}_t^t$ is clearly that value of x_t which minimizes V.

By the principle of optimality [4], we have the following functional equation for V:

$$V(x_t, t) = \min_{\{w_\tau, t-\delta t \leqslant \tau \leqslant t\}} \left\{ V(x_t - \delta x, t - \delta t) \right.$$

$$\left. + \frac{1}{2} \int_{t-\delta t}^{t} [(y_\tau - Mx_\tau)^T R^{-1}(y_\tau - Mx_\tau) + w_\tau^T Q^{-1} w_\tau]\, d\tau \right\}. \tag{7.127}$$

Now for δt small, the integral in (7.127) is

$$[(y_t - Mx_t)^T R^{-1}(y_t - Mx_t) + w_t^T Q^{-1} w_t]\, \delta t + o(\delta t)$$

and

$$V(x_t - \delta x, t - \delta t) = V(x_t, t) - V_t(x_t, t)\, \delta t - V_x^T(x_t, t)\, \delta x + o(\delta t),$$

so that (7.127) becomes

$$V_t = \min_{\{w_\tau, t-\delta t \leqslant \tau \leqslant t\}} \{-V_x{}^\mathrm{T}(\delta x/\delta t) + \tfrac{1}{2}(y_t - Mx_t)^\mathrm{T} R^{-1}(y_t - Mx_t)$$
$$+ \tfrac{1}{2}w_t{}^\mathrm{T}Q^{-1}w_t + o(\delta t)/\delta t\}.$$

Substituting for $\delta x/\delta t$ from (7.114) and passing to the $\delta t = 0$ limit, we formally get the Bellman partial differential equation

$$V_t = \min_{w_t}\{-V_x{}^\mathrm{T}(Fx_t + Gw_t) + \tfrac{1}{2}(y_t - Mx_t)^\mathrm{T} R^{-1}(y_t - Mx_t) + \tfrac{1}{2}w_t{}^\mathrm{T}Q^{-1}w_t\}.$$
$$(7.128)$$

The minimizing w_t is easy to compute by setting the derivative of the term in braces equal to zero:

$$w_t = QG^\mathrm{T}V_x. \tag{7.129}$$

Comparing (7.129) with (7.118), we see that V_x plays the role of the Lagrange multiplier. Thus we finally have

$$V_t = -V_x{}^\mathrm{T}Fx_t - \tfrac{1}{2}V_x{}^\mathrm{T}GQG^\mathrm{T}V_x + \tfrac{1}{2}(y_t - Mx_t)^\mathrm{T} R^{-1}(y_t - Mx_t). \tag{7.130}$$

The boundary condition is clearly

$$V(x_{t_0}, t_0) = \tfrac{1}{2}(x_{t_0} - \bar{x}_{t_0})^\mathrm{T} P_{t_0}^{-1}(x_{t_0} - \bar{x}_{t_0}).$$

The solution of the estimation problem is reduced to the solution of (7.130). Once $V(x(\tau), \tau)$ is found, then, in view of (7.129), backward integration of (7.114) produces $\hat{x}_\tau{}^t$.

To get a formal filtering solution, assume V of the form

$$V(x_t, t) = \tfrac{1}{2}(x_t - \hat{x}_t{}^t)^\mathrm{T} P_t^{t-1}(x_t - \hat{x}_t{}^t) + a(t). \tag{7.131}$$

$\hat{x}_t{}^t$ clearly minimizes V with respect to x_t. From (7.131),

$$V_x = P_t^{t-1}(x_t - \hat{x}_t{}^t),$$

$$V_t = \dot{a} - (x_t - \hat{x}_t{}^t)^\mathrm{T} P_t^{t-1} d\hat{x}_t{}^t/dt + \tfrac{1}{2}(x_t - \hat{x}_t{}^t)^\mathrm{T} [d(P_t^{t-1})/dt] (x_t - \hat{x}_t{}^t).$$

Substituting these expressions into (7.130) and equating like powers of x_t, we get

$$d(P_t^{t-1})/dt = -P_t^{t-1}F - F^\mathrm{T}P_t^{t-1} - P_t^{t-1}GQG^\mathrm{T}P_t^{t-1} + M^\mathrm{T}R^{-1}M,$$
$$d\hat{x}_t{}^t/dt = F\hat{x}_t{}^t + P_t{}^t M^\mathrm{T}R^{-1}(y_t - M\hat{x}_t{}^t). \tag{7.132}$$

Since

$$PP^{-1} = I,$$
$$dP/dt = -P[d(P^{-1})/dt]\, P.$$

Using this fact, (7.132) gives the variance equation. ▲

Example 7.13 (Invariant Imbedding). We return to the variational two-point boundary-value problem (7.119)–(7.122) of Example 7.11, and obtain the filtering solution by invariant imbedding (see Bellman *et al.* [5] and Detchmendy and Sridhar [14]).

Suppose we solve the problem (7.119)–(7.122) and obtain the missing terminal condition on x_t, call it $\hat{x}_t^t$. Then, for a new time $t' > t$, the two-point boundary-value problem has to be re-solved to produce $\lambda(t') = 0$ and obtain $\hat{x}_t^{t'}$. If the solution of the original problem is continued from t to t', then, in general, $\lambda(t') = c \neq 0$. Clearly, x_t is a function of t and c:

$$x_t = r(c, t) \tag{7.133}$$

and

$$\hat{x}_t^t = r(0, t). \tag{7.134}$$

We have imbedded the original problem in a family of problems parameterized by c. $r(0, t)$ is the solution of our filtering problem.

Now, on the one hand,

$$r(c + \delta c, t + \delta t) = r(c, t) + (GQG^T c + Fr)\, \delta t + o(\delta t) \tag{7.135}$$

in view of (7.120), and on the other, using a Taylor series expansion,

$$r(c + \delta c, t + \delta t) = r(c, t) + r_c\, \delta c + r_t\, \delta t + o(\delta t), \tag{7.136}$$

where, from (7.119),

$$\delta c = [-F^T c + M^T R^{-1} M r - M^T R^{-1} y_t]\, \delta t\,.$$

Equating (7.135) and (7.136), dividing by δt, and passing to the $\delta t = 0$ limit, we formally get

$$r_t + r_c(-F^T c + M^T R^{-1} M r - M^T R^{-1} y_t) = GQG^T c + Fr. \tag{7.137}$$

This is a partial differential equation for r.

Assume a solution of the form

$$r(c, t) = \hat{x}_t^t + P(t)\, c. \tag{7.138}$$

This is motivated by (7.134). Then, from (7.138),

$$r_c = P(t), \qquad r_t = \hat{x}_t{}^t + \dot{P}(t)\, c. \tag{7.139}$$

Substituting (7.138) and (7.139) into (7.137), and regarding the result as an identity in c (equate coefficients of c^0 and c to zero), produces the continuous Kalman–Bucy filter. ▲

In the next three examples, we obtain the continuous filter in the case of correlated system and measurement noise, colored measurement noise, and the continuous linear smoother by using the limiting procedure of Example 7.10 in Examples 7.5–7.8.

Example 7.14 (Correlated System and Measurement Noise). Suppose in the system (7.95), (7.96) that

$$\mathcal{E}\{d\beta_t\, d\eta_t{}^{\mathrm{T}}\} = C(t)\, dt.$$

Then, using the limiting procedure of Example 7.10 in the filter of Example 7.5, where

$$\mathcal{E}\{w_{t_k} v_{t_k}{}^{\mathrm{T}}\} = C(t_k)/\Delta t,$$

we easily obtain

$$d\hat{x}_t{}^t/dt = F(t)\, \hat{x}_t{}^t + [P_t{}^t M^{\mathrm{T}}(t) + G(t)\, C(t)]\, R^{-1}(t)(y_t - M(t)\, \hat{x}_t{}^t),$$

$$dP_t{}^t/dt = F(t)\, P_t{}^t + P_t{}^t F^{\mathrm{T}} + G(t)Q(t)\, G^{\mathrm{T}}(t) \tag{7.140}$$

$$\qquad - [P_t{}^t M^{\mathrm{T}}(t) + G(t)\, C(t)]\, R^{-1}(t)[M(t)\, P_t{}^t + C^{\mathrm{T}}(t)\, G^{\mathrm{T}}(t)].$$

The reader can verify for himself that the same result is obtained from the discrete filter of Example 7.6. The filter (7.140) was first derived by Kalman [36], and also by Cox [13]. ▲

Example 7.15 (Colored Measurement Noise). Suppose v_t in (7.96′) is generated by

$$dv_t/dt = L(t)\, v_t + u_t, \qquad v_{t_0} \sim N(0, V), \qquad \mathcal{E}\{x_{t_0} v_{t_0}{}^{\mathrm{T}}\} = 0. \tag{7.141}$$

$\{u_t\}$ is white, Gaussian, $u_t \sim N[0, S(t)]$, independent of $\{w_t\}$. We apply the limiting procedure of Example 7.10 to the discrete filter of Example 7.7. We have

$$v_{t_{k+1}} - v_{t_k} = L(t_k)\, v_{t_k}\, \Delta t + \Delta t\, u_{t_k} + o(\Delta t),$$

where

$$\mathscr{E}\{u_{t_k}u_{t_k}^{\mathrm{T}}\} = S(t_k)/\Delta t,$$

so that

$$\Psi(t_{k+1}, t_k) = I + \Delta t\, L(t_k).$$

Note that u_k in (7.71) is replaced by $\Delta t\, u_{t_k}$.
Then, $t_k \leqslant t \leqslant t_{k+1}$,

$$\xi_{k+1} = [dy_t/dt - L(t)\, y_t]\, \Delta t + o(\Delta t),$$

$$H(k) = \bar{H}(t)\, \Delta t + o(\Delta t),$$

$$\bar{H}(t) \triangleq dM/dt + [M(t)F(t) - L(t)\, M(t)], \qquad (7.142)$$

$$C_k = Q(t)\, G^{\mathrm{T}}(t)\, M^{\mathrm{T}}(t) + O(\Delta t),$$

$$\mathbb{R}_k = \bar{R}(t)\, \Delta t + o(\Delta t),$$

$$\bar{R}(t) \triangleq M(t)\, G(t)\, Q(t)\, G^{\mathrm{T}}(t)\, M^{\mathrm{T}}(t) + S(t). \qquad (7.143)$$

Therefore, from (7.78),

$$(\hat{x}_{t_{k+1}}^{t_{k+1}} - \hat{x}_{t_k}^{t_k})/\Delta t = F\hat{x}_{t_k}^{t_k} + [P_{t_k}^{t_k}\bar{H}^{\mathrm{T}} + GQG^{\mathrm{T}}M^{\mathrm{T}} + O(\Delta t)]$$

$$\times\, [\bar{R}\,\Delta t + o(\Delta t)]^{-1}[(dy/dt)\,\Delta t - Ly\,\Delta t - \bar{H}\hat{x}_{t_k}^{t_k}\,\Delta t + o(\delta t)],$$

where we have omitted the time argument t. In the $\Delta t = 0$ limit, we get

$$d\hat{x}_t^{\,t}/dt = F(t)\,\hat{x}_t^{\,t} + [P_t^{\,t}\bar{H}^{\mathrm{T}}(t) + G(t)\,Q(t)\,G^{\mathrm{T}}(t)\,M^{\mathrm{T}}(t)]$$

$$\times\, \bar{R}^{-1}(t)[dy_t/dt - L(t)\, y_t - \bar{H}(t)\,\hat{x}_t^{\,t}]. \qquad (7.144)$$

The filter equation (7.144) involves the differentiation of observations. This can be removed for practical applications in the following way. Define

$$B(t) \triangleq [P_t^{\,t}\bar{H}^{\mathrm{T}}(t) + G(t)\,Q(t)\,G^{\mathrm{T}}(t)\,M^{\mathrm{T}}(t)]\, \bar{R}^{-1}(t). \qquad (7.145)$$

Since

$$\frac{d}{dt}\,(\hat{x}_t^{\,t} - B(t)\, y_t) = d\hat{x}_t^{\,t}/dt - B(t)\, dy_t/dt - \dot{B}(t)\, y_t,$$

(7.144) becomes

$$\frac{d}{dt}\,(\hat{x}_t^{\,t} - B(t)\, y_t) = F(t)\,\hat{x}_t^{\,t} + B(t)[-L(t)\, y_t - \bar{H}(t)\,\hat{x}_t^{\,t}] - \dot{B}(t)\, y_t. \qquad (7.146)$$

Similarly, from (7.78) we get the variance equation

$$dP_t{}^t/dt = F(t)\,P_t{}^t + P_t{}^t F^{\mathrm{T}}(t) + G(t)\,Q(t)\,G^{\mathrm{T}}(t) - B(t)\,\bar{R}(t)\,B^{\mathrm{T}}(t). \quad (7.147)$$

The initial conditions for (7.146), (7.147) follow from (7.76):

$$\hat{x}_{t_0}^{t_0} = \hat{x}_{t_0} + P_{t_0} M^{\mathrm{T}}(t_0)[M(t_0)\,P_{t_0} M^{\mathrm{T}}(t_0) + V]^{-1}\,(y_{t_0} - M(t_0)\,\hat{x}_{t_0}),$$

$$\qquad\qquad\qquad\qquad\qquad\qquad\qquad\qquad (7.148)$$

$$P_{t_0}^{t_0} = P_{t_0} - P_{t_0} M^{\mathrm{T}}(t_0)[M(t_0)\,P_{t_0} M^{\mathrm{T}}(t_0) + V]^{-1}\,M(t_0)\,P_{t_0}.$$

Thus, there is a discontinuity in the estimate and the covariance matrix at the initial time.

This filter was first derived by Bryson and Johansen [8], and later by Mehra and Bryson [39], who used variational techniques. The latter reference also derives the smoother for the colored measurement noise case. ▲

Example 7.16 (Continuous Linear Smoother). Using again the limiting procedure of Example 7.10, applied to the discrete smoother of Example 7.8, we derive the continuous smoother for system (7.95), (7.96). Let $t_k \leqslant t \leqslant t_{k+1} \leqslant t_l$, t_l fixed. From (7.86),

$$\hat{x}_{t_{k+1}}^{t_l} - \hat{x}_{t_k}^{t_l} = \Phi(t_{k+1}, t_k)\,\hat{x}_{t_k}^{t_k} - \hat{x}_{t_k}^{t_l} + S_k^{-1}(\hat{x}_{t_k}^{t_l} - \hat{x}_{t_k}^{t_k})$$

$$= (\hat{x}_{t_k}^{t_k} - \hat{x}_{t_k}^{t_l}) + \Delta t\,F(t_k)\,\hat{x}_{t_k}^{t_k} + P_{t_k}^{t_k}[I + \Delta t\,F^{\mathrm{T}}(t_k)]^{-1}\,P_{t_k}^{t_k^{-1}}(\hat{x}_{t_k}^{t_l} - \hat{x}_{t_k}^{t_k})$$

$$\quad + \Delta t[F(t_k)\,P_{t_k}^{t_k} + P_{t_k}^{t_k}F^{\mathrm{T}}(t_k) + G(t_k)\,Q(t_k)\,G^{\mathrm{T}}(t_k)]$$

$$\quad \times [I + O(\Delta t)]^{-1}\,P_{t_k}^{t_k^{-1}}(\hat{x}_{t_k}^{t_l} - \hat{x}_{t_k}^{t_k}) + o(\Delta t).$$

But

$$[I + \Delta t\,F^{\mathrm{T}}(t_k)]^{-1} = I - \Delta t\,F^{\mathrm{T}}(t_k)[I + \Delta t\,F^{\mathrm{T}}(t_k)]^{-1},$$

and so, in the $\Delta t = 0$ limit, we get

$$d\hat{x}_t^{t_l}/dt = F(t)\,\hat{x}_t^{t_l} + G(t)\,Q(t)\,G^{\mathrm{T}}(t)\,P_t^{t^{-1}}(\hat{x}_t^{t_l} - \hat{x}_t^{t}). \quad (7.149)$$

Similarly,

$$dP_t^{t_l}/dt = [F(t) + G(t)\,Q(t)\,G^{\mathrm{T}}(t)\,P_t^{t^{-1}}]\,P_t^{t_l} + P_t^{t_l}[F^{\mathrm{T}}(t) + P_t^{t^{-1}}G(t)\,Q(t)\,G^{\mathrm{T}}(t)]$$

$$\quad - G(t)\,Q(t)\,G^{\mathrm{T}}(t). \quad (7.150)$$

Equations (7.149) and (7.150) are to be integrated backwards from t_l with initial conditions

$$\hat{x}_{t_l}^{t_l}, \qquad P_{t_l}^{t_l}.$$

This smoothing algorithm was first derived by Rauch, Tung, and Striebel [45] (see also Bryson and Frazier [7]). ▲

5. OBSERVABILITY AND INFORMATION

Assuming that a dynamical system (dynamics and measurement equation) has been defined, the preceding sections have dealt with derivations of the optimal filter for that system. Now it is important to ask what, if anything, can be gained from filtering the data. How much information about the state of the system is contained in the data? Can the state be determined from the data? Intuitively, it would seem that answers to such questions are related to the system model itself, and indeed this is so. The importance of such questions is obvious. If little is to be gained from filtering, then we should consider remodeling the system. This might involve taking additional or alternate measurements, or redesigning the dynamics of the system.

How well the state is known is measured by the estimation error covariance matrix $P_t{}^t$. But $P_t{}^t$ depends on its initial condition P_{t_0} (the initial data) and does not reflect the uncertainty in the estimate by virtue of filtering the data alone. Consider the discrete least squares problem of Example 7.2. Set $P_0^{-1} = 0$, which means that no weight is attached to the prior estimate $\bar{x}_0$; there is no prior information. Then from (7.41), it is clear that, in order to determine x_k, the *information matrix*

$$\mathscr{I}(k, 1) \triangleq \sum_{i=1}^{k} \Phi^{\mathrm{T}}(i, k) \, M^{\mathrm{T}}(i) \, R_i^{-1} M(i) \, \Phi(i, k) \qquad (7.151)$$

must be positive definite. If $\mathscr{I}(k, 1)$ is singular, then certain linear combinations of the elements of x_k cannot be determined; there is no information about them in the data $\{y_1, ..., y_k\}$. Notice that the information matrix depends on Φ and M, the system model, and *not* on the data themselves. Also note that the dynamical system in Example 7.2 is noise-free. From (7.44), we see that uncertainty is inversely proportional to information.

This discussion motivates the following definitions, which are due to Kalman [29, 34]. The discrete dynamical system (7.27), (7.2) is said to be *completely observable* if, and only if,

$$\mathscr{I}(k, 0) > 0 \qquad \text{for some} \quad k > 0. \text{[6]} \qquad (7.152)$$

---

[6] We assume that the first observation is taken at t_0 rather than at t_1.

It is *uniformly completely observable* if there exist a positive integer N and positive constants α, β such that

$$0 < \alpha I \leqslant \mathscr{I}(k, k - N) \leqslant \beta I, \qquad (7.153)$$

for *all* $k \geqslant N$.[7] Finally, we say that x_k *is completely observable with respect to* $\{y_l, y_{l+1}, ..., y_k\}$ if, and only if,

$$\mathscr{I}(k, l) > 0 \qquad (k \geqslant l). \qquad (7.154)$$

We next note some properties of the information matrix, already developed in Example 7.2, for a noise-free system ($w_k \equiv 0$). We saw that the information matrix satisfies the following difference equation:

$$\mathscr{I}(k + 1, 0) = \Phi^{\mathrm{T}}(k, k + 1)\, \mathscr{I}(k, 0)\, \Phi(k, k + 1) + M^{\mathrm{T}}(k + 1)\, R_{k+1}^{-1} M(k + 1),$$
$$(7.155)$$

and is related to the covariance matrix by

$$P_k^{k^{-1}} = \Phi^{\mathrm{T}}(0, k)\, P_0^{-1} \Phi(0, k) + \mathscr{I}(k, 0). \qquad (7.156)$$

A concept dual to that of observability is the concept of controllability, also introduced by Kalman [29, 34, 35]. Consider the effect of the forcing term in (7.27). At t_k, the difference between x_k and $\Phi(k, 0)x_0$ is

$$\Delta_k = \sum_{i=0}^{k-1} \Phi(k, i + 1)\, \Gamma(i)\, w_{i+1}$$

and

$$\mathscr{E}\{\Delta_k\} \equiv 0, \qquad \mathscr{E}\{\Delta_k \Delta_k^{\mathrm{T}}\} = \sum_{i=0}^{k-1} \Phi(k, i + 1)\, \Gamma(i) Q_{i+1} \Gamma^{\mathrm{T}}(i)\, \Phi^{\mathrm{T}}(k, i + 1).$$

We define the *controllability matrix*

$$\mathscr{C}(k, 0) \triangleq \sum_{i=0}^{k-1} \Phi(k, i + 1)\, \Gamma(i) Q_{i+1} \Gamma^{\mathrm{T}}(i)\, \Phi^{\mathrm{T}}(k, i + 1). \qquad (7.157)$$

The discrete dynamical system (7.27), (7.2) is said to be *completely controllable* if, and only if,

$$\mathscr{C}(k, 0) > 0 \qquad \text{for some} \quad k > 0. \qquad (7.158)$$

---

[7] For two symmetric matrices A and B, $A > B$ ($A \geqslant B$) means that $A - B > 0$ ($A - B \geqslant 0$) is positive definite (positive semidefinite).

It is *uniformly completely controllable* if there exist a positive integer N and positive constants α, β such that

$$0 < \alpha I \leqslant \mathscr{C}(k, k - N) \leqslant \beta I \qquad (7.159)$$

for *all* $k \geqslant N$.

Uniform complete observability and controllability imply stochastic observability and controllability as defined by Aoki [1]. We prefer the present definitions because they involve intrinsic properties of the system.

Completely analogous definitions are made for the continuous dynamical system (7.95), (7.96). In the limit of continuous observations [replace R_i by $R(t_i)/\Delta t$, Q_i by $Q(t_i)/\Delta t$, and $\Gamma(i)$ by $\Delta t\, G(t_i)$; see Example 7.10], we have the *information matrix*

$$\mathscr{I}(t, t_0) \triangleq \int_{t_0}^{t} \Phi^{\mathrm{T}}(\tau, t)\, M^{\mathrm{T}}(\tau)\, R^{-1}(\tau)\, M(\tau)\, \Phi(\tau, t)\, d\tau \qquad (7.160)$$

and the *controllability matrix*

$$\mathscr{C}(t, t_0) \triangleq \int_{t_0}^{t} \Phi(t, \tau)\, G(\tau)\, Q(\tau)\, G^{\mathrm{T}}(\tau)\, \Phi^{\mathrm{T}}(t, \tau)\, d\tau. \qquad (7.161)$$

The continuous dynamical system (7.95), (7.96) is said to be *completely observable* (*controllable*) if, and only if,

$$\mathscr{I}(t, t_0) > 0 \qquad (\mathscr{C}(t, t_0) > 0) \qquad \text{for some} \quad t > t_0. \qquad (7.162)$$

It is *uniformly completely observable* (*controllable*) if there exist positive constants σ, α, β such that

$$0 < \alpha I \leqslant \mathscr{I}(t, t - \sigma) \leqslant \beta I \qquad (0 < \alpha I \leqslant \mathscr{C}(t, t - \sigma) \leqslant \beta I)$$

$$\text{for all} \quad t \geqslant t_0 + \sigma. \qquad (7.163)$$

See Kalman [29, 34, 35] for equivalent and simpler conditions for controllability and observability of constant (stationary) continuous systems.

By the limiting procedure of Example 7.10 applied to (7.155) (or differentiate 7.160), we get the differential equation for the information matrix in the continuous case:

$$d\mathscr{I}/dt = -F^{\mathrm{T}}(t)\,\mathscr{I} - \mathscr{I}F(t) + M^{\mathrm{T}}(t)\, R^{-1}(t)\, M(t). \qquad (7.164)$$

$M^{\mathrm{T}}(t)\, R^{-1}(t)\, M(t)$ is appropriately called the *information rate* matrix. Comparing (7.164) with (7.132), we see that $\mathscr{I}$ is analogous to P_t^{-1}. $\mathscr{I}^{-1}$ satisfies the variance equation in the case of no system noise.

In the next several sections, we use the concepts of observability and controllability in a study of qualitative properties of linear filters.

6. BOUNDS AND STABILITY—DISCRETE FILTER

This section is devoted to the development of qualitative properties of the discrete filter (Theorem 7.2). Our main objective is to prove, under appropriate conditions, the uniform asymptotic stability of the discrete filter. In this connection, we note that optimality does not imply stability. The stability needs to be proved. In the course of proving stability, we develop upper and lower bounds on the covariance matrix $P_k{}^k$ and study its asymptotic properties. Our development is based on the work of Kalman [36], Deyst and Price [15], and Sorenson [52].

Before proceeding, it will be useful to write the discrete filter in the equivalent form [see (7.11)]

$$\hat{x}_k{}^k = P(k \mid k)[P^{-1}(k \mid k-1)\, \Phi(k, k-1)\, \hat{x}_{k-1}^{k-1} + M^{\mathrm{T}}(k)\, R_k^{-1} y_k], \quad (7.165)$$

$$P(k \mid k) = [P^{-1}(k \mid k-1) + M^{\mathrm{T}}(k)\, R_k^{-1} M(k)]^{-1},$$

$$P(k+1 \mid k) = \Phi(k+1, k)\, P(k \mid k)\, \Phi^{\mathrm{T}}(k+1, k) + \Gamma(k)\, Q_{k+1}\Gamma^{\mathrm{T}}(k). \quad (7.166)$$

The filter equation (7.165) can also be written as

$$\hat{x}_k{}^k = [I - K(k)\, M(k)]\, \Phi(k, k-1)\, \hat{x}_{k-1}^{k-1} + K(k)\, y_k. \quad (7.167)$$

$[I - K(k)\, M(k)]\, \Phi(k, k-1) \triangleq \psi(k, k-1)$ is the state transition matrix of the filter, and, in view of (7B.1), it is nonsingular.

Lemma 7.1. *If the dynamical system* (7.27), (7.2) *is uniformly completely observable and uniformly completely controllable, and if* $P_0 \geqslant 0$, *then* $P(k \mid k)$ *is uniformly bounded from above for all* $k \geqslant N$,

$$P(k \mid k) \leqslant \mathscr{I}^{-1}(k, k - N) + \mathscr{C}(k, k - N) \leqslant \left(\frac{1 + \alpha\beta}{\alpha}\right) I, \quad k \geqslant N.$$

PROOF: Consider the estimate

$$\bar{x}_k{}^N = \mathscr{I}^{-1}(k, k - N) \sum_{i=k-N}^{k} \Phi^{\mathrm{T}}(i, k)\, M^{\mathrm{T}}(i)\, R_i^{-1} y_i, \quad k \geqslant N.$$

In view of Eq. (7.41) of Example 7.2, this is the least squares estimate of x_k based on the most recent N observations, ignoring the system noise w_k . This estimate is suboptimal (not minimum variance) and, as a consequence,

$$P(k \mid k) \leqslant \mathscr{E}\{(x_k - \bar{x}_k{}^N)(x_k - \bar{x}_k{}^N)^{\mathrm{T}}\}. \tag{7.168}$$

We compute the covariance matrix in (7.168). Since, from (7.27),

$$x_i = \Phi(i, k)\, x_k - \Phi(i, k) \sum_{j=1}^{k-1} \Phi(k, j+1)\, \Gamma(j)\, w_{j+1},$$

the measurement y_i can be written as

$$y_i = M(i)\, \Phi(i, k)\, x_k + v_i - M(i)\, \Phi(i, k) \sum_{j=1}^{k-1} \Phi(k, j+1)\, \Gamma(j)\, w_{j+1}$$

Therefore,

$$x_k - \bar{x}_k{}^N = -\mathscr{I}^{-1}(k, k-N) \sum_{i=k-N}^{k} \Phi^{\mathrm{T}}(i, k)\, M^{\mathrm{T}}(i)\, R_i^{-1} v_i$$

$$+ \mathscr{I}^{-1}(k, k-N) \sum_{i=k-N}^{k} \Phi^{\mathrm{T}}(i, k)\, M^{\mathrm{T}}(i)\, R_i^{-1} M(i)\, \Phi(i, k)$$

$$\times \sum_{j=i}^{k-1} \Phi(k, j+1)\, \Gamma(j)\, w_{j+1},$$

and

$$\mathscr{E}\{(x_k - \bar{x}_k{}^N)(x_k - \bar{x}_k{}^N)^{\mathrm{T}}\} = \mathscr{I}^{-1}(k, k-N)$$

$$+ \operatorname{cov}\left\{\mathscr{I}^{-1}(k, k-N) \sum_{i=k-N}^{k} \Phi^{\mathrm{T}}(i, k)\, M^{\mathrm{T}}(i)\, R_i^{-1} M(i)\, \Phi(i, k)\right.$$

$$\left. \times \sum_{j=i}^{k-1} \Phi(k, j+1)\, \Gamma(j)\, w_{j+1}\right\}$$

$$\leqslant \mathscr{I}^{-1}(k, k-N)$$

$$+ \operatorname{cov}\left\{\mathscr{I}^{-1}(k, k-N) \sum_{i=k-N}^{k} \Phi^{\mathrm{T}}(i, k)\, M^{\mathrm{T}}(i)\, R_i^{-1} M(i)\, \Phi(i, k)\right.$$

$$\left. \times \sum_{j=k-N}^{k-1} \Phi(k, j+1)\, \Gamma(j)\, w_{j+1}\right\}$$

$$= \mathscr{I}^{-1}(k, k-N) + \mathscr{C}(k, k-N).$$

This, together with (7.168), proves the lemma. ∎

By considering the estimate

$$\bar{x}_k{}^k = \mathscr{I}^{-1}(k, 0) \sum_{i=0}^{k} \Phi^{\mathrm{T}}(i, k) M^{\mathrm{T}}(i) R_i^{-1} y_i$$

based on all the observations, and imitating the proof of the lemma, we can derive the bound

$$P(k \mid k) \leqslant \mathscr{I}^{-1}(k, 0) + \mathscr{C}(k, 0), \qquad k \geqslant N. \tag{7.169}$$

Suppose our dynamical system is *uniformly completely observable* and *noise-free* ($\mathscr{C} \equiv 0$). Then

$$P(k \mid k) \leqslant \mathscr{I}^{-1}(k, 0), \qquad k \geqslant N. \tag{7.170}$$

[This also follows directly from (7.156).] Defining

$$\tilde{\mathscr{I}}(k, 0) \triangleq \Phi^{\mathrm{T}}(k, 0) \mathscr{I}(k, 0) \Phi(k, 0) = \sum_{i=0}^{k} \Phi^{\mathrm{T}}(i, 0) M^{\mathrm{T}}(i) R_i^{-1} M(i) \Phi(i, 0), \tag{7.171}$$

(7.170) becomes

$$P(k \mid k) \leqslant \Phi(k, 0) \tilde{\mathscr{I}}^{-1}(k, 0) \Phi^{\mathrm{T}}(k, 0). \tag{7.172}$$

Now

$$\tilde{\mathscr{I}}(k, 0) = \tilde{\mathscr{I}}(N, 0) + \tilde{\mathscr{I}}(2N, N) + \cdots,$$

the sum of positive definite matrices. Thus, if $\| \tilde{\mathscr{I}}^{-1}(k, 0) \| \to 0$ faster than $\| \Phi(k, 0) \|^2$ increases, as $k \to \infty$, $P(k \mid k) \to 0$ ($k \to \infty$). The state of the system can be estimated as accurately as desired. This also shows that the initial statistic P_0 is gradually forgotten. The system noise, in the general system (7.27), prevents us from estimating the state exactly.

Returning to the *noise-free* system, we see from (7.156) that, if the system is *completely observable*, then $P(k \mid k)$ is positive definite for some k. Furthermore, it is evident that, once $P(k \mid k) > 0$, it remains positive definite thereafter. [This last property is also true for the general system, in view of (7.166).] In contrast with this, we saw in Example 7.7 that, if the measurements are noise-free, $P(k \mid k)$ is *never* positive definite.

Lemma 7.2. *If the dynamical system (7.27), (7.2) is uniformly completely observable and uniformly completely controllable, and if $P_0 > 0$, then $P(k \mid k)$ is uniformly bounded from below for all $k \geqslant N$,*

$$P(k \mid k) \geqslant [\mathscr{I}(k, k - N) + \mathscr{C}^{-1}(k, k - N)]^{-1} \geqslant \left(\frac{\alpha}{1 + \alpha\beta} \right) I, \qquad k \geqslant N.$$

PROOF: It is evident from (7.166) that $P(k \mid k) > 0$ for all k, since $P_0 > 0$. Define

$$S(k \mid k) \triangleq P^{-1}(k \mid k), \tag{7.173}$$

$$\bar{S}(k \mid k) \triangleq S(k \mid k) - M^{\mathrm{T}}(k) R_k^{-1} M(k), \tag{7.174}$$

$$\bar{S}(k+1 \mid k) \triangleq \Phi^{-\mathrm{T}}(k+1, k) S(k \mid k) \Phi^{-1}(k+1, k). \tag{7.175}$$

Then, using (7.166),

$$\begin{aligned}
\bar{S}(k \mid k) &= [\bar{S}^{-1}(k \mid k-1) + \Gamma(k-1) Q_k \Gamma^{\mathrm{T}}(k-1)]^{-1}, \\
\bar{S}(k+1 \mid k) &= \Phi^{-\mathrm{T}}(k+1, k) \bar{S}(k \mid k) \Phi^{-1}(k+1, k) \\
&\quad + \Phi^{-\mathrm{T}}(k+1, k) M^{\mathrm{T}}(k) R_k^{-1} M(k) \Phi^{-1}(k+1, k).
\end{aligned} \tag{7.176}$$

Comparing (7.176) with (7.166), we see that $\bar{S}(k \mid k)$ is the estimation error covariance matrix of the system

$$\bar{x}_{k+1} = \Phi^{-\mathrm{T}}(k+1, k) \bar{x}_k + \Phi^{-\mathrm{T}}(k+1, k) M^{\mathrm{T}}(k) \bar{w}_k, \qquad \mathscr{E}\{\bar{w}_k \bar{w}_k^{\mathrm{T}}\} = R_k^{-1},$$

$$\bar{y}_k = \Gamma^{\mathrm{T}}(k-1) \bar{x}_k + \bar{v}_k, \qquad \mathscr{E}\{\bar{v}_k \bar{v}_k^{\mathrm{T}}\} = Q_k^{-1}.^8 \tag{7.177}$$

It is easy to verify that (7.177) is uniformly completely observable and uniformly completely controllable, since the dynamical system (7.27), (7.2) is. Therefore, applying Lemma 7.1 to (7.177), we have that

$$\begin{aligned}
\bar{S}(k \mid k) &\leqslant \left[\sum_{i=k-N}^{k} \Phi^{-1}(i, k) \Gamma(i-1) Q_i \Gamma^{\mathrm{T}}(i-1) \Phi^{-\mathrm{T}}(i, k) \right]^{-1} \\
&\quad + \sum_{i=k-N}^{k-1} \Phi^{-\mathrm{T}}(k, i+1) \Phi^{-\mathrm{T}}(i+1, i) \\
&\qquad \times M^{\mathrm{T}}(i) R_i^{-1} M(i) \Phi^{-1}(i+1, i) \Phi^{-1}(k, i+1) \\
&= \left[\sum_{i=k-N-1}^{k-1} \Phi(k, i+1) \Gamma(i) Q_{i+1} \Gamma^{\mathrm{T}}(i) \Phi^{\mathrm{T}}(k, i+1) \right]^{-1} \\
&\quad + \sum_{i=k-N}^{k-1} \Phi^{\mathrm{T}}(i, k) M^{\mathrm{T}}(i) R_i^{-1} M(i) \Phi(i, k).
\end{aligned}$$

Now, in view of (7.174),

$$S(k \mid k) \leqslant \mathscr{C}^{-1}(k, k-N) + \mathscr{I}(k, k-N),$$

---

⁸ The assumption that $Q_k > 0$ is no loss of generality.

and, in view of (7.173),

$$P(k \mid k) \geqslant [\mathscr{C}^{-1}(k, k - N) + \mathscr{I}(k, k - N)]^{-1}. \quad \blacksquare$$

Lemma 7.3. *If the dynamical system* (7.27), (7.2) *is uniformly completely controllable and* $P_0 \geqslant 0$, *then* $P(k \mid k) > 0$ *for all* $k \geqslant N$.

PROOF: In view of (7.166), if $P(N \mid N) > 0$, then $P(k \mid k) > 0$ for all $k \geqslant N$. So it suffices to prove that $P(N \mid N)$ is nonsingular. Suppose $P(N \mid N)$ is singular. Then there exists a vector $v \neq 0$ such that

$$v^{\mathrm{T}} P(N \mid N) v = 0. \quad (7.178)$$

Let

$$S(k) \triangleq \Psi(N, k) P(k \mid k) \Psi^{\mathrm{T}}(N, k), \quad (7.179)$$

where

$$\Psi(k, k - 1) = [I - K(k) M(k)] \Phi(k, k - 1)$$

is the filter state transition matrix [see (7.167)]. It is easy to compute

$$S(k) - S(k - 1) = \Psi(N, k)[[I - K(k) M(k)] \Gamma(k - 1) Q_k \Gamma^{\mathrm{T}}(k - 1)$$
$$\times [I - K(k) M(k)]^{\mathrm{T}} + K(k) R_k K^{\mathrm{T}}(k)] \Psi^{\mathrm{T}}(N, k), \quad (7.180)$$

using Theorem 7.2 and (7.16). Now $S(0) \geqslant 0$, $S(k) - S(k - 1) \geqslant 0$, and, in view of (7.178), $v^{\mathrm{T}} S(N) v = 0$. Therefore, $v^{\mathrm{T}} S(k) v = 0$ for all $k \leqslant N$. That is,

$$v^{\mathrm{T}} \Psi(N, k) P(k \mid k) \Psi^{\mathrm{T}}(N, k) v = 0, \quad \text{all} \quad k \leqslant N, \quad (7.181)$$

or, equivalently,

$$P(k \mid k) \Psi^{\mathrm{T}}(N, k) v = 0, \quad \text{all} \quad k \leqslant N. \quad (7.182)$$

Now let

$$w(k) \triangleq \Psi^{\mathrm{T}}(N, k) v, \quad w(N) = v. \quad (7.183)$$

We easily compute

$$w(k - 1) = \Psi^{\mathrm{T}}(k, k - 1) w(k) \quad (7.184)$$

and

$$\Phi^{\mathrm{T}}(k - 1, k) w(k - 1) = (I - K(k) M(k))^{\mathrm{T}} \Psi^{\mathrm{T}}(N, k) v. \quad (7.185)$$

Since $v^{\mathrm{T}}[S(k) - S(k - 1)]v = 0$, $1 \leqslant k \leqslant N$, using (7.185) in (7.180), we have that

$$w^{\mathrm{T}}(k - 1) \Phi(k - 1, k) \Gamma(k - 1) Q_k \Gamma^{\mathrm{T}}(k - 1) \Phi^{\mathrm{T}}(k - 1, k) w(k - 1) = 0,$$
$$1 \leqslant k \leqslant N. \quad (7.186)$$

Now (7.184) can be written as

$$w(k - 1) = \Phi^T(k, k - 1)\, w(k) - \Phi^T(k, k - 1)\, M^T(k)\, K^T(k)\, w(k). \quad (7.187)$$

But

$$K^T(k)\, w(k) = [M(k)\, P(k \mid k - 1)\, M^T(k) + R_k]^{-1}\, M(k)\, P(k \mid k - 1)\, \Psi^T(N, k)\, v,$$

and, using (7.29) and (7.182), it is easy to show that

$$P(k \mid k - 1)\, \Psi^T(N, k)\, v = 0, \qquad \text{all} \quad k \leqslant N.$$

Therefore,

$$w(k - 1) = \Phi^T(k, k - 1)\, w(k),$$

which has the solution

$$w(k) = \Phi^T(N, k)\, v. \quad (7.188)$$

As a result, (7.186) becomes

$$v^T\Phi(N, k)\, \Gamma(k - 1)\, Q_k\Gamma^T(k - 1)\, \Phi^T(N, k)\, v = 0, \qquad 1 \leqslant k \leqslant N,$$

and

$$v^T\left[\sum_{k=1}^{N} \Phi(N, k)\, \Gamma(k - 1)\, Q_k\Gamma^T(k - 1)\, \Phi^T(N, k)\right] v = 0,$$

which is

$$v^T\mathscr{C}(N, 0)\, v = 0.$$

But this contradicts the hypothesis of uniform complete controllability, and the lemma is proved. ∎

We are now ready to consider the stability of the filter. For the definitions of various types of stability, and stability theory in general, the reader is referred to Kalman and Bertram [30, 31] and Hahn [23]. We are concerned here only with the linear discrete system (7.165) or (7.167).

The linear system

$$z_{k+1} = \Psi(t_{k+1}, t_k)\, z_k + u_{k+1}, \qquad t_k \geqslant t_0, \quad (7.189)$$

is said to be *stable*[9] if

$$\| \Psi(t_k, t_0)\| \leqslant c_1 \qquad \text{for all} \quad t_k \geqslant t_0. \quad (7.190)$$

[9] When speaking of the stability of (7.189), we mean the stability of the homogeneous part of (7.189).

The system is *asymptotically stable* if, in addition,

$$\| \Psi(t_k, t_0)\| \to 0 \qquad (t_k \to \infty). \tag{7.191}$$

It is *uniformly asymptotically stable* if

$$\| \Psi(t_k, t_0)\| \leqslant c_2 \exp(-c_3(t_k - t_0)) \qquad \text{for all} \quad t_k \geqslant t_0. \tag{7.192}$$

(The c_i are positive constants.) We note in passing that, in uniformly asymptotically stable systems, bounded inputs u_k produce bounded outputs z_k.

Kalman and Bertram [30, 31] show that (7.189) is uniformly asymptotically stable if there exist scalar functions $V(z_k, t_k), \gamma_1(\| z_k \|), \gamma_2(\| z_k \|)$, and $\gamma_3(\| z_k \|)$, such that, for some integers N, $M > 0$,

$$0 < \gamma_1(\| z_k \|) \leqslant V(z_k, t_k) \leqslant \gamma_2(\| z_k \|), \qquad z_k \neq 0, \qquad \text{all} \quad k \geqslant M, \tag{7.193}$$

where

$$\gamma_1(0) = \gamma_2(0) = 0, \qquad \gamma_1(\| z \|) \to \infty \qquad (\| z \| \to \infty);$$

and

$$V(z_k, t_k) - V(z_{k-N}, t_{k-N}) \leqslant \gamma_3(\| z_k \|) < 0, \qquad z_k \neq 0, \qquad \text{all} \quad k \geqslant M, \tag{7.194}$$

where

$$\gamma_3(0) = 0;$$

and $V(0, t_k) \equiv 0$; γ_1, γ_2, and γ_3 are continuous, and γ_1 and γ_2 are nondecreasing. The function V is called a *Lyapunov* function for the system (7.189).

We shall prove the uniform asymptotic stability of the linear filter by producing a Lyapunov function (and $\gamma_1, \gamma_2, \gamma_3$) for it.

Theorem 7.4. *If the dynamical system* (7.27), (7.2) *is uniformly completely observable and uniformly completely controllable, and if* $P_0 \geqslant 0$, *then the discrete linear filter of Theorem 7.2 is uniformly asymptotically stable.*

PROOF: We shall prove that

$$V(z_k, k) = z_k^{\mathrm{T}} P^{-1}(k \mid k) z_k \tag{7.195}$$

is a Lyapunov function for (7.165), or for

$$z_k = P(k \mid k) P^{-1}(k \mid k-1) \Phi(k, k-1) z_{k-1}. \tag{7.196}$$

In view of Lemma 7.3, $P(k \mid k) > 0$ for $k \geqslant N$. Then, by Lemmas 7.1 and 7.2,

$$\left(\frac{\alpha}{1 + \alpha\beta}\right) I \leqslant P^{-1}(k \mid k) \leqslant \left(\frac{1 + \alpha\beta}{\alpha}\right) I, \qquad k \geqslant 2N,$$

so that

$$\left(\frac{\alpha}{1 + \alpha\beta}\right) \| z_k \|^2 \leqslant V(z_k, k) \leqslant \left(\frac{1 + \alpha\beta}{\alpha}\right) \| z_k \|^2, \qquad k \geqslant 2N.$$

This establishes property (7.193). It remains to establish (7.194). Following Deyst and Price [15], we define

$$
\begin{aligned}
u_k &\triangleq [P(k \mid k) P^{-1}(k \mid k - 1) - I] \, \Phi(k, k - 1) \, z_{k-1}, \\
\bar{z}_k &\triangleq \Phi(k, k - 1) \, z_{k-1}.
\end{aligned}
\tag{7.197}
$$

Then the system (7.196) can be written as

$$z_k = \bar{z}_k + [P(k \mid k) P^{-1}(k \mid k - 1) - I] \, \bar{z}_k = \bar{z}_k + u_k. \tag{7.198}$$

Using (7.166), we have

$$
\begin{aligned}
& z_k{}^T P^{-1}(k \mid k) \, z_k \\
&= z_k{}^T [P^{-1}(k \mid k - 1) + M^T(k) \, R_k^{-1} M(k)] \, z_k \\
&= z_k{}^T P^{-1}(k \mid k - 1) \, z_k + 2 z_k{}^T [P^{-1}(k \mid k) - P^{-1}(k \mid k - 1)] \, z_k \\
&\quad - z_k{}^T M^T(k) \, R_k^{-1} M(k) \, z_k + \bar{z}_k{}^T P^{-1}(k \mid k - 1) \, \bar{z}_k \\
&\quad - \bar{z}_k{}^T P^{-1}(k \mid k - 1) \, \bar{z}_k \\
&= \bar{z}_k{}^T P^{-1}(k \mid k - 1) \, \bar{z}_k - z_k{}^T M^T(k) \, R_k^{-1} M(k) \, z_k - u_k{}^T P^{-1}(k \mid k - 1) \, u_k \\
&= z_{k-1}^T [P(k - 1 \mid k - 1) \\
&\quad + \Phi(k - 1, k) \, \Gamma(k - 1) \, Q_k \Gamma^T(k - 1) \, \Phi^T(k - 1, k)]^{-1} \, z_{k-1} \\
&\quad - z_k{}^T M^T(k) \, R_k^{-1} M(k) \, z_k - u_k{}^T P^{-1}(k \mid k - 1) \, u_k \\
&\leqslant z_{k-1}^T P(k - 1 \mid k - 1) \, z_{k-1} \\
&\quad - z_k{}^T M^T(k) \, R_k^{-1} M(k) \, z_k - u_k{}^T P^{-1}(k \mid k - 1) \, u_k.
\end{aligned}
$$

Therefore,

$$V(z_k, t_k) - V(z_{k-1}, t_{k-1}) \leqslant -z_k{}^T M^T(k) \, R_k^{-1} M(k) \, z_k - u_k{}^T P^{-1}(k \mid k - 1) \, u_k,$$

and

$$V(z_k, t_k) - V(z_{k-N-1}, t_{k-N-1})$$

$$\leqslant - \sum_{i=k-N}^{k} [z_i^{\mathrm{T}} M^{\mathrm{T}}(i) R_i^{-1} M(i) z_i + u_i^{\mathrm{T}} P^{-1}(i \mid i - 1) u_i] \triangleq -J,$$

$$k \geqslant 2N + 1. \quad (7.199)$$

Let J^0 be the minimum value of J. Then we have

$$V(z_k, t_k) - V(z_{k-N-1}, t_{k-N-1}) \leqslant -J \leqslant -J^0, \qquad k \geqslant 2N + 1. \quad (7.200)$$

We are thus led to the problem of minimizing J with respect to $\{u_k\}$ for the system (7.198):

$$z_i = \Phi(i, i - 1) z_{i-1} + u_i; \qquad i = k - N,..., k; \qquad z_{k-N-1} \text{ given.} \quad (7.201)$$

This is an ordinary minimum problem once the z_i, $i \geqslant k - N$, are eliminated from J in favor of z_{k-N-1} via (7.201).

This problem is readily solved for $\{u_k^0\}$, the minimizing $\{u_k\}$ as a function of z_{k-N-1}, which results in J^0 as a function of z_{k-N-1}. Then, using the hypothesis of uniform complete observability, it can be shown that

$$J^0 \geqslant c_4 \| z_{k-N-1} \|^2, \qquad c_4 > 0. [10]$$

Then, in view of (7.196),

$$J^0 \geqslant c_5 \| z_k \|^2, \qquad c_5 > 0,$$

and (7.200) becomes

$$V(z_k, t_k) - V(z_{k-N-1}, t_{k-N-1}) \leqslant -c_5 \| z_k \|^2, \qquad k \geqslant 2N + 1,$$

which establishes property (7.194). ■

An immediate consequence of Theorem 7.4 is

Theorem 7.5. *Let the dynamical system* (7.27), (7.2) *be uniformly completely observable and uniformly completely controllable. Suppose* $P^1(k \mid k)$ *and* $P^2(k \mid k)$ *are any two solutions of* (7.166) *with respective initial conditions* P_0^1 *and* P_0^2; $P_0^1, P_0^2 \geqslant 0$. *Let* $\delta P(t_k) = P^1(k \mid k) - P^2(k \mid k)$. *Then*

$$\| \delta P(t_k) \| \leqslant c_2^2 e^{-2c_3(t_k - t_0)} \| P_0^1 - P_0^2 \| \to 0 \qquad (k \to \infty)$$

$(c_2, c_3 > 0)$.

---

[10] We assume, of course, that all the matrices in the system description are bounded.

PROOF: It is easy to verify that $\delta P(t_k)$ satisfies the difference equation

$$\delta P(t_k) = [I - K^1(k)\,M(k)]\,\Phi(k, k-1)\,\delta P(t_{k-1})\,\Phi^{\mathrm{T}}(k, k-1)[I - K^2(k)\,M(k)]^{\mathrm{T}}, \tag{7.202}$$

where K^1 and K^2 are the Kalman gains corresponding to $P^1(k \mid k)$ and $P^2(k \mid k)$, respectively. But

$$[I - K(k)\,M(k)]\,\Phi(k, k-1) = \Psi(k, k-1)$$

is the filter state transition matrix [see (7.167)]. Therefore, the solution of (7.202) is

$$\delta P(t_k) = \Psi_1(t_k, t_0)(P_0{}^1 - P_0{}^2)\,\Psi_2{}^{\mathrm{T}}(t_k, t_0). \tag{7.203}$$

According to Theorem 7.4, the filter is uniformly asymptotically stable, and, by definition, its state transition matrix has property (7.192). Taking norms in (7.203) and using (7.192) proves the theorem. ∎

Theorem 7.5 has two important applications. It essentially states that the effect of the initial statistic P_0 is forgotten as more and more data are processed. This is important, since P_0 is often poorly known and sometimes rather arbitrarily set. Furthermore, this means that the computation of $P(k \mid k)$ is stable. Numerical errors in $P(k \mid k)$ are also forgotten.

7. BOUNDS AND STABILITY—CONTINUOUS FILTER

The qualitative properties of the continuous filter (Theorem 7.3) are identical to those already developed for the discrete filter in the preceding section. That is, Lemmas 7.1–7.3 and Theorems 7.4 and 7.5 hold if the dynamical system (7.95), (7.96) is uniformly completely observable and uniformly completely controllable, and $P(t \mid t)$ is the solution of the variance equation (7.99). The homogeneous part of the filter equation is, of course [see (7.98)],

$$d\hat{x}_t{}^t/dt = [F(t) - P(t \mid t)\,M^{\mathrm{T}}(t)\,R^{-1}(t)\,M(t)]\,\hat{x}_t{}^t.$$

The proofs of these lemmas and theorems are given by Kalman [36]. They are by and large analogous to those already given for the discrete case. Because of this, and because they are available in the book already cited, we shall not present these proofs here. In fact, the reader himself can adapt the proofs in the preceding section to the continuous case.

In the case of the continuous filter, one can also speak of the *steady-*

state solution of the variance equation. The existence of the steady-state solution and its properties for constant (stationary) systems are treated in detail by Kalman and Bucy [33] and Kalman [36].

8. ERROR SENSITIVITY—DISCRETE FILTER

In applying the linear filter (Theorem 7.2) to a specific system, the dynamical model parameters (Φ, Γ, M), noise statistics (Q, R), and *a priori* data $[\hat{x}_0, P(0)]$ must be specified. Since the system model is usually an approximation to a physical situation, the model parameters and noise statistics are seldom exact. That is, the system model used in constructing the filter differs from the (real) system that generates the observations. Sometimes such an approximation is intentional. For example, it may be desirable to use a system model of lower dimension than the dimension of the real system in order to gain computational speed and simplicity. It has already been remarked in several places that the prior data are seldom precisely known and are generally rather arbitrarily specified to start the filtering process. We have seen (Theorem 7.5) that the prior data are eventually forgotten, after sufficient observations have been processed.

It is clear that an inexact filter model will degrade the filter performance. In fact, such an inexact model may cause the filter to diverge (see Chapter 8). In designing a filter model, it is therefore important to evaluate the effect on performance of the various approximations made. This section is a study of the qualitative effects of such approximations.

Suppose that the *real* or *actual* system is described by

$$x_{k+1} = \Phi(k+1, k)\, x_k + \varphi(k) + \Gamma(k)\, w_{k+1}\, ,$$

$$y_k = M(k)\, x_k + v_k\, , \qquad\qquad\qquad\qquad (7.204)$$

$$w_k \sim N[0, Q(k)], \qquad v_k \sim N[0, R(k)], \qquad x_0 \sim N[\hat{x}(0), P(0)],$$

whereas we *model* this system by

$$\bar{x}_{k+1} = \Phi_c(k+1, k)\, \bar{x}_k + \varphi_c(k) + \Gamma_c(k)\, \bar{w}_{k+1}\, ,$$

$$\bar{y}_k = M_c(k)\, \bar{x}_k + \bar{v}_k\, , \qquad\qquad\qquad\qquad (7.205)$$

$$\bar{w}_k \sim N[0, Q_c(k)], \qquad \bar{v}_k \sim N[0, R_c(k)], \qquad \bar{x}_0 \sim N[\hat{x}_c(0), P_c(0)].$$

We include the fixed bias $\varphi(k)$ in the dynamics to account in an approximate, yet convenient, way for errors due to nonlinearities, reduction in system dimension, and so on. In what follows, $\{\varphi(k)\}$ could readily be treated as a random sequence. We assume that $\Phi(k + 1, k)$, $\Phi_c(k + 1, k)$, $\Gamma(k)$, $\Gamma_c(k)$, $M(k)$, $M_c(k)$, $Q(k)$, $Q_c(k)$, $R(k)$, $R_c(k)$ are uniformly bounded; that Φ and Φ_c are nonsingular; that $R(k)$, $R_c(k) > 0$; and that $P(0)$, $P_c(0) \geqslant 0$.

Using the model (7.205), our filter design is

$$\hat{x}_{k+1}^k = \Phi_c(k + 1, k)\,\hat{x}_k^{\ k} + \varphi_c(k), \tag{7.206}$$

$$\hat{x}_k^{\ k} = [I - K_c(k)\,M_c(k)]\,\hat{x}_k^{k-1} + K_c(k)\,y_k\,,$$

where

$$K_c(k) = P_c(k \mid k - 1)\,M_c^{\mathrm{T}}(k)[M_c(k)\,P_c(k \mid k - 1)\,M_c^{\mathrm{T}}(k) + R_c(k)]^{-1}, \tag{7.207}$$

and

$$P_c(k + 1 \mid k) = \Phi_c(k + 1, k)\,P_c(k \mid k)\,\Phi_c^{\mathrm{T}}(k + 1, k) + \Gamma_c(k)\,Q_c(k + 1)\,\Gamma_c^{\mathrm{T}}(k),$$

$$P_c(k \mid k) = [I - K_c(k)\,M_c(k)]\,P_c(k \mid k - 1)$$

$$= [I - K_c(k)\,M_c(k)]\,P_c(k \mid k - 1)[I - K_c(k)\,M_c(k)]^{\mathrm{T}} \tag{7.208}$$

$$+ K_c(k)\,R_c(k)\,K_c^{\mathrm{T}}(k),$$

$$P_c(0 \mid 0) = P_c(0).$$

Note that the filter operates on the *real* data $\{y_k\}$.

Now the *computed* matrix $P_c(k \mid k)$ *is not* the estimation error covariance matrix, since the filter model differs from the real model. Neither is this filter the minimum variance filter for the actual system (7.204). The *actual* estimation errors are

$$\tilde{x}_{k+1}^k \triangleq x_{k+1} - \hat{x}_{k+1}^k = \Phi_c(k + 1, k)\,\tilde{x}_k^{\ k} + \Delta\Phi(k + 1, k)\,x_k + \Delta\varphi(k) + \Gamma(k)w_{k+1}\,,$$

$$\tag{7.209}$$

$$\tilde{x}_k^{\ k} \triangleq x_k - \hat{x}_k^{\ k} = [I - K_c(k)\,M_c(k)]\,\tilde{x}_k^{k-1} - K_c(k)\,\Delta M(k)\,x_k - K_c(k)\,v_k\,,$$

where

$$\Delta\Phi(k + 1, k) \triangleq \Phi(k + 1, k) - \Phi_c(k + 1, k),$$

$$\Delta\varphi(k) \triangleq \varphi(k) - \varphi_c(k), \tag{7.210}$$

$$\Delta M(k) \triangleq M(k) - M_c(k).$$

A measure of the filter performance is provided by the *actual estimation error covariance matrix*

$$P(k \mid k) \triangleq \mathscr{E}\{\tilde{x}_k^{\ k}\tilde{x}_k^{k\mathrm{T}}\}, \qquad P(k + 1 \mid k) \triangleq \mathscr{E}\{\tilde{x}_{k+1}^k\tilde{x}_{k+1}^{k\mathrm{T}}\}. \tag{7.211}$$

We next develop a recursion for this matrix.

Using (7.209), we get

$$
\begin{aligned}
P(k+1 \mid k) = {} & \Phi_c(k+1, k) \, P(k \mid k) \, \Phi_c^{\mathrm{T}}(k+1, k) + \Gamma(k) \, Q(k+1) \, \Gamma^{\mathrm{T}}(k) \\
& + \Delta\Phi(k+1, k) \, X(k) \, \Delta\Phi^{\mathrm{T}}(k+1, k) \\
& + \Delta\Phi(k+1, k) \, C(k \mid k) \, \Phi_c^{\mathrm{T}}(k+1, k) \\
& + \Phi_c(k+1, k) \, C^{\mathrm{T}}(k \mid k) \, \Delta\Phi^{\mathrm{T}}(k+1, k) \\
& + \Delta\Phi(k+1, k) \, m(k) \, \Delta\varphi^{\mathrm{T}}(k) \\
& + \Delta\varphi(k) \, m^{\mathrm{T}}(k) \, \Delta\Phi^{\mathrm{T}}(k+1, k) + \Delta\varphi(k) \, \Delta\varphi^{\mathrm{T}}(k) \\
& + \Phi_c(k+1, k) \, \Delta m(k \mid k) \, \Delta\varphi^{\mathrm{T}}(k) \\
& + \Delta\varphi(k) \, \Delta m^{\mathrm{T}}(k \mid k) \, \Phi_c^{\mathrm{T}}(k+1, k),
\end{aligned} \tag{7.212}
$$

$$
\begin{aligned}
P(k \mid k) = {} & [I - K_c(k) \, M_c(k)] \, P(k \mid k-1)[I - K_c(k) \, M_c(k)]^{\mathrm{T}} \\
& + K_c(k) \, R(k) \, K_c^{\mathrm{T}}(k) \\
& - K_c(k) \, \Delta M(k) \, C(k \mid k-1)[I - K_c(k) \, M_c(k)]^{\mathrm{T}} \\
& - [I - K_c(k) \, M_c(k)] \, C^{\mathrm{T}}(k \mid k-1) \, \Delta M^{\mathrm{T}}(k) K_c^{\mathrm{T}}(k) \\
& + K_c(k) \, \Delta M(k) \, X(k) \, \Delta M^{\mathrm{T}}(k) \, K_c^{\mathrm{T}}(k),
\end{aligned} \tag{7.213}
$$

where

$$
X(k) \triangleq \mathscr{E}\{x_k x_k^{\mathrm{T}}\},
$$

$$
C(k \mid k) \triangleq \mathscr{E}\{x_k \tilde{x}_k^{k\mathrm{T}}\}, \qquad C(k+1 \mid k) \triangleq \mathscr{E}\{x_{k+1} \tilde{x}_{k+1}^{k\mathrm{T}}\}, \tag{7.214}
$$

$$
m(k) \triangleq \mathscr{E}\{x_k\}, \qquad \Delta m(k \mid k) \triangleq \mathscr{E}\{\tilde{x}_k^k\}.
$$

Difference equations for these latter quantities are similarly computed to be

$$
\begin{aligned}
X(k+1) = {} & \Phi(k+1, k) \, X(k) \, \Phi^{\mathrm{T}}(k+1, k) + \varphi(k) \, \varphi^{\mathrm{T}}(k) \\
& + \Gamma(k) \, Q(k+1) \, \Gamma^{\mathrm{T}}(k) + \Phi(k+1, k) \, m(k) \, \varphi^{\mathrm{T}}(k) \\
& + \varphi(k) \, m^{\mathrm{T}}(k) \, \Phi^{\mathrm{T}}(k+1, k),
\end{aligned} \tag{7.215}
$$

$$
\begin{aligned}
C(k+1 \mid k) = {} & \Phi(k+1, k) \, C(k \mid k) \, \Phi_c^{\mathrm{T}}(k+1, k) \\
& + \Phi(k+1, k) \, X(k) \, \Delta\Phi^{\mathrm{T}}(k+1, k) \\
& + \Gamma(k) \, Q(k+1) \, \Gamma^{\mathrm{T}}(k) \\
& + \Phi(k+1, k) \, m(k) \, \Delta\varphi^{\mathrm{T}}(k) \\
& + \varphi(k) \, m^{\mathrm{T}}(k) \, \Delta\Phi^{\mathrm{T}}(k+1, k) \\
& + \varphi(k) \, \Delta m^{\mathrm{T}}(k \mid k) \, \Phi_c^{\mathrm{T}}(k+1, k) + \varphi(k) \, \Delta\varphi^{\mathrm{T}}(k),
\end{aligned} \tag{7.216}
$$

$$
C(k \mid k) = C(k \mid k-1)[I - K_c(k) \, M_c(k)]^{\mathrm{T}} - X(k) \, \Delta M^{\mathrm{T}}(k) \, K_c^{\mathrm{T}}(k),
$$

$$
m(k+1) = \Phi(k+1, k) \, m(k) + \varphi(k), \tag{7.217}
$$

$$\Delta m(k + 1 \mid k + 1) = [I - K_c(k + 1) M_c(k + 1)][\Phi_c(k + 1, k) \Delta m(k \mid k)$$
$$+ \Delta\Phi(k + 1, k) m(k) + \Delta\varphi(k)]$$
$$- K_c(k + 1) \Delta M(k + 1) m(k + 1).$$
$$\hspace{6cm}(7.218)$$

Initial conditions for (7.212)–(7.218) are

$$P(0 \mid 0) = P(0),$$
$$X(0) = \mathscr{E}\{x_0 x_0^\mathrm{T}\} = P(0) + \hat{x}(0)\,\hat{x}^\mathrm{T}(0),$$
$$C(0 \mid 0) = \mathscr{E}\{x_0[x_0 - \hat{x}(0)]^\mathrm{T}\} = P(0),$$
$$m(0) = \mathscr{E}\{x_0\} = \hat{x}(0),$$
$$\Delta m(0 \mid 0) = \mathscr{E}\{x_0 - \hat{x}(0)\} = 0.$$
$$\hspace{6cm}(7.219)$$

Various special cases of the actual estimation error covariance matrix equations have been developed by several authors. Battin [3, p. 334] considered the effect of incorrect measurement error statistics. Fagin [16] was concerned with errors in Φ and Q. Heffes [24] considered errors in $P(0)$, Q, and R. Heffes also presents some numerical simulations that show that the actual estimation error variance is greater than or equal to the minimum variance (errors degrade the filter performance). His simulations also indicate that, under appropriate conditions, the actual error variance is less than or equal to the computed variance $[P(k \mid k) \leqslant P_c(k \mid k)]$. Nishimura [41, 42] considers initial errors only [errors in $P(0)$] and develops an upper bound on $P(k \mid k)$. His result is a special case of Theorem 7.6. Nishimura also presents applications and numerical simulations. Price [44] treats errors in $\varphi(k)$, Q, and R and proves, under certain conditions, that $P(k \mid k)$ is bounded. This result is contained in our Theorem 7.7. The most complete development is given by Griffin and Sage [22]. They consider all the errors except $\Delta\varphi$. They do not, however, prove any qualitative results but merely develop equations. Griffin and Sage also define and develop equations for local error sensitivities,

$$[\partial P(k \mid k)/\partial \xi_i]_{\bar{\xi}},$$

where ξ is a vector of parameters in the actual system, and $\bar{\xi}$ is its value used in the filter model. Such equations[11] are developed trivially by differentiating (7.212) and (7.213). Griffin and Sage also present some numerical results. Of some interest are their simulations of a model dimension error (via $\Delta\Phi$).

[11] These equations, as given by Griffin and Sage, appear to be in error in that they neglect to differentiate X and C.

The equations just developed enable the *a priori* determination of the effect of various approximations on filter performance. As a result, they are a useful tool for filter design. Analogous equations for the linear smoother are presented in Griffin and Sage [22].

Consider first the special case of errors in $P(0)$, $Q(k)$, and $R(k)$ only. In this case, Eqs. (7.212) and (7.213) reduce to

$$P(k + 1 \mid k) = \Phi(k + 1, k) P(k \mid k) \Phi^{\mathrm{T}}(k + 1, k) + \Gamma(k) Q(k + 1) \Gamma^{\mathrm{T}}(k),$$

$$P(k \mid k) = [I - K_c(k) M(k)] P(k \mid k - 1)[I - K_c(k) M(k)]^{\mathrm{T}} \qquad (7.220)$$
$$+ K_c(k) R(k) K_c^{\mathrm{T}}(k),$$

whereas

$$P_c(k + 1 \mid k) = \Phi(k + 1, k) P_c(k \mid k) \Phi^{\mathrm{T}}(k + 1, k) + \Gamma(k) Q_c(k + 1) \Gamma^{\mathrm{T}}(k),$$

$$P_c(k \mid k) = [I - K_c(k) M(k)] P_c(k \mid k - 1)[I - K_c(k) M(k)]^{\mathrm{T}} \qquad (7.221)$$
$$+ K_c(k) R_c(k) K_c^{\mathrm{T}}(k).$$

Define

$$E(k \mid k) \triangleq P(k \mid k) - P_c(k \mid k),$$
$$E(k + 1 \mid k) = P(k + 1 \mid k) - P_c(k + 1 \mid k). \qquad (7.222)$$

Then

$$E(k + 1 \mid k) = \Phi(k + 1, k) E(k \mid k) \Phi^{\mathrm{T}}(k + 1, k)$$
$$+ \Gamma(k)[Q(k + 1) - Q_c(k + 1)] \Gamma^{\mathrm{T}}(k),$$
$$E(k \mid k) = [I - K_c(k) M(k)] E(k \mid k - 1)[I - K_c(k) M(k)]^{\mathrm{T}} \qquad (7.223)$$
$$+ K_c(k)[R(k) - R_c(k)] K_c^{\mathrm{T}}(k).$$

Suppose $Q(k) \leqslant Q_c(k)$, $R(k) \leqslant R_c(k)$, all k. Then, if $E(k \mid k - 1) \leqslant 0$, it is evident from (7.223) that $E(k \mid k) \leqslant 0$ and $E(k + 1 \mid k) \leqslant 0$. Therefore, by induction, we have

Theorem 7.6. *If $P(0) \leqslant P_c(0)$ and $Q(k) \leqslant Q_c(k)$, $R(k) \leqslant R_c(k)$, all k, then $P(k \mid k) \leqslant P_c(k \mid k)$ and $P(k + 1 \mid k) \leqslant P_c(k + 1 \mid k)$ for all k.*

The implications of this result are the following. Although the actual values $P(0)$, $Q(k)$, and $R(k)$ are not generally available, conservative estimates of these quantities can often be made. In that case, the *actual* estimation error variances are bounded by the *computed* "variances," which are generated as part of the filter solution. In fact, $P_c(k \mid k)$ may be precomputed to determine whether the conservative estimates of

$P(0)$, $Q(k)$, and $R(k)$ give satisfactory filter performance, or whether somewhat less conservative estimates are desirable. Theorem 7.6 has the following corollary.

Corollary. *Hypothesis of Theorem 7.6. Further suppose that the system model (7.205) is uniformly completely observable and uniformly completely controllable. Then there is an integer $N > 0$ such that $P(k \mid k)$ is uniformly bounded (from above) for all $k \geqslant N$.*

PROOF: By Lemma 7.1, we have

$$P_c(k \mid k) \leqslant \gamma I \quad (\gamma > 0), \quad k \geqslant N.$$

Combining this with Theorem 7.6 proves the corollary. ■

Consider now the case of errors in $P(0)$, $Q(k)$, $R(k)$, and $\varphi(k)$ only. As was mentioned earlier, $\varphi(k)$ accounts, in an approximate way, for possible neglected nonlinearities, reduction in system dimension, and so on. In this case, Eqs. (7.212) and (7.213) become

$$P(k + 1 \mid k) = \Phi(k + 1, k) P(k \mid k) \Phi^T(k + 1, k) + \Gamma(k) Q(k + 1) \Gamma^T(k)$$
$$+ \Delta\varphi(k) \Delta\varphi^T(k) + \Phi(k + 1, k) \Delta m(k \mid k) \Delta\varphi^T(k)$$
$$+ \Delta\varphi(k) \Delta m^T(k \mid k) \Phi^T(k + 1, k),$$

$$P(k \mid k) = [I - K_c(k) M(k)] P(k \mid k - 1)[I - K_c(k) M(k)]^T$$
$$+ K_c(k) R(k) K_c^T(k),$$

and, combining these equations, we have

$$P(k + 1 \mid k + 1) = [I - K_c(k + 1) M(k + 1)] \Phi(k + 1, k) P(k \mid k) \Phi^T(k + 1, k)$$
$$\times [I - K_c(k + 1) M(k + 1)]^T + F(k), \quad (7.224)$$

where

$$F(k) = [I - K_c(k + 1) M(k + 1)][\Gamma(k) Q(k + 1) \Gamma^T(k) + \Delta\varphi(k) \Delta\varphi^T(k)$$
$$+ \Phi(k + 1, k) \Delta m(k \mid k) \Delta\varphi^T(k) + \Delta\varphi(k) \Delta m^T(k \mid k) \Phi^T(k + 1, k)]$$
$$\times [I - K_c(k + 1) M(k + 1)]^T + K_c(k + 1) R(k + 1) K_c^T(k + 1),$$
$$(7.225)$$

and where [from (7.218)]

$$\Delta m(k + 1 \mid k + 1) = [I - K_c(k + 1) M(k + 1)] \Phi(k + 1, k) \Delta m(k \mid k) + f(k),$$
$$(7.226)$$

where

$$f(k) = [I - K_c(k + 1) M(k + 1)] \Delta\varphi(k). \qquad (7.227$$

Note that

$$[I - K_c(k + 1) M(k + 1)] \Phi(k + 1, k) = \Psi_c(k + 1, k) \qquad (7.228)$$

is the state transition matrix of the filter [see (7.167)]. We have the following theorem due to Price [44].

Theorem 7.7. *If the system model* (7.205) *is uniformly completely observable and uniformly completely controllable, and if F(k) is uniformly bounded and P(0) is bounded, then P(k | k) is uniformly bounded for all k.*

PROOF: We can write the solution of (7.224) as

$$P(k \mid k) = \Psi_c(k, 0) P(0) \Psi_c^{\mathrm{T}}(k, 0) + \sum_{i=0}^{k-1} \Psi_c(k, i + 1) F(i) \Psi_c^{\mathrm{T}}(k, i + 1). \qquad (7.229)$$

Now in view of Theorem 7.4, the filter (7.206) is uniformly asymptotically stable; that is,

$$\| \Psi_c(k, j) \| \leqslant c_1 e^{-c_2(k-j)} \qquad (c_1, c_2 > 0), \qquad \text{all} \quad k \geqslant j. \qquad (7.230)$$

Using this and the uniform bound on $F(k)$ in (7.229), it is not difficult to show that $(c_3, c_4 > 0)$

$$\| P(k \mid k) \| \leqslant c_3 \sum_{i=0}^{k} e^{-c_4 i} \leqslant c_3 \sum_{i=0}^{\infty} e^{-c_4 i} < \infty.$$

(This is an application of the well-known theorem that, in uniformly asymptotically stable systems, a uniformly bounded input produces a uniformly bounded response.) ∎

Consider the problem of bounding $F(k)$, which is a hypothesis of the theorem. In view of the assumed uniform complete observability and controllability, and Lemma 7.1, $P_c(k \mid k)$ and $P_c(k + 1 \mid k)$ are uniformly bounded, at least for $k \geqslant N$. It has been assumed that all the matrices in the description of system (7.205) are bounded (except φ and φ_c). It therefore remains to bound $\Delta\varphi(k)$ and $\Delta m(k \mid k)$ in $F(k)$. Suppose $\Delta\varphi(k)$ is uniformly bounded. Then $f(k)$ in (7.226) is uniformly bounded. Then Theorem 7.7, applied to $\Delta m(k \mid k)$, proves that it is uniformly bounded.

As a consequence, we have the

Corollary. *If the system model* (7.205) *is uniformly completely observable and uniformly completely controllable, and if* $\Delta\varphi(k)$ *is uniformly bounded and* $P(0)$ *is bounded, then* $P(k \mid k)$ *is uniformly bounded.*

Now if $\varphi(k)$ is known to be bounded, then it is easy to bound $\Delta\varphi(k)$ [set $\varphi_c(k) \equiv 0$, for example]. The difficult problem occurs if $\varphi(k)$ is unbounded. An unbounded $\varphi(k)$ can result, for example, if $\varphi(k)$ stands for neglected states (due, for example, to a reduction of system dimension), and these states are unstable. In that case, it appears that our infinite memory filter may diverge [unbounded $P(k \mid k)$], and we may have to resort to the limited memory filter techniques described in Section 10. Suppose a bound on $\Delta\varphi(k)$ is available. Since all the hypotheses of Theorem 7.7 involve bounds on the parameters of the system model (7.205), it is possible, at least in principle, to design the filter in such a way as to produce the desired bound on $P(k \mid k)$. In particular, $R_c(k)$ and $Q_c(k)$ can be so chosen. The hypothesis of uniform complete controllability points out the requirement of having *some* system noise (Q_c) in the filter.

Consider now the possibility of bounding $P(k \mid k)$ in the general situation described by Eqs. (7.212)–(7.218). Assume that *all* the matrices, including φ and φ_c, in systems (7.204) and (7.205) are uniformly bounded. If Eqs. (7.212) and (7.213) are combined, we see that the forcing term in the difference equation for $P(k \mid k)$ depends on $X(k)$, $C(k \mid k)$, and $m(k)$, as well as on $\Delta m(k \mid k)$. An inspection of the equations for X, C, and m reveals that these matrices can be bounded *if* the *actual system* (7.204) is uniformly asymptotically stable, that is, if $\Phi(k + 1, k)$ has property (7.192). Thus, in order to bound $P(k \mid k)$ in the general case, we need the uniform asymptotic stability of the actual system in addition to the hypotheses of Theorem 7.7. This is a very stringent requirement. We can expect the filter to diverge [unbounded $P(k \mid k)$] in cases where the actual system is not uniformly asymptotically stable.

9. ERROR SENSITIVITY—CONTINUOUS FILTER

As in the discrete case, assume that the *actual* system is described by

$$dx_t/dt = F(t) x_t + f(t) + G(t) w_t,$$

$$y_t = M(t) x_t + v_t, \qquad (7.231)$$

$$w_t \sim N[0, Q(t)], \qquad v_t \sim N[0, R(t)], \qquad x_{t_0} \sim N[\hat{x}(t_0), P(t_0)],$$

whereas its *model* is

$$d\bar{x}_t/dt = F_c(t)\,\bar{x}_t + f_c(t) + G_c(t)\,\bar{w}_t,$$

$$\bar{y}_t = M_c(t)\,\bar{x}_t + \bar{v}_t, \tag{7.232}$$

$$\bar{w}_t \sim N[0, Q_c(t)], \qquad \bar{v}_t \sim N[0, R_c(t)], \qquad \bar{x}_{t_0} \sim N[\hat{x}_c(t_0), P_c(t_0)].$$

Boundedness requirements, analogous to those of Sect. 8, are imposed on the matrices in the description of systems (7.231) and (7.232). Our filter design is

$$d\hat{x}_t{}^t/dt = F_c(t)\,\hat{x}_t{}^t + K_c(t)(y_t - M_c(t)\,\hat{x}_t{}^t),$$

$$dP_c(t \mid t)/dt = F_c(t)\,P_c(t \mid t) + P_c(t \mid t)\,F_c{}^{\mathrm{T}}(t)$$
$$+ G_c(t)\,Q_c(t)\,G_c{}^{\mathrm{T}}(t) - K_c(t)\,R_c(t)\,K_c{}^{\mathrm{T}}(t), \tag{7.233}$$

$$K_c(t) = P_c(t \mid t)\,M_c{}^{\mathrm{T}}(t)\,R_c^{-1}(t),$$

$$P_c(t_0 \mid t_0) = P_c(t_0).$$

Define

$$\Delta F(t) \triangleq F(t) - F_c(t),$$
$$\Delta f(t) \triangleq f(t) - f_c(t), \tag{7.234}$$
$$\Delta M(t) \triangleq M(t) - M_c(t),$$

and

$$P(t \mid t) \triangleq \mathscr{E}\{(x_t - \hat{x}_t{}^t)(x_t - \hat{x}_t{}^t)^{\mathrm{T}}\},$$
$$X(t) \triangleq \mathscr{E}\{x_t x_t{}^{\mathrm{T}}\},$$
$$C(t \mid t) \triangleq \mathscr{E}\{x_t(x_t - \hat{x}_t{}^t)^{\mathrm{T}}\}, \tag{7.235}$$
$$m(t) \triangleq \mathscr{E}\{x_t\},$$
$$\delta m(t \mid t) \triangleq \mathscr{E}\{x_t - \hat{x}_t{}^t\}.$$

Then, applying the limiting procedure of Example 7.10 to Eqs. (7.212)–(7.218), we (tediously) get the following differential equations for the *actual* estimation error covariance matrix $P(t \mid t)$, and for $X(t)$, $C(t \mid t)$, $m(t)$, and $\delta m(t \mid t)$:

$$dP(t \mid t)/dt = [F_c(t) - K_c(t)\,M_c(t)]\,P(t \mid t)$$
$$+ P(t \mid t)[F_c(t) - K_c(t)\,M_c(t)]^{\mathrm{T}}$$
$$+ G(t)\,Q(t)\,G^{\mathrm{T}}(t) + K_c(t)\,R(t)\,K_c{}^{\mathrm{T}}(t)$$
$$+ \delta m(t \mid t)\,\Delta f^{\mathrm{T}}(t) + \Delta f(t)\,\delta m^{\mathrm{T}}(t \mid t)$$
$$+ [\Delta F(t) - K_c(t)\,\Delta M(t)]\,C(t \mid t)$$
$$+ C^{\mathrm{T}}(t \mid t)[\Delta F(t) - K_c(t)\,\Delta M(t)]^{\mathrm{T}}, \tag{7.236}$$

$$dX(t)/dt = F(t)\,X(t) + X(t)\,F^{\mathrm{T}}(t) + G(t)\,Q(t)\,G^{\mathrm{T}}(t)$$
$$+ m(t)\,f^{\mathrm{T}}(t) + f(t)\,m^{\mathrm{T}}(t), \tag{7.237}$$

$$dC(t\mid t)/dt = F(t)\,C(t\mid t) + C(t\mid t)[F_c(t) - K_c(t)\,M_c(t)]^{\mathrm{T}}$$
$$+ G(t)\,Q(t)\,G^{\mathrm{T}}(t) + f(t)\,\delta m^{\mathrm{T}}(t\mid t) + m(t)\,\Delta f^{\mathrm{T}}(t)$$
$$+ X(t)[\Delta F(t) - K_c(t)\,\Delta M(t)]^{\mathrm{T}}, \tag{7.238}$$

$$dm(t)/dt = F(t)\,m(t) + f(t), \tag{7.239}$$

$$d\,\delta m(t\mid t)/dt = [F_c(t) - K_c(t)\,M_c(t)]\,\delta m(t\mid t) + \Delta f(t)$$
$$+ [\Delta F(t) - K_c(t)\,\Delta M(t)]\,m(t). \tag{7.240}$$

Initial conditions are those given in (7.219).

Several authors have developed special cases of the actual estimation error covariance equation (7.236). Nishimura [43], Friedland [18], and Mehra [38] considered the case of errors in $P(t_0)$, $Q(t)$, and $R(t)$. Fitzgerald [17] and Griffin and Sage [21] consider all the errors except $\Delta f(t)$. Nishimura, Fitzgerald, and Griffin and Sage present examples and numerical simulations. Analogous equations for the linear smoother are developed by Griffin and Sage and by Mehra. Fitzgerald discusses the case of constant coefficient systems in detail and determines conditions under which divergence will occur.

The analogs of Theorems 7.6 and 7.7 hold in the continuous case. We shall develop these theorems below, but our treatment will be sketchy, since it essentially duplicates the discrete results. The reader should refer to Section 8 for discussion, since everything carries over from the discrete to the continuous case.

In case of errors in $P(t_0)$, $Q(t)$, and $R(t)$ only, (7.236) reduces to

$$dP(t\mid t)/dt = [F(t) - K_c(t)\,M(t)]\,P(t\mid t) + P(t\mid t)[F(t) - K_c(t)\,M(t)]^{\mathrm{T}}$$
$$+ G(t)\,Q(t)\,G^{\mathrm{T}}(t) + K_c(t)\,R(t)\,K_c^{\mathrm{T}}(t), \tag{7.241}$$

whereas

$$dP_c(t\mid t)/dt = F(t)\,P_c(t\mid t) + P_c(t\mid t)\,F^{\mathrm{T}}(t)$$
$$+ G(t)\,Q_c(t)\,G^{\mathrm{T}}(t) - K_c(t)\,R_c(t)\,K_c^{\mathrm{T}}(t). \tag{7.242}$$

Defining

$$E(t) \triangleq P(t\mid t) - P_c(t\mid t), \tag{7.243}$$

we have

$$dE(t)/dt = [F(t) - K_c(t)\,M(t)]\,E(t) + E(t)[F(t) - K_c(t)\,M(t)]^{\mathrm{T}}$$
$$+ G(t)[Q(t) - Q_c(t)]\,G^{\mathrm{T}}(t) + K_c(t)[R(t) - R_c(t)]\,K_c^{\mathrm{T}}(t). \tag{7.244}$$

Equation (7.244) is linear and has the solution

$$E(t) = \Psi_c(t, t_0) E(t_0) \Psi_c^T(t, t_0) + \int_{t_0}^{t} \Psi_c(t, \tau) G(\tau)[Q(\tau) - Q_c(\tau)] G^T(\tau) \Psi_c^T(t, \tau) d\tau$$

$$+ \int_{t_0}^{t} \Psi_c(t, \tau) K_c(\tau)[R(\tau) - R_c(\tau)] K_c^T(\tau) \Psi_c^T(t, \tau) d\tau, \qquad (7.245)$$

where $\psi_c(t, \tau)$ is the state transition matrix of the filter (see Section 7); that is,

$$d\Psi_c(t, \tau)/dt = [F(t) - K_c(t) M(t)] \Psi_c(t, \tau), \qquad \Psi_c(\tau, \tau) = I. \qquad (7.246)$$

Inspection of (7.245) produces the theorem due to Nishimura.

Theorem 7.8. *If $P(t_0) \leqslant P_c(t_0)$ and $Q(t) \leqslant Q_c(t)$, $R(t) \leqslant R_c(t)$, all t, then $P(t \mid t) \leqslant P_c(t \mid t)$ for all t.*

The following corollary is immediate if we apply the continuous analog of Lemma 7.1 (see Section 7).

Corollary. *Hypotheses of Theorem 7.8. Further suppose that the system model (7.232) is uniformly completely observable and uniformly completely controllable. Then there is a $\sigma > 0$ such that $P(t \mid t)$ is uniformly bounded for all $t \geqslant t_0 + \sigma$.*

Now consider the case of errors in $P(t_0)$, $Q(t)$, $R(t)$, and $f(t)$. Equation (7.236) becomes

$$dP(t \mid t)/dt = [F(t) - K_c(t) M(t)] P(t \mid t) + P(t \mid t)[F(t) - K_c(t) M(t)]^T + D(t), \qquad (7.247)$$

where

$$D(t) = G(t) Q(t) G^T(t) + K_c(t) R(t) K_c^T(t) + \delta m(t \mid t) \Delta f^T(t) + \Delta f(t) \delta m^T(t \mid t), \qquad (7.248)$$

and where, from (7.240),

$$d\, \delta m(t \mid t)/dt = [F(t) - K_c(t) M(t)] \delta m(t \mid t) + \Delta f(t). \qquad (7.249)$$

Theorem 7.9. *If the system model (7.232) is uniformly completely observable and uniformly completely controllable, and if $\Delta f(t)$ is uniformly bounded and $P(t_0)$ is bounded, then $P(t \mid t)$ is uniformly bounded.*

PROOF: Note that both (7.249) and (7.247) are linear, and their solutions may be written down using the filter state transition matrix defined in (7.246). By Theorem 7.4 for the continuous filter, (7.249) is uniformly asymptotically stable, and, since $\Delta f(t)$ is uniformly bounded, so is $\delta m(t \mid t)$. As a result, $D(t)$ is uniformly bounded. But then, by the same arguments, $P(t \mid t)$ is uniformly bounded. ■

10. LINEAR LIMITED MEMORY FILTER

The results and discussion of Sections 8 and 9 indicate that, if the dynamical model parameters are not sufficiently well known, the "optimal" filter may diverge. Even if the actual estimation error covariance matrix is bounded, the actual estimation errors may be too large and the estimate unusable. In that case, for all practical purposes, the filter is diverging. The onset of divergence is, of course, determined by statistical inconsistency of the actual estimation errors with the computed error covariance matrix.

The problem of divergence is discussed in detail in the next chapter, where various remedies for divergence are outlined. One of these is the idea of limiting the memory of the filter, thus curtailing the growth of the actual estimation error covariance matrix. This has already been discussed in Section 6.11, as well as in Section 8. The limited memory filter appears to be the only device for preventing divergence in the presence of unbounded $\Delta\varphi(k)$ or $\Delta f(t)$ (see Theorems 7.7 and 7.9).

Since the limited memory filter has a sound theoretical basis (most other remedies are *ad hoc*), we develop its theory here. We note that the idea of limiting the filter memory by various means has been advocated by a number of authors. Schweppe [49] develops a form of the limited memory filter. Similar ideas are contained in Cosaert and Gottzein [12]. Schmidt [48] advocates overweighting the most recent observations, and limiting information by other means. Our development here follows that of Jazwinski [28].

We assume that our discrete dynamical system is noise-free:

$$x_{k+1} = \Phi(k+1, k)\, x_k, \qquad x_0 \sim N(\hat{x}_0, P_0), \qquad P_0 > 0, \qquad (7.250)$$

with observations

$$y_k = M(k)\, x_k + v_k, \qquad v_k \sim N[0, R(k)], \qquad R(k) > 0. \qquad (7.251)$$

We propose to limit the filter memory sufficiently so that (7.250), (7.251) is an adequate approximation to the real system over the limited time arcs. The absence of noise in (7.250) is fundamental to our theory. We now specialize the results of Section 6.11 to this linear case.

Using the notation (Y_m, Y_n, Y_N) defined in Section 6.11, let

$$\hat{x}_n^n \triangleq \mathscr{E}\{x_n \mid Y_n\},$$

$$P_n^n \triangleq \mathscr{E}\{(x_n - \hat{x}_n^n)(x_n - \hat{x}_n^n)^{\mathrm{T}} \mid Y_n\},$$

$$\hat{x}_n^m \triangleq \mathscr{E}\{x_n \mid Y_m\} = \Phi(n, m)\,\hat{x}_m^m,$$

$$P_n^m \triangleq \mathscr{E}\{(x_n - \hat{x}_n^m)(x_n - \hat{x}_n^m)^{\mathrm{T}} \mid Y_m\} = \Phi(n, m)\,P_m^m\Phi^{\mathrm{T}}(n, m).$$

(7.252)

Theorem 7.10. *If x_n is completely observable with respect to Y_N, then $\hat{x}_n^{N|}$, $P_n^{N|}$ (given below) are, respectively, the conditional mean of x_n and the conditional covariance matrix, where the conditioning is on Y_N only:*

$$\hat{x}_n^{N|} = P_n^{N|}(P_n^{n-1}\hat{x}_n^n - P_n^{m-1}\hat{x}_n^m),$$

$$P_n^{N|-1} = P_n^{n-1} - P_n^{m-1}.$$

(7.253)

$\hat{x}_n^{N|}$ is also the maximum likelihood estimate of x_n based on Y_N.

PROOF: Since $P_0 > 0$, and in view of the discussion following Lemma 7.1, P_n^n, $P_n^m > 0$, so that the densities appearing in Lemma 6.5 are well-defined Gaussian densities. According to Example 7.2, the parameters of $p(x_n \mid Y_N)$ (that is, its mean and covariance matrix) are, respectively,

$$\mathscr{E}\{x_n \mid Y_N, p(x_0)\} = \mathrm{cov}\{x_n \mid Y_N, p(x_0)\}$$

$$\times \left[\sum_{i=m+1}^{n} \Phi^{\mathrm{T}}(i, n)\,M^{\mathrm{T}}(i)\,R^{-1}(i)\,y_i + \Phi^{\mathrm{T}}(0, n)\,P_0^{-1}\hat{x}_0 \right],$$

(7.254)

$$\mathrm{cov}^{-1}\{x_n \mid Y_N, p(x_0)\} = \Phi^{\mathrm{T}}(0, n)\,P_0^{-1}\Phi(0, n) + \mathscr{I}(n, m+1),$$

where

$$\mathscr{I}(n, m+1) = \sum_{i=m+1}^{n} \Phi^{\mathrm{T}}(i, n)\,M^{\mathrm{T}}(i)\,R^{-1}(i)\,M(i)\,\Phi(i, n) > 0 \quad (7.255)$$

by hypothesis. Therefore, the $P_0^{-1} = 0$ limits in (7.254) exist, so that $p(x_n \mid Y_N)$ is a well-defined Gaussian density [defined according to (6.133)].

We now compute the parameters of $p(x_n \mid Y_N)$ using Theorem 6.8.

Now $p(x_n \mid Y_N)$ has parameters $\hat{x}_n{}^n$, $P_n{}^n$; and $p(x_n \mid Y_m)$ has parameters $\hat{x}_n{}^m$, $P_n{}^m$. Therefore, by Theorem 6.8,

$$p(x_n \mid Y_N \mid)$$
$$= c_2 \exp\{-\tfrac{1}{2}[(x_n - \hat{x}_n{}^n)^{\mathrm{T}} P_n^{n^{-1}}(x_n - \hat{x}_n{}^n) - (x_n - \hat{x}_n{}^m)^{\mathrm{T}} P_n^{m^{-1}}(x_n - \hat{x}_n{}^m)]\}.$$
$$(7.256)$$

Since this density is Gaussian, its mean coincides with its mode. The mode minimizes the quantity in brackets in (7.256). Setting the gradient (with respect to x_n) of that quantity to zero, we get

$$P_n^{n^{-1}}(x_n - \hat{x}_n{}^n) - P_n^{m^{-1}}(x_n - \hat{x}_n{}^m) = 0. \qquad (7.257)$$

$\hat{x}_n^{N\mid}$ is the solution of (7.257). $P_n^{N\mid^{-1}}$ is the matrix of second derivatives

$$P_n^{n^{-1}} - P_n^{m^{-1}} \equiv P_n^{N\mid^{-1}}.$$

Note that $P_n^{N\mid} = \mathcal{I}^{-1}(n,\, m + 1)$, which is positive definite by hypothesis, so that $P_n^{N\mid}$ is nonsingular. This establishes (7.253) and proves the theorem. ∎

Several remarks concerning Theorem 7.10 are in order. We note that $\hat{x}_n{}^n$, $P_n{}^n$ are the outputs of the discrete Kalman filter of Theorem 7.2. $\hat{x}_n{}^m$, $P_n{}^m$ are predicted values computed via (7.28). We may solve (7.253) for $\hat{x}_n{}^n$ and $P_n{}^n$ and use the resulting equations to process batches of preprocessed data (preprocessed via a least-squares procedure) and characterized by $\hat{x}_n^{N\mid}$ and $P_n^{N\mid}$. If $m = 0$, Eqs. (7.253) can be used to remove the conditioning of a Kalman estimate on any initial data. In fact, we can remove $\hat{x}_0$, P_0 and *replace* it by $\hat{x}_0'$, P_0'. We may note that, with $n = m$ ($N = 0$), $P_n^{N\mid^{-1}} = 0$ (no information), and $\hat{x}_n^{N\mid}$ is arbitrary. For N large, $P_n^{N\mid^{-1}} \approx P_n^{n^{-1}}$ and $\hat{x}_n^{N\mid} \approx \hat{x}_n{}^n$, which are the intuitively correct properties.

Equations (7.253) generate the minimum variance estimate based on a "moving window" of the most recent N observations. Two Kalman filters and a predictor are required in their implementation, and N observations have to be stored in memory. Three indicated matrix inversions are required. Now, aside from the large computational load, this scheme is clearly no solution to the problem of divergence. The two required Kalman filters would diverge!

Based on Theorem 7.10, we propose the following limited memory filter, which, at the same time, decreases the computational load and eliminates the storage requirement. The conditioning of the estimate on old data is discarded in batches of N. We filter N observations and also predict over the same time arc. This produces $\hat{x}_N{}^N$, $P_N{}^N$, $\hat{x}_N{}^0$, $P_N{}^0$. Equations (7.253) are used to compute $\hat{x}_N^{N|}$, $P_N^{N|}$ (we have just eliminated the prior data). We next set $\hat{x}_N^{N|}$, $P_N^{N|}$ as initial conditions for both the filter and predictor, and repeat the procedure.

The limited memory filter just described produces limited memory estimates with memory varying between N and $2N$. Clearly, variations on this are possible, and N itself may be varied in the process. N should be chosen so that the model (7.250) represents an adequate approximation to the real system over time arcs of $2N$.

We next discuss some alternate computational procedures. The second of Eqs. (7.253) involves two matrix inversions and a subtraction, and could be ill-conditioned computationally. Recognizing that $P_n^{N|} = \mathscr{I}^{-1}(n, m + 1)$ (see 7.255), we may compute this matrix by the recursion

$$\mathscr{I}(k, m + 1) = \Phi^{\mathrm{T}}(k - 1, k)\,\mathscr{I}(k - 1, m + 1)\,\Phi(k - 1, k)$$
$$+ M^{\mathrm{T}}(k)\,R^{-1}(k)\,M(k), \quad k = m + 1,..., n, \quad (7.258)$$

with

$$\mathscr{I}(m, m + 1) = 0.$$

Then the first of Eqs. (7.253) can be written as

$$\hat{x}_n^{N|} = \hat{x}_n{}^n + \mathscr{I}^{-1}(n, m + 1)\,P_n^{m^{-1}}(\hat{x}_n{}^n - \hat{x}_n{}^m). \quad (7.259)$$

Equation (7.258) requires the inverse of the state transition matrix, but this is sometimes available in closed form.[12] In that case, $P_n^{m^{-1}}$ can also be computed recursively:

$$P_k^{m^{-1}} = \Phi^{\mathrm{T}}(k - 1, k)\,P_{k-1}^{m^{-1}}\Phi(k - 1, k), \quad k = m + 1,..., n. \quad (7.260)$$

Recursions (7.258) and (7.260) can be carried along with the Kalman filter.

As was mentioned earlier, Schweppe [49] develops another form of the limited memory filter with a constant memory length. That algorithm, however, requires the storage of N observations together with $M(k)$ and $R(k)$, $k = m + 1,..., n$; the inverse of the state transition matrix; prediction over arcs of length N every time a new observation is acquired; and inversion of an information matrix every time a new observation is acquired. Although a constant memory length is desirable, the computational load associated with the filter just described is very large.

---

[12] In orbit determination, closed-form approximations to the state transition matrix and its inverse are available.

APPENDIX 7A

CLASSICAL PARAMETER ESTIMATION

In this Appendix, we point out some interesting connections between filtering theory and classical parameter estimation theory. We shall see that estimation and filtering are based on essentially the same theory. Consider the problem of estimating the (fixed) parameter x_k from the observations

$$y_i = M(i)\, \Phi(i, k)\, x_k + v_i\,, \qquad i = 1,\dots, k, \qquad (7A.1)$$

where $\{v_i\}$ is a white, but *not necessarily* Gaussian, vector sequence with

$$\mathscr{E}\{v_i\} = 0, \qquad \mathscr{E}\{v_i v_j^{\mathrm{T}}\} = R_i\, \delta_{ij}\,, \qquad R_i > 0.$$

The parameter x_k may be viewed as the state, at time t_k, of the dynamical system

$$x_{i+1} = \Phi(i + 1, i)\, x_i\,, \qquad y_i = M(i)\, x_i + v_i\,.$$

We assume that x_k is completely observable with respect to $\{y_1,\dots, y_k\}$. We wish to find the *linear, unbiased, minimum variance* estimator for x_k. That is, we seek that estimate x_k^* in the class of estimates $\{\xi_k\}$,

$$\xi_k = \sum_{i=1}^{k} A_i y_i\,, \qquad (7A.2)$$

which is unbiased,

$$\mathscr{E}\{x_k^*\} = \mathscr{E}\{x_k\} = x_k\,, \qquad (7A.3)$$

and for which

$$\mathrm{tr}\, \mathscr{E}\{(x_k^* - x_k)(x_k^* - x_k)^{\mathrm{T}}\}$$

is a minimum. The answer is provided in the classical

Theorem 7A.1 (Gauss–Markov). *Let x_k be completely observable with respect to $\{y_1,\dots, y_k\}$. Let $\{v_i\}$ be a zero-mean, white, not necessarily Gaussian vector sequence, $R_i > 0$. Then the linear, unbiased, minimum variance estimate of x_k is*

$$x_k^* = \mathscr{I}_{k,1}^{-1} \sum_{i=1}^{k} \Phi^{\mathrm{T}}(i, k)\, M^{\mathrm{T}}(i)\, R_i^{-1} y_i\,,$$

where

$$\mathscr{I}_{k,1} = \sum_{i=1}^{k} \Phi^{\mathrm{T}}(i, k)\, M^{\mathrm{T}}(i)\, R_i^{-1} M(i)\, \Phi(i, k)$$

is the information matrix. The covariance matrix of x_k^* is $\mathscr{I}_{k,1}^{-1}$.

PROOF: Consider any linear, unbiased estimate

$$\xi_k = \sum_{i=1}^{k} A_i y_i .$$

We have

$$\mathscr{E}\{\xi_k\} = \mathscr{E}\left\{\sum_{i=1}^{k} A_i[M(i)\,\Phi(i, k)\,x_k + v_i]\right\} = \sum_{i=1}^{k} A_i M(i)\,\Phi(i, k)\,x_k$$

and

$$\xi_k - x_k = \xi_k - \mathscr{E}\{\xi_k\}.$$

Therefore,

$$\mathscr{E}\{(\xi_k - x_k)(\xi_k - x_k)^{\mathrm{T}}\}$$

$$= \sum_{i=1}^{k} A_i R_i A_i^{\mathrm{T}}$$

$$= \sum_{i=1}^{k} [\mathscr{I}_{k,1}^{-1}L_i + (A_i - \mathscr{I}_{k,1}^{-1}L_i)]\, R_i[\mathscr{I}_{k,1}^{-1}L_i + (A_i - \mathscr{I}_{k,1}^{-1}L_i)]^{\mathrm{T}}, \quad (7A\ 4)$$

where

$$L_i \triangleq \Phi^{\mathrm{T}}(i, k)\, M^{\mathrm{T}}(i)\, R_i^{-1}.$$

Now

$$\sum_{i=1}^{k} (A_i - \mathscr{I}_{k,1}^{-1}L_i)\, R_i L_i^{\mathrm{T}}\mathscr{I}_{k,1}^{-1} = \left[\sum_{i=1}^{k} A_i M(i)\,\Phi(i, k) - I\right]\mathscr{I}_{k,1}^{-1} = 0,$$

since ξ_k is unbiased. [From (7A.3),

$$0 = \mathscr{E}\{\xi_k\} - x_k = \left[\sum_{i=1}^{k} A_i M(i)\,\Phi(i, k) - I\right]x_k$$

for all x_k]. As a result, (7A.4) becomes

$$\mathscr{E}\{(\xi_k - x_k)(\xi_k - x_k)^{\mathrm{T}}\}$$

$$= \sum_{i=1}^{k} [\mathscr{I}_{k,1}^{-1}L_i R_i L_i^{\mathrm{T}}\,\mathscr{I}_{k,1}^{-1} + (A_i - \mathscr{I}_{k,1}^{-1}L_i)\, R_i(A_i - \mathscr{I}_{k,1}^{-1}L_i)^{\mathrm{T}}]. \quad (7A.5)$$

Since $R_i > 0$, the trace of this matrix is minimum when $A_i = \mathscr{I}_{k,1}^{-1}L_i$, that is, when $\xi_k = x_k{}^*$. From (7A.5), it then follows that

$$\mathscr{E}\{(x_k{}^* - x_k)(x_k{}^* - x_k)^{\mathrm{T}}\} = \mathscr{I}_{k,1}^{-1}. \quad \blacksquare$$

Compare the linear, unbiased, minimum variance estimate (Theorem 7A.1) with the estimate (7.41) of Example 7.2, which derives the discrete Kalman–Bucy filter. We see that with $P_0^{-1} = 0$ they are identical. Thus we see again that the absence of *a priori* information about x_k is expressed formally by setting $P_0^{-1} = 0$. We also have

Theorem 7A.2. *The minimum variance filter in the Gaussian case is also the linear minimum variance filter in the general case, without any assumption of Gaussianness.*

This conclusion can also be understood in the following way. If we seek the best *linear* estimate, then only the first two moments of the $\{y_k\}$ process need be known. These moments define a unique Gaussian process. Therefore, the linear minimum variance filter in the Gaussian case must be the same as in the general case. But in the Gaussian case, the variance is minimized by the conditional mean which *is* a linear functional of the $\{y_k\}$ process. Therefore, in the Gaussian case, the minimum variance and linear minimum variance filters are the same. Thus the conclusion follows.

Recall that, in the derivation of Example 7.2, no statistical significance was attached to the R_i, that is, the errors v_i were not considered random variables with a given distribution. Now if one is willing to stipulate that the v_i are *zero-mean*, one can conclude that the estimate (of Theorem 7A.1) is *unbiased*. If one is further willing to specify the *second moments* of v_i, that is, R_i, one can conclude by the Gauss–Markov theorem that the estimate is a *linear, minimum variance, unbiased* estimate. If the v_i are furthermore *Gaussian*, then Theorem 7A.2 asserts that the estimate is the *minimum variance* estimate. Thus there is a direct relationship between the assumptions made about the noise and the conclusions that can be drawn about the estimate.

APPENDIX 7B

SOME MATRIX EQUALITIES

Let P, R, and M be $n \times n$, $m \times m$, and $m \times n$ matrices, respectively. Suppose $P \geqslant 0$ and $R > 0$. Then the following equalities hold:

$$(I + PM^{\mathrm{T}}R^{-1}M)^{-1} = I - PM^{\mathrm{T}}(MPM^{\mathrm{T}} + R)^{-1}M, \qquad (7B.1)$$

$$(I + PM^{\mathrm{T}}R^{-1}M)^{-1}P = P - PM^{\mathrm{T}}(MPM^{\mathrm{T}} + R)^{-1}MP, \qquad (7B.2)$$

$$(I + PM^{\mathrm{T}}R^{-1}M)^{-1}PM^{\mathrm{T}}R^{-1} = PM^{\mathrm{T}}(MPM^{\mathrm{T}} + R)^{-1}. \qquad (7B.3)$$

If, in addition, $P > 0$, then

$$(I + PM^{T}R^{-1}M)^{-1}P = (P^{-1} + M^{T}R^{-1}M)^{-1}. \quad (7B.4)$$

Using (7B.4) in (7B.2) and (7B.3), we get (for P, $R > 0$)

$$(P^{-1} + M^{T}R^{-1}M)^{-1} = P - PM^{T}(MPM^{T} + R)^{-1}MP, \quad (7B.5)$$

$$(P^{-1} + M^{T}R^{-1}M)^{-1}M^{T}R^{-1} = PM^{T}(MPM^{T} + R)^{-1}. \quad (7B.6)$$

These matrix equalities are relatively well known in numerical analysis (e.g., Bodewig [6, p. 30]). Equation (7B.5) was introduced to the filtering field by Ho [26].

Equation (7B.1) is easily proved by direct verification:

$$(I + PM^{T}R^{-1}M)(I - PM^{T}(MPM^{T} + R)^{-1}M)$$
$$= I + PM^{T}R^{-1}M - PM^{T}R^{-1}R(MPM^{T} + R)^{-1}M$$
$$\quad - PM^{T}R^{-1}MPM^{T}(MPM^{T} + R)^{-1}M$$
$$= I.$$

Equation (7B.2) follows trivially. To prove (7B.3), multiply (7B.2) on the right by $M^{T}R^{-1}$:

$$(I + PM^{T}R^{-1}M)^{-1}PM^{T}R^{-1}$$
$$= PM^{T}R^{-1} - PM^{T}(MPM^{T} + R)^{-1}MPM^{T}R^{-1}$$
$$\quad + [PM^{T}(MPM^{T} + R)^{-1} - PM^{T}(MPM^{T} + R)^{-1}RR^{-1}]$$
$$= PM^{T}(MPM^{T} + R)^{-1}.$$

Equation (7B.4) follows from

$$[(I + PM^{T}R^{-1}M)^{-1}P]^{-1} = P^{-1}(I + PM^{T}R^{-1}M) = P^{-1} + M^{T}R^{-1}M.$$

REFERENCES

1. M. Aoki, "Optimization of Stochastic Systems." Academic Press, New York, 1967.
2. R. H. Battin, Statistical Optimizing Navigation Procedure for Space Flight, *ARS J.* 32, 1681–1696 (1962).
3. R. H. Battin, "Astronautical Guidance." McGraw-Hill, New York, 1964.
4. R. E. Bellman and S. E. Dreyfus, "Applied Dynamic Programming." Princeton Univ. Press, Princeton, New Jersey, 1962.
5. R. E. Bellman, H. H. Kagiwada, R. E. Kalaba, and R. Sridhar, Invariant Imbedding and Nonlinear Filtering Theory, *J. Astronaut. Sci.* 13, 110–115 (1966).
6. E. Bodewig, "Matrix Calculus." Wiley (Interscience), New York, 1959.

7. A. E. Bryson and M. Frazier, Smoothing for Linear and Nonlinear Dynamic Systems, *Proc. Optimum Systems Synthesis Conf.* Wright-Patterson Air Force Base, Ohio, February 1963. U. S. Air Force Tech. Rept. ASD-TDR-063-119, 1963.

8. A. E. Bryson and D. E. Johansen, Linear Filtering for Time-Varying Systems Using Measurements Containing Colored Noise, *IEEE Trans. Automatic Control* 10, 4–10 (1965).

9. A. E. Bryson, Jr. and L. J. Henrikson, Estimation Using Sampled-Data Containing Sequentially Correlated Noise, *AIAA Guidance, Control and Flight Dynamics Conf., Huntsville, Alabama.* Paper 67-541, (1967).

10. A. E. Bryson, Jr. and Y. C. Ho, "Applied Optimal Control." Ginn (Blaisdell), Waltham, Massachusetts, 1969.

11. E. A. Coddington and N. Levinson, "Theory of Ordinary Differential Equations." McGraw-Hill, New York, 1955.

12. R. Cosaert and E. Gottzein, A Decoupled Shifting Memory Filter Method for Radio Tracking of Space Vehicles, *Intern. Astronaut. Congr. 18th, Belgrade, Yugoslavia.* Presented paper (1967).

13. H. Cox, Estimation of State Variables via Dynamic Programming, *Proc. 1964 Joint Automatic Control Conf., Stanford, California,* pp. 376–381, (1964).

14. D. M. Detchmendy and R. Sridhar, Sequential Estimation of States and Parameters in Noisy Non-linear Dynamical Systems, *Proc. 1965 Joint Automatic Control Conf.,* Troy, New York, pp. 56–63 (1965).

15. J. J. Deyst and C. F. Price, Conditions for Asymptotic Stability of the Discrete, Minimum Variance, Linear Estimator, *IEEE Trans. Automatic Control* (to appear).

16. S. L. Fagin, Recursive Linear Regression Theory, Optimal Filter Theory, and Error Analysis of Optimal Systems, *IEEE Intern. Conv. Record* 12, 216–240 (1964).

17. R. J. Fitzgerald, Error Divergence in Optimal Filtering Problems, *Second IFAC Symp. Automatic Control in Space, Vienna, Austria, September 1967.* Presented paper (1967).

18. B. Friedland, On the Effect of Incorrect Gain in the Kalman Filter, *IEEE Trans. Automatic Control* 12, 610 (1967).

19. P. A. Gainer, A Method for Computing the Effect of an Additional Observation on a Previous Least-Squares Estimate, NASA Technical Note D-1599, 1963.

20. I. M. Gelfand and S. V. Fomin, "Calculus of Variations." Prentice-Hall, Englewood Cliffs, New Jersey, 1963.

21. R. E. Griffin and A. P. Sage, Large and Small Scale Sensitivity Analysis of Optimum Estimation Algorithms, *Proc. Joint Automatic Control Conf., Ann Arbor, Michigan, June 1968,* pp. 977–988 (1968).

22. R. E. Griffin and A. P. Sage, Sensitivity Analysis of Discrete Filtering and Smoothing Algorithms, *AIAA Guidance, Control, and Flight Dynamics Conf., Pasadena, California, August 1968.* Paper No. 68–824 (1968).

23. W. Hahn, "Theory and Application of Liapunov's Direct Method." Prentice-Hall, Englewood Cliffs, New Jersey, 1963.

24. H. Heffes, The Effect of Erroneous Models on the Kalman Filter Response, *IEEE Trans. Automatic Control* 11, 541–543 (1966).

25. Y. C. Ho, The Method of Least Squares and Optimal Filtering Theory, The RAND Corporation, Memorandum RM-3329-PR, 1962.

26. Y. C. Ho, On the Stochastic Approximation Method and Optimal Filtering Theory, *J. Math. Anal. Appl.* 6, 152–154 (1963).

27. Y. C. Ho and R. C. K. Lee, A Bayesian Approach to Problems in Stochastic Estimation and Control, *IEEE Trans. Automatic Control* 9, 333–339 (1964).

28. A. H. Jazwinski, Limited Memory Optimal Filtering, *Proc. 1968 Joint Automatic Control Conf. Ann Arbor, Michigan, June 1968,* pp. 383–393 [Also in *IEEE Trans. Automatic Control* 13, 558–563 (1968)].

29. R. E. Kalman, Contributions to the Theory of Optimal Control, *Bol. Soc. Mat. Mexicana* 5, 102–119 (1960).

30. R. E. Kalman and J. E. Bertram, Control System Analysis and Design via the "Second Method" of Lyapunov, I. Continuous-Time Systems, *Trans. ASME, Ser. D: J. Basic Eng.* 82, 371–393 (1960).

31. R. E. Kalman and J. E. Bertram, Control System Analysis and Design via the "Second Method" of Lyapunov, II. Discrete-Time Systems, *Trans. ASME, Ser. D: J. Basic Eng.* 82, 394–400 (1960).

32. R. E. Kalman, A New Approach to Linear Filtering and Prediction Problems, *Trans. ASME, Ser. D: J. Basic Eng.* 82, 35–45 (1960).

33. R. E. Kalman and R. S. Bucy, New Results in Linear Filtering and Prediction Theory, *J. Basic Eng.* 83, 95–108, 1961.

34. R. E. Kalman, On the General Theory of Control Systems, *Proc. IFAC Congr 1st.* 1, 481–491 (1961).

35. R. E. Kalman, Y. C. Ho, and K. S. Narenda, Controllability of Linear Dynamical Systems, *in* "Contributions to Differential Equations" (J. P. Lasalle and J. B. Diaz, eds.), Vol. 1. Wiley (Interscience), New York, 1963.

36. R. E. Kalman, New Methods in Wiener Filtering Theory, *Proc. Symp. Eng. Appl. Random Function Theory and Probability,* (J. L. Bogdanoff and F. Kozin, eds.), Wiley, New York, 1963.

37. R. C. K. Lee, "Optimal Estimation, Identification, and Control," Res. Monograph No. 28. M.I.T. Press, Cambridge, Massachusetts, 1964.

38. R. K. Mehra, On Optimal and Suboptimal Linear Smoothing, Tech. Rept. No. 559, Division of Eng. and Appl. Phys., Harvard Univ., Cambridge, Massachusetts, March 1968.

39. R. K. Mehra and A. E. Bryson, Smoothing for Time-Varying Systems Using Measurements Containing Colored Noise, *Proc. 1968 Joint Automatic Control Conf., Ann Arbor, Michigan, June 1968,* pp. 871–883 (1968).

40. V. O. Mowery, Least Squares Recursive Differential-Correction Estimation in Nonlinear Problems, *IEEE Trans. Automatic Control* 10, 399–407 (1965).

41. T. Nishimura, On the a priori Information in Sequential Estimation Problems, *IEEE Trans. Automatic Control* 11, 197–204 (1966).

42. T. Nishimura, Correction to and Extension of "On the a priori Information in Sequential Estimation Problems, *IEEE Trans. Automatic Control* 12, 123 (1967).

43. T. Nishimura, Error Bounds of Continuous Kalman Filters and the Application to Orbit Determination Problems, *IEEE Trans. Automatic Control* 12, 268–275 (1967).

44. C. F. Price, An Analysis of the Divergence Problem in the Kalman Filter, *IEEE Trans. Automatic Control* (to appear).

45. H. E. Rauch, F. Tung, and C. T. Striebel, Maximum Likelihood Estimates of Linear Dynamic Systems, *AIAA J.* 3, 1445–1450 (1965).

46. R. W. Rishel, Convergence of Sampled Discrete Optimum Filters, *Proc. 1968 Joint Automatic Control Conf. Ann Arbor, Michigan, June 1968,* pp. 863–870 (1968).

47. S. F. Schmidt, Application of State-Space Methods to Navigation Problems, *Advan. Control Syst.* 3, 293–340 (1966).

48. S. F. Schmidt, Compensation for Modeling Errors in Orbit Determination Problems, Analytical Mechanics Associates, Inc., Seabrook, Maryland. Report No. 67-16, November 1967.

49. F. C. Schweppe, Algorithms for Estimating a Re-entry Body's Position, Velocity and Ballistic Coefficient in Real Time or from Post Flight Analysis, M.I.T., Lincoln Lab. Lexington, Massachusetts. Rept. ESD-TDR-64-583, December 1964.

50. G. L. Smith, S. F. Schmidt, and L. A. McGee, Applications of Statistical Filter Theory to the Optimal Estimation of Position and Velocity on Board a Circumlunar Vehicle, NASA TR R-135, 1962.

51. H. W. Sorenson, Kalman Filtering Techniques, *Advan. Control Syst.* 3 (1966).

52. H. W. Sorenson, On the Error Behavior in Linear Minimum Variance Estimation Problems, *IEEE Trans. Automatic Control* 12, 557–562 (1967).

53. P. Swerling, First Order Error Propagation in a Stagewise Smoothing Procedure for Satellite Observations, *J. Astronaut. Sci.* 6, 46–52 (1959).

8

Applications of Linear Theory

1. INTRODUCTION

The practical implications of linear filtering theory were soon recognized by the engineering community. Whereas classical least squares methods involve simultaneous processing of batches of observations with attendant data storage requirements, the filter operates on the data sequentially, requiring no data storage. The filter generates new state estimates as new observations become available, thus opening the possibility of real-time estimation. As a by-product, the filter generates the estimation error covariance matrix, which measures the uncertainty in the estimate.

The importance of having realistic estimation error variances cannot be overemphasized. Of what value is an estimate if we do not really know how good it is? Knowledge of the error variance is especially important if some decision is to be made on the basis of the estimate. Suppose we are journeying to the moon and we detect an error in our target destination. Should we correct our trajectory now or wait for more observations to be taken? How much should we correct? Should we correct now and again later? Answers to such questions as these depend not only on our estimate of the error, but also on the certainty of the error estimate.

Other important advantages accrue from the filter theory. It is possible, for example, to perform a complete error analysis without actually simulating the filter. This is accomplished by solving the covariance matrix equations that are independent of the observations. As we shall subsequently see (Sections 3 and 7), the filter structure offers advantages over least squares in applications to nonlinear problems. Thus, although the linear filter is completely equivalent to least squares when the latter is properly interpreted (see Example 7.2), it offers numerous advantages in applications.

In the following paragraphs, we briefly sketch the applications of linear filter theory which have been made to aerospace engineering systems. Our sketch is no doubt incomplete, and the author apologizes in advance for any errors of omission. Although linear filter theory has been applied in other fields, the author is most familiar with aerospace applications. It is probably fair to say that most of the applications have been made by the aerospace community, and that this community has contributed significantly to the nontrivial task of reducing the theory to practice.

Perhaps the first practical application of linear filter theory was made by Smith, Schmidt, and their colleagues (see [22, 30, 35, 36]) to celestial navigation. Simultaneously, a similar application was made by Battin [4]. Pines et al. [23] applied the theory to orbit determination where, hitherto, least squares techniques had been in use. Controlled simulations indicated that the filter theory can be successfully applied to these problems. Such simulations are also reported by Smith [37]. We note that orbit determination problems are nonlinear, and standard linearization techniques (Section 3) were used in these applications.

Several problem areas were encountered by these investigators. One was the requirement of high precision in the filter computations. (See Section 13 for a technique that reduces numerical precision requirements.) The other, more fundamental problem encountered was that of modeling errors that can lead to divergence of the filter. Several investigators, Gunckel [14] and Pines et al. [24] among them, found that the filter did not perform as well under actual conditions (real observations) as it did under controlled conditions (simulated observations). This problem was already analyzed in Chapter 7, and will occupy much of our effort in the present chapter.

Following these early pioneering investigations, a number of authors report interesting applications of linear filter theory. These include the work of Rauch [28], Farrell [11], and Fitzgerald [12] in orbit determination and navigation, and the work of Wagner [39] in reentry trajectory estimation. By now the linear filter theory is relatively well known in the aerospace community, being the subject of survey papers and articles (e.g., Schmidt [31] and Sorenson [38]). The Kalman–Bucy filter is now being used for ascent guidance at Cape Kennedy, and for various phases of the Apollo lunar mission [5]. It is also used in various places for error analyses.

The reduction of theory to practice is generally a long and difficult process. Such is the situation in the case of linear filter theory. Yet reports of applications often deal only with the successful final results and are thus, in a sense, very misleading, especially to the uninitiated. Schmidt *et al.* [34] recently reported the application of the theory to the C-5 aircraft navigation problem. This study goes into considerable detail in reporting the many problem areas encountered along the way in this application, and is recommended reading for the engineer interested in applications.

The Goddard study[1] of applications in orbit determination [23, 24] still continues. A recent status report on this work is given by Wolf [41]. After six years of effort (the author has been associated with this work for the past three years), the fundamental modeling problem still remains. Although several successful applications have been made, they have by no means been routine. Each new satellite presents new versions of the same problem. A dynamical model adequate for one satellite is inadequate for another one. This problem, of course, exists in all applications, but is particularly acute in tracking a satellite for many revolutions (orbits). These problems, and several solutions, are discussed in Sections 8–12.

This chapter deals with applications of linear filter theory to orbit determination and trajectory estimation problems. Since these problems are generally discrete (or continuous-discrete, which is essentially the same thing), we focus our attention here on the discrete filter. We review the discrete filter in Section 2 and outline its application to nonlinear problems in Section 3. The filter is then extended in Section 4 to systems with uncertain parameters which may, but need not, be estimated. In order to give the reader a "feel" for the filter, we treat a simple academic example in Section 5, giving results of numerical computations. The next two sections (Sections 6 and 7) present some applications of the linear filter in orbit determination and reentry trajectory estimation. The intent here is to show how the filter can perform in these systems under controlled conditions (simulated data). The next five sections deal with the important remaining problem area in filter applications, namely, the problem of filter divergence due to modeling errors. The

---

[1] This study is being conducted by the Special Projects Branch of the NASA Goddard Space Flight Center, under the direction of Mr. R. K. Squires.

problem is introduced and exhibited in Section 8. Sections 9 and 10 summarize various techniques that have been found useful in preventing divergence. In Sections 11 and 12, we show the applications of more sophisticated and promising techniques to the divergence problem. Finally, Section 13 briefly describes various miscellaneous developments in linear filter theory which are not treated in detail in this book.

2. REVIEW OF THE DISCRETE FILTER

Let us outline the computations involved in simulating the discrete linear filter. Our dynamical system model is (7.27, 7.2)

$$x_{k+1} = \Phi(k+1, k)x_k + \Gamma(k)w_{k+1},$$

$$y_k = M(k)x_k + v_k,$$

$$(8.1)$$

with

$$\mathscr{E}\{w_k\} = 0, \qquad \mathscr{E}\{w_k w_l^{\mathrm{T}}\} = Q(k)\,\delta_{kl},$$

$$\mathscr{E}\{v_k\} = 0, \qquad \mathscr{E}\{v_k v_l^{\mathrm{T}}\} = R(k)\,\delta_{kl}, \qquad R(k) > 0.$$

Refer to Section 7.3 for a more complete description of (8.1). Recall that $Y_k = \{y_1, ..., y_k\}$.

TABLE 8.1

FILTER VARIABLES

Variable	Definition	Dimension
$\hat{x}(k \mid k)$	State estimate at t_k given Y_k	$n \times 1$
$P(k \mid k)$	Covariance matrix of the error in $\hat{x}(k \mid k)$	$n \times n$
$\Phi(k+1 \mid k)$	State transition matrix (from t_k to t_{k+1})	$n \times n$
$\Gamma(k)$	System noise coefficient matrix	$n \times r$
$Q(k+1)$	System noise covariance matrix	$r \times r$
$\hat{x}(k+1 \mid k)$	State estimate at t_{k+1} given Y_k	$n \times 1$
$P(k+1 \mid k)$	Covariance matrix of the error in $\hat{x}(k+1 \mid k)$	$n \times n$
$M(k+1)$	Measurement matrix	$m \times n$
$R(k+1)$	Measurement noise covariance matrix	$m \times m$
$K(k+1)$	Filter (Kalman) gain matrix at t_{k+1}	$n \times m$
y_{k+1}	Measurement (observation) at t_{k+1}	$m \times 1$

With the variables of the filter described in Table 8.1, the filter computations can proceed as follows (see Theorem 7.2):

(1) Store the filter state $[\hat{x}(k \mid k), P(k \mid k)]$;

(2)   Compute the predicted state

$$\hat{x}(k + 1 \mid k) = \Phi(k + 1, k)\,\hat{x}(k \mid k); \tag{8.2}$$

(3)   Compute the predicted error covariance matrix

$$P(k + 1 \mid k) = \Phi(k + 1, k)\,P(k \mid k)\,\Phi^{\mathrm{T}}(k + 1, k) + \Gamma(k)Q(k + 1)\,\Gamma^{\mathrm{T}}(k); \tag{8.3}$$

(4)   Compute the filter gain matrix

$$\begin{aligned} K(k + 1) &= P(k + 1 \mid k)\,M^{\mathrm{T}}(k + 1) \\ &\quad \times [M(k + 1)\,P(k + 1 \mid k)\,M^{\mathrm{T}}(k + 1) + R(k + 1)]^{-1}; \end{aligned} \tag{8.4}$$

(5)   Process the observation y_{k+1}

$$\hat{x}(k + 1 \mid k + 1) = \hat{x}(k + 1 \mid k) + K(k + 1)[\,y_{k+1} - M(k + 1)\,\hat{x}(k + 1 \mid k)]; \tag{8.5}$$

(6)   Compute the new error covariance matrix by

$$P(k + 1 \mid k + 1) = [I - K(k + 1)\,M(k + 1)]\,P(k + 1 \mid k), \tag{8.6}$$

or by

$$\begin{aligned} P(k + 1 \mid k + 1) &= [I - K(k + 1)\,M(k + 1)] \\ &\quad \times P(k + 1 \mid k)[I - K(k + 1)\,M(k + 1)]^{\mathrm{T}} \\ &\quad + K(k + 1)\,R(k + 1)\,K^{\mathrm{T}}(k + 1); \end{aligned} \tag{8.7}$$

(7)   Set $k = k + 1$, and return to step (1).

We have listed two different, although equivalent, equations for updating the error covariance matrix at an observation, namely, (8.6) and (8.7). As Aoki [2] points out, Eq. (8.7) is preferable for two reasons. First, the right-hand side of (8.7) is the sum of two symmetric, positive definite matrices, and, when these are added, the sum will be positive definite. On the other hand, (8.6) is, at best, the difference of two positive definite matrices. As a consequence, (8.7) is better conditioned for numerical computations, and will tend to retain more faithfully the positive definiteness and symmetry of $P(k + 1 \mid k + 1)$. As a second consideration, suppose a small error δK is made in the computation of the filter gain K. Then, to first order, we have from (8.6)

$$\delta P(k + 1 \mid k + 1) = -\delta K(k + 1)\,M(k + 1)\,P(k + 1 \mid k),$$

whereas from (8.7), to first order,

$$\delta P(k + 1 \mid k + 1) = 0.$$

Thus, to first order, (8.7) is insensitive to errors in the filter gain and is to be preferred in numerical computations.

We mentioned, in passing, in Section 7.2 that

$$r(k+1 \mid k) \triangleq y_{k+1} - \mathscr{E}\{y_{k+1} \mid Y_k\} = y_{k+1} - M(k+1)\,\hat{x}(k+1 \mid k) \quad (8.8)$$

is called the *predicted residual* (more precisely, the predicted measurement residual error). The correction to the state estimate [see (8.5)] is proportional to the predicted residual, the constant of proportionality being the filter gain. It is easy to compute

$$\mathscr{E}\{r(k+1 \mid k)\} = 0, \quad (8.9)$$

and

$$Y(k+1 \mid k) \triangleq \mathscr{E}\{r(k+1 \mid k)\,r^{\mathrm{T}}(k+1 \mid k)\}$$
$$= M(k+1)\,P(k+1 \mid k)\,M^{\mathrm{T}}(k+1) + R(k+1). \quad (8.10)$$

We see that the filter gain is proportional to the inverse of Y. Predicted residuals provide us with a useful tool for judging the performance of the filter in actual practice. These residuals are available to us, as are their statistics (8.9), (8.10). By checking whether residuals indeed possess their (theoretical) statistical properties, we are able to assess the performance of the filter. This is exploited in the adaptive filter, which we develop in Section 11.

The filter computations, as outlined in the foregoing, involve the inversion of the $m \times m$ residual covariance matrix Y. Recall that m is the dimension of the measurement vector. We have already indicated in Section 7.2 that this inversion can be avoided, under certain conditions, by processing the components of the measurement vector one at a time. More generally, suppose the measurement vector is partitioned as

$$y_{k+1} = \begin{bmatrix} \overset{(m_1 \times 1)}{y^1_{k+1}} \\ \vdots \\ \overset{(m_l \times 1)}{y^l_{k+1}} \end{bmatrix} = \begin{bmatrix} \overset{(m_1 \times n)}{M^1(k+1)}\,x_{k+1} + \overset{(m_1 \times 1)}{v^1_{k+1}} \\ \vdots \\ \overset{(m_l \times n)}{M^l(k+1)}\,x_{k+1} + \overset{(m_l \times 1)}{v^l_{k+1}} \end{bmatrix},$$

and that

$$\mathscr{E}\left\{ \begin{bmatrix} v^1_{k+1} \\ \vdots \\ v^l_{k+1} \end{bmatrix} [v^{1\mathrm{T}}_{k+1}, \dots, v^{l\mathrm{T}}_{k+1}] \right\} = \begin{bmatrix} \overset{(m_1 \times m_1)}{R^1(k+1)} & 0 & \cdots & \cdots & 0 \\ 0 & \overset{(m_2 \times m_2)}{R^2(k+1)} & 0 & \cdots & 0 \\ \vdots & \vdots & \vdots & & \vdots \\ 0 & \cdots & & 0 & \overset{(m_l \times m_l)}{R^l(k+1)} \end{bmatrix};$$

that is, $\{v_{k+1}^1, \ldots, v_{k+1}^l\}$ are uncorrelated. Then, instead of processing the measurement vector y_{k+1} at once, we may equivalently process $y_{k+1}^1, \ldots, y_{k+1}^l$ one at a time. This is accomplished by iterating the computational procedure [steps (1)–(7)] l times with $\Phi \equiv I$, $\Gamma \equiv Q \equiv 0$. On the ith iteration, we process y_{k+1}^i, where $M(k + 1)$ is replaced by $M^i(k + 1)$, $R(k + 1)$ by $R^i(k + 1)$, and y_{k+1} by y_{k+1}^i. Thus on the ith iteration we invert

$$Y^i(k + 1 \mid k) = M^i(k + 1)\, P(k + 1 \mid k)\, M^{i^T}(k + 1) + R^i(k + 1),$$

which is an $m_i \times m_i$ matrix. [During the iterations, we clearly do not increase the index k in step (7).] In this procedure, the inversion of one $m \times m$ matrix is replaced by the inversion of l matrices of dimensions $m_1 \times m_1$, $m_2 \times m_2, \ldots, m_l \times m_l$. In case $l = m$ (the components of v_{k+1} are uncorrelated), we need only invert m scalars.

In concluding this section, we note that the equivalent "inverse" form of the filter, given in (7.11), can be used. This form, however, involves the inversion of $n \times n$ covariance matrices and is generally computationally inferior to the Kalman filter. If $m > n$, however, and the measurement noise components are correlated, then the inverse form can offer computational advantages.

3. EXTENSION TO NONLINEAR PROBLEMS

As was noted in the Introduction, orbit determination (and most other) problems are nonlinear and must be linearized (approximated) before the linear filter theory can be applied. We proceed to this task now. We shall deal with continuous nonlinear systems and discrete nonlinear observations, as this is the structure of the orbit determination problem (see Appendix 8A).

Suppose our nonlinear system is described by the stochastic n-vector differential equation

$$dx_t/dt = f(x_t, t) + G(t)w_t, \qquad t \geqslant t_0, \qquad x_{t_0} \sim N(\hat{x}_{t_0}, P_{t_0}), \qquad (8.11)$$

where $\{w_t\}$ is a zero-mean, white Gaussian noise process with

$$\mathscr{E}\{w_t w_\tau^T\} = Q(t)\, \delta(t - \tau).$$

Now suppose that we generate a *reference* (or *nominal*) deterministic trajectory $\bar{x}(t)$, with given $\bar{x}(t_0)$, satisfying

$$d\bar{x}(t)/dt = f(\bar{x}(t), t), \qquad t \geqslant t_0. \qquad (8.12)$$

Define

$$\delta x_t \triangleq x_t - \bar{x}(t), \qquad (8.13)$$

the *deviation* (or *perturbation*, or *variation*) from the reference trajectory. Then we see that $\{\delta x_t\}$ is a stochastic process satisfying the differential equation

$$d(\delta x_t)/dt = f(x_t, t) - f(\bar{x}(t), t) + G(t)w_t, \qquad \delta x_{t_0} \sim N(\hat{x}_{t_0} - \bar{x}(t_0), P_{t_0}). \quad (8.14)$$

If the deviations from the reference trajectory are "small" (say in the mean square sense), then a Taylor series expansion gives

$$f(x_t, t) - f(\bar{x}(t), t) \cong F[t; \bar{x}(t_0)] \, \delta x_t, \qquad (8.15)$$

where

$$F[t; \bar{x}(t_0)] \triangleq \left[\frac{\partial f_i(\bar{x}(t), t)}{\partial x_j} \right] \qquad (8.16)$$

is the matrix of partial derivatives evaluated along the reference trajectory. Thus we obtain the *approximate linear* equation (called the *perturbation* or *variational* equation)

$$d(\delta x_t)/dt = F[t; \bar{x}(t_0)] \, \delta x_t + G(t)w_t, \qquad \delta x_{t_0} \sim N(\hat{x}_{t_0} - \bar{x}(t_0), P_{t_0}). \quad (8.17)$$

We have emphasized the dependence of δx_t on the choice of the nominal trajectory by including $\bar{x}(t_0)$ in the argument of F. It should be understood, however, that F is a function of time only, being evaluated at the values $\bar{x}(t)$.

Now, as we did in Eqs. (7.19)–(7.26) of Section 7.2, we can discretize (8.17) as

$$\delta x_{t_{k+1}} = \Phi(t_{k+1}, t_k; \bar{x}(t_k)) \, \delta x_{t_k} + w_{t_{k+1}}, \qquad \delta x_{t_0} \sim N(\hat{x}_{t_0} - \bar{x}(t_0), P_{t_0}), \quad (8.18)$$

where Φ is the state transition matrix [defined in (7.20)], and $\{w_{t_k}\}$ is a zero-mean, white Gaussian sequence, $w_{t_{k+1}} \sim N[0, Q(k+1)]$, where

$$Q(k+1) = \int_{t_k}^{t_{k+1}} \Phi(t_{k+1}, \tau) \, G(\tau) \, Q(\tau) \, G^{\mathrm{T}}(\tau) \, \Phi^{\mathrm{T}}(t_{k+1}, \tau) \, d\tau.$$

We, of course, now have the problem of computing the state transition matrix. We do not deal with that problem in this book. A good account of exact and approximate methods of computing the state transition matrix in orbit determination problems may be found in Lee *et al.* [21]. Our reason for discretizing the dynamics is that we prefer to compute the state transition matrix, rather than solving the variance equation

(7.99). It is noted in passing that the dynamics of satellite motion are usually considered to be deterministic ($w_{t_k} \equiv 0$). The noise input is, however, a useful, although artificial, method of accounting for computational errors, neglected nonlinearities [in (8.15)], and model errors. We shall have much more to say on this subject in Sections 8, 9, and 11.

We now turn to the measurement equation, which is assumed of the form

$$y_{t_k} = h(x_{t_k}, t_k) + v_k . \tag{8.19}$$

Defining the *nominal measurement* as

$$\bar{y}(t_k) \triangleq h(\bar{x}(t_k), t_k), \tag{8.20}$$

and

$$\delta y_{t_k} \triangleq y_{t_k} - \bar{y}(t_k), \tag{8.21}$$

and performing a similar linearization, we get the *linearized measurement equation*

$$\delta y_{t_k} = M[t_k ; \bar{x}(t_k)] \, \delta x_{t_k} + v_k , \tag{8.22}$$

where

$$M[t_k ; \bar{x}(t_k)] \triangleq \left[\frac{\partial h_i(\bar{x}(t_k), t_k)}{\partial x_j} \right]. \tag{8.23}$$

Let us summarize. We started with the *nonlinear* system

$$\begin{aligned} dx_t/dt &= f(x_t, t) + G(t)w_t , \qquad t \geq t_0 , \qquad x_{t_0} \sim N(\hat{x}_{t_0}, P_{t_0}), \\ y_{t_k} &= h(x_{t_k}, t_k) + v_k , \end{aligned} \tag{8.24}$$

linearized about $\bar{x}(t)$, the solution of (8.12), and obtained the *linearized* discrete system

$$\begin{aligned} \delta x_{t_{k+1}} &= \varPhi[t_{k+1}, t_k ; \bar{x}(t_k)] \, \delta x_{t_k} + w_{t_{k+1}} , \qquad \delta x_{t_0} \sim N(\hat{x}_{t_0} - \bar{x}(t_0), P_{t_0}), \\ \delta y_{t_k} &= M[t_k ; \bar{x}(t_k)] \, \delta x_{t_k} + v_k . \end{aligned} \tag{8.25}$$

Comparing (8.25) with (8.1), we see that the linear filter is directly applicable to the linearized system. Instead of "state" and "measurement," we now speak of the state deviation and measurement deviation. Given a reference trajectory and measurements y_{t_k}, we can compute δy_{t_k} [via (8.21)] and process the measurement deviations through the linear filter to estimate the state deviations. Then, from (8.13),

$$\hat{x}(t_k \mid t_k) = \bar{x}(t_k) + \widehat{\delta x}(t_k \mid t_k) \qquad (\widehat{\delta x} = \delta \hat{x}),$$

and

$$\tilde{x}(t_k \mid t_k) \triangleq x_{t_k} - \hat{x}(t_k \mid t_k) = \delta x_{t_k} - \delta\hat{x}(t_k \mid t_k) = \delta\tilde{x}(t_k \mid t_k),$$

so that

$$P(k \mid k) = \operatorname{cov}\{\tilde{x}(t_k \mid t_k)\} = \operatorname{cov}\{\delta\tilde{x}(t_k \mid t_k)\}.$$

The remaining problems are the choice of the reference trajectory and the question of the validity of the linearized equations. The latter question will occupy us in Chapter 9. Here we mention only that we expect the linearized dynamics to represent a good approximation if the system noise covariance (Q) and the initial uncertainty (P_{t_0}) are sufficiently small. For then, with the choice $\bar{x}(t_0) = \hat{x}_{t_0}$, the state deviations will remain small with a high probability, or in the mean square sense [see (3.132) and (4.164)]. Of course, what we have said is really incomplete, since it is clear from (8.25) that the size of the state deviations is also dependent upon the stability properties of the linearized system. Discussion of stochastic stability is beyond the scope of this book, and the interested reader is referred to Kushner [20]. Foreseeing the results of Chapter 9, we expect the linearized measurement equation to represent a good approximation if the measurement noise covariance (R) is sufficiently large. For then the neglected nonlinearities will be small compared to the noise, provided, of course, that the state deviations are small.

The obvious choice of the reference trajectory is made with $\bar{x}(t_0) = \hat{x}_{t_0}$, the prior estimate of the state. Then

$$\delta x_{t_0} \sim N(0, P_{t_0}),$$

and, as is evident from (8.25),

$$\mathscr{E}\{\delta x_{t_k}\} = 0, \quad \text{all} \quad t_k.$$

But then, as we process the observations, obviously

$$\delta\hat{x}(t_k \mid t_k) = \mathscr{E}\{\delta x_{t_k} \mid Y_k\} \neq 0, \quad k > 0,$$

in general, and our initially wise choice of a reference trajectory may turn out to be a poor one. This will be especially true if P_{t_0} is large (so that $\hat{x}_{t_0}$ is not a good estimate), and/or if the system noise (Q) is large. In that case, the estimates of the state deviations can become large, violating our linearity assumptions.

Suppose that our dynamical system is unforced by noise[2] $(w_t \equiv 0)$.

_____

[2] The methods we are about to develop are probably good even in the presence of small system noise, but our intuitive arguments fail in that case.

Then we can imagine that there exists a "true" trajectory of the system, which we could learn if we had enough observations available (see Lemma 7.1 and the discussion following). Suppose that, having chosen $\bar{x}(t_0) = \hat{x}_{t_0}$, we filter the batch of observations

$$\{y_1 ,..., y_k\}$$

and obtain $\delta\hat{x}(t_k \mid t_k)$. Then, predicting backward to t_0,

$$\delta\hat{x}(t_0 \mid t_k) = \Phi[t_0 , t_k ; \bar{x}(t_0)] \, \delta\hat{x}(t_k \mid t_k).$$

If our batch of observations is sufficiently large, then we can expect

$$\bar{x}'(t_0) = \bar{x}(t_0) + \delta\hat{x}(t_0 \mid t_k) \tag{8.26}$$

to be closer to the "true" initial state (x_{t_0}) than $\bar{x}(t_0)$ is. Taking $\bar{x}'(t_0)$ now as the reference initial condition, we can reprocess our batch of observations, that is, linearize about $\bar{x}'(t)$ and run the filter for the linearized system over the same batch of observations.[3] This procedure can be repeated several times (iterated) until (8.26) produces no change in the reference initial condition. Presumably, we shall have then converged to the true trajectory, or as close to the true trajectory as the information in our batch of observations and neglected nonlinear effects allow.

We shall call the procedure just described _global iteration_ of the Kalman filter. This type of iteration is commonly used in batch (least squares) estimation techniques. In fact, in view of the equivalence of least squares and the (recursive) filter shown in Example 7.2, the global iteration we have described is equivalent to iterated least squares. It should be noted that global iteration destroys the advantages of the filter as described in the introductory section. Furthermore, there is no guarantee that the iteration will converge. In fact, the iteration may diverge, requiring alternate guesses of the reference trajectory.

With the linear filter recursive structure, we can go a step further. We can relinearize about each new estimate as new estimates become available. This procedure goes as follows. At t_0, linearize about $\hat{x}_{t_0}$. Once y_1 is processed, relinearize about $\hat{x}(t_1 \mid t_1)$, and so on. The point of this is to use a better reference trajectory as soon as one is available. As a consequence of relinearization, large initial estimation errors are not allowed to propagate through time, and, therefore, our linearity assumptions are less likely to be violated. We note that this advantage is not available in least squares batch estimation techniques.

---

[3] When running the filter for the second time, we _do not_ change the statistics of x_{t_0}. We only change the reference trajectory.

Now if we initially linearize about $\hat{x}_{t_0}$, then

$$\delta\hat{x}(t_0 \mid t_0) = 0,$$

and, in view of (8.2) (applied to the deviations),

$$\delta\hat{x}(t_1 \mid t_0) = 0.$$

Since we subsequently linearize about $\hat{x}(t_1 \mid t_1)$,

$$\delta\hat{x}(t_1 \mid t_1) = 0,$$

so that again

$$\delta\hat{x}(t_2 \mid t_1) = 0;$$

and, in general,

$$\delta\hat{x}(t \mid t_k) = 0, \qquad t_k \leqslant t \leqslant t_{k+1}, \qquad \text{all} \quad k. \tag{8.27}$$

As a result, between observations the best estimate of the state is the reference state, and, accordingly,

$$d\hat{x}(t \mid t_k)/dt = f(\hat{x}(t \mid t_k), t), \qquad t_k \leqslant t \leqslant t_{k+1}. \tag{8.28}$$

Prediction is accomplished using the *nonlinear* system differential equation (8.24) with $w_t \equiv 0$. Compare this with Example 7.12. We shall have more to say about this in Chapter 9.

Now write the equation for the correction to the estimate at an observation (8.5) for the linearized system

$$\delta\hat{x}(k+1 \mid k+1) = \delta\hat{x}(k+1 \mid k) + K(k+1)[\delta y_{k+1} - M(k+1)\,\delta\hat{x}(k+1 \mid k)].$$

Since

$$\delta\hat{x}(k+1 \mid k+1) = \hat{x}(k+1 \mid k+1) - \hat{x}(k+1 \mid k)$$

in view of relinearization, and using (8.27), this becomes

$$\hat{x}(k+1 \mid k+1) = \hat{x}(k+1 \mid k) + K(k+1)[y_{k+1} - h(\hat{x}(k+1 \mid k), t_{k+1})]. \tag{8.29}$$

We have just derived what is known as the *extended* (or *modified*) Kalman filter for the nonlinear system (8.24). This filter, the result of the relinearization procedure described, consists of the computations outlined in steps (1)–(7) of Section 2, but with (8.2) replaced by (8.28), and (8.5) replaced by (8.29). The extended Kalman filter is commonly used in orbit determination. We summarize it in the following theorem.

Theorem 8.1 (Extended Kalman Filter). *The extended Kalman filter for the nonlinear system* (8.24) *consists of prediction via*

$$\hat{x}(t_{k+1} \mid t_k) = \hat{x}(t_k \mid t_k) + \int_{t_k}^{t_{k+1}} f(\hat{x}(t \mid t_k), t)\, dt, \tag{8.30}$$

$$P(t_{k+1} \mid t_k) = \Phi[t_{k+1}, t_k \,;\, \hat{x}(t_k \mid t_k)]\, P(t_k \mid t_k)\, \Phi^{\mathrm{T}}[t_{k+1}, t_k \,;\, \hat{x}(t_k \mid t_k)] + Q(t_{k+1}); \tag{8.31}$$

and at an observation

$$\hat{x}(t_{k+1} \mid t_{k+1}) = \hat{x}(t_{k+1} \mid t_k) + K[t_{k+1} \,;\, \hat{x}(t_{k+1} \mid t_k)]$$
$$\times \, [y_{t_{k+1}} - h(\hat{x}(t_{k+1} \mid t_k), t_{k+1})], \tag{8.32}$$

$$P(t_{k+1} \mid t_{k+1}) = [I - K\{t_{k+1} \,;\, \hat{x}(t_{k+1} \mid t_k)\}\, M\{t_{k+1} \,;\, \hat{x}(t_{k+1} \mid t_k)\}]\, P(t_{k+1} \mid t_k)$$
$$\times \, [I - K\{t_{k+1} \,;\, \hat{x}(t_{k+1} \mid t_k)\}\, M\{t_{k+1} \,;\, \hat{x}(t_{k+1} \mid t_k)\}]^{\mathrm{T}}$$
$$+ \, K[t_{k+1} \,;\, \hat{x}(t_{k+1} \mid t_k)]\, R(k+1)\, K^{\mathrm{T}}[t_{k+1} \,;\, \hat{x}(t_{k+1} \mid t_k)]. \tag{8.33}$$

The Kalman gain is

$$K[t_{k+1} \,;\, \hat{x}(t_{k+1} \mid t_k)] = P(t_{k+1} \mid t_k)\, M^{\mathrm{T}}[t_{k+1} \,;\, \hat{x}(t_{k+1} \mid t_k)]$$
$$\times \, [M\{t_{k+1} \,;\, \hat{x}(t_{k+1} \mid t_k)\}\, P(t_{k+1} \mid t_k)$$
$$\times \, M^{\mathrm{T}}\{t_{k+1} \,;\, \hat{x}(t_{k+1} \mid t_k)\} + R(k+1)]^{-1}. \tag{8.34}$$

The matrices Φ and M are those of the linearized system (8.25). We have emphasized the fact that these matrices depend on $\hat{x}(t_k \mid t_k)$, or, equivalently, on $\hat{x}(t_{k+1} \mid t_k)$, by including these estimates as arguments. We shall often omit these arguments for convenience of notation.

We next describe some *local* iteration algorithms that are based largely on the extended Kalman filter. By local iteration, we mean iteration at a point t_{k+1}, or on an interval $[t_k, t_{k+1}]$. The purpose of the iterations is to improve the reference trajectory, and thus the estimate, in the presence of significant nonlinearities. Since the iteration is local, the recursive filter structure is retained; new estimates are computed as new observations become available.

Consider the estimator, Eq. (8.32), in the extended Kalman filter. We obtained (8.32) by evaluating (8.5) (for the linearized system),

$$\delta\hat{x}(t_{k+1} \mid t_{k+1}) = \delta\hat{x}(t_{k+1} \mid t_k) + K[t_{k+1} \,;\, \bar{x}(t_{k+1})]$$
$$\times \, [\delta y_{k+1} - M\{t_{k+1} \,;\, \bar{x}(t_{k+1})\}\, \delta\hat{x}(t_{k+1} \mid t_k)], \tag{8.35}$$

about $\bar{x}(t_{k+1}) = \hat{x}(t_{k+1} \mid t_k)$ as reference. Then, processing the observation y_{k+1} via (8.32), we get $\hat{x}(t_{k+1} \mid t_{k+1})$. Now $\hat{x}(t_{k+1} \mid t_{k+1})$ is a "better" estimate (presumably closer to the true trajectory) than $\hat{x}(t_{k+1} \mid t_k)$. Let us relinearize about $\hat{x}(t_{k+1} \mid t_{k+1})$ then, and recompute the estimate. In general, evaluate (8.35) about η_i, and call the output of (8.35) η_{i+1} to get the iterator

$$\eta_{i+1} = \hat{x}(t_{k+1} \mid t_k) + K(t_{k+1} ; \eta_i)[y_{k+1} - h(\eta_i , t_{k+1}) - M(t_{k+1} ; \eta_i)$$
$$\times \{\hat{x}(t_{k+1} \mid t_k) - \eta_i\}], \tag{8.36}$$

$i = 1, 2,...; \eta_1 = \hat{x}(t_{k+1} \mid t_k)$. Note that η_2 is just the result of the extended Kalman filter (8.32). Note also that the gain K is reevaluated on each iteration, as are the measurement function h and the matrix M. The iteration is terminated when there is no significant change in consecutive iterates. Then the last iterate, η_l say, is taken for the estimate $\hat{x}(t_{k+1} \mid t_{k+1})$. This iterated extended Kalman filter is apparently due to J. V. Breakwell. It is presented and analyzed by Denham and Pines [7], who show that it can be useful in reducing the effect of measurement function (h) nonlinearity, thus improving the linear filter performance. We shall have more to say about this iterator in Chapter 9. For now, we summarize it in the following theorem.

Theorem 8.2 (Iterated Extended Kalman Filter). *The iterated extended Kalman filter consists of Theorem 8.1 with (8.32) replaced by the iterator:*

$$\eta_{i+1} = \hat{x}(t_{k+1} \mid t_k) + K(t_{k+1} ; \eta_i)[y_{k+1} - h(\eta_i , t_{k+1}) - M(t_{k+1} ; \eta_i)$$
$$\times \{\hat{x}(t_{k+1} \mid t_k) - \eta_i\}], \qquad i = 1,..., l, \tag{8.36}$$

$$\hat{x}(t_{k+1} \mid t_{k+1}) = \eta_l . \tag{8.37}$$

The iteration starts with $\eta_1 = \hat{x}(t_{k+1} \mid t_k)$, and terminates when there is no significant difference between consecutive iterates. The covariance matrix in (8.33) is then computed based on η_l, that is, based on linearization about η_l.

The preceding iterator is designed for measurement nonlinearities and does not improve the previous reference trajectory on $[t_k , t_{k+1})$. Thus, $\hat{x}(t_{k+1} \mid t_k)$ can be in error due to system nonlinearities acting on the interval $[t_k , t_{k+1})$. Of course, the reference trajectory on $[t_{k+1} , t_{k+2})$ will be improved. We can, in the following way, include the reference trajectory $[t_k , t_{k+1})$ in the iteration loop. We start at t_k with $\hat{x}(t_k \mid t_k)$ and $P(t_k \mid t_k)$. We linearize and predict to t_{k+1} and apply one iteration of (8.36). Based on that new estimate, we smooth back to t_k. The smoothing closes the loop, providing an improved reference for prediction to t_{k+1}. Let us derive the equations for this iterator.

We are at t_k with $\hat{x}(t_k \mid t_k)$ and $P(t_k \mid t_k)$. Linearize about ξ_i. Then, from (8.12),

$$\bar{x}(t_{k+1} ; \xi_i) = \xi_i + \int_{t_k}^{t_{k+1}} f(\bar{x}(\tau), \tau) \, d\tau, \tag{8.38}$$

and, from (8.2) for the linearized system,

$$\delta\hat{x}(t_{k+1} \mid t_k) = \Phi(t_{k+1} , t_k ; \xi_i) \, \delta\hat{x}(t_k \mid t_k),$$

or

$$\hat{x}(t_{k+1} \mid t_k) = \bar{x}(t_{k+1} ; \xi_i) + \Phi(t_{k+1} , t_k ; \xi_i)[\hat{x}(t_k \mid t_k) - \xi_i]. \tag{8.39}$$

Now apply iterator (8.36) to process the observation and get η_{i+1}. We now have the following choice. Linearize about η_{i+1} to smooth back to t_k [this involves solving (8.12) backward in time], or keep $\bar{x}(t, \xi_i)$ as reference for the smoothing. We make the latter choice, since we shall improve $\bar{x}(t, \xi_i)$ after the smoothing anyway. From Example 7.8, the linearized smoother is

$$\delta\hat{x}(t_k \mid t_{k+1}) = \delta\hat{x}(t_k \mid t_k) + S(t_k ; \xi_i)[\delta\hat{x}(t_{k+1} \mid t_{k+1}) - \delta\hat{x}(t_{k+1} \mid t_k)],$$

with

$$S(t_k ; \xi_i) = P(t_k \mid t_k) \, \Phi^{\mathrm{T}}(t_{k+1} , t_k ; \xi_i) \, P^{-1}(t_{k+1} \mid t_k); \tag{8.40}$$

or, using the smoothed estimate as ξ_{i+1},

$$\xi_{i+1} = \hat{x}(t_k \mid t_k) + S(t_k ; \xi_i)[\eta_{i+1} - \hat{x}(t_{k+1} \mid t_k)]. \tag{8.41}$$

[Note that $P(t_{k+1} \mid t_k)$ in (8.40) depends on ξ_i.] ξ_{i+1} is our improved reference for the prediction in (8.38). This iteration starts with $\xi_1 = \hat{x}(t_k \mid t_k)$ and $\eta_1 = \hat{x}(t_{k+1} \mid t_k)$. Thus η_2 is the output of the extended Kalman filter.

This last iterator was apparently first derived by Wishner *et al.* [40], but in an entirely different way. We shall call it the *iterated linear filter-smoother*. Actually, the iterators we have derived are closely related in theory; both have a similar probabilistic interpretation that we shall pursue in Chapter 9. We summarize as follows:

Theorem 8.3 (Iterated Linear Filter-Smoother). *Given $\hat{x}(t_k \mid t_k)$ and $P(t_k \mid t_k)$, the iterated linear filter-smoother consists of the iteration*

$$\eta_{i+1} = \hat{x}(t_{k+1} \mid t_k) + K(t_{k+1} ; \eta_i , \xi_i)[y_{k+1} - h(\eta_i , t_{k+1}) - M(t_{k+1} ; \eta_i)$$
$$\times \{\hat{x}(t_{k+1} \mid t_k) - \eta_i\}], \tag{8.42}$$

$$\xi_{i+1} = \hat{x}(t_k \mid t_k) + S(t_k ; \xi_i)[\eta_{i+1} - \hat{x}(t_{k+1} \mid t_k)], \qquad i = 1,..., l, \tag{8.43}$$

$$\eta_1 = \hat{x}(t_{k+1} \mid t_k), \qquad \xi_1 = \hat{x}(t_k \mid t_k),$$

where

$$\bar{x}(t_{k+1} \,; \xi_i) = \xi_i + \int_{t_k}^{t_{k+1}} f(\bar{x}(\tau), \tau) \, d\tau, \tag{8.44}$$

$$\hat{x}(t_{k+1} \mid t_k) = \bar{x}(t_{k+1} \,; \xi_i) + \Phi(t_{k+1}, t_k \,; \xi_i)(\hat{x}(t_k \mid t_k) - \xi_i), \tag{8.45}$$

$$K(t_{k+1} \,; \eta_i \,, \xi_i) = P(t_{k+1} \mid t_k) \, M^{\mathrm{T}}(t_{k+1} \,; \eta_i)[M(t_{k+1} \,; \eta_i) \, P(t_{k+1} \mid t_k)$$
$$\times \, M^{\mathrm{T}}(t_{k+1} \,; \eta_i) + R(k+1)]^{-1}, \tag{8.46}$$

$$P(t_{k+1} \mid t_k) = \Phi(t_{k+1}, t_k \,; \xi_i) \, P(t_k \mid t_k) \, \Phi^{\mathrm{T}}(t_{k+1}, t_k \,; \xi_i) + Q(t_{k+1}), \tag{8.47}$$

$$S(t_k \,; \xi_i) = P(t_k \mid t_k) \, \Phi^{\mathrm{T}}(t_{k+1}, t_k \,; \xi_i) \, P^{-1}(t_{k+1} \mid t_k). \tag{8.48}$$

Then

$$\hat{x}(t_{k+1} \mid t_{k+1}) = \eta_l \,, \tag{8.49}$$

$$P(t_{k+1} \mid t_{k+1}) = [I - K(t_{k+1} \,; \eta_l \,, \xi_l) \, M(t_{k+1} \,; \eta_l)] \, P(t_{k+1} \mid t_k \,; \xi_l)$$
$$\times \, [I - K(t_{k+1} \,; \eta_l \,, \xi_l) \, M(t_{k+1} \,; \eta_l)]^{\mathrm{T}}$$
$$+ K(t_{k+1} \,; \eta_l \quad \xi_l) \, R(k+1) \, K^{\mathrm{T}}(t_{k+1} \,; \eta_l \,, \xi_l). \tag{8.50}$$

We note in closing this section that the extended Kalman filter may also be *iterated globally*. Having processed $\{y_1, ..., y_k\}$, starting with $\hat{x}_{t_0}$ and P_{t_0}, we have $\hat{x}(t_k \mid t_k)$ and $P(t_k \mid t_k)$. Then, assuming $w_t \equiv 0$, we solve (8.12) backward in time with $\hat{x}(t_k \mid t_k)$ as initial condition. Thus we get the smoothed estimate $\hat{x}(t_0 \mid t_k)$. We can then reprocess the observations, starting with $\hat{x}(t_0 \mid t_k)$, P_{t_0}. Note that in doing this we are changing the initial statistic $\mathscr{E}\{x_{t_0}\}$, which is, strictly speaking, part of the problem statement. This is, however, often justified by the fact that $\hat{x}_{t_0}$ is only an educated guess anyway.

4. UNCERTAIN PARAMETERS

Frequently, in applications, the dynamical system depends on certain parameters whose values are imprecisely known. Such parameters may be regarded as random variables with known *a priori* statistics. We model such a system containing uncertain parameters by

$$x_{k+1} = \Phi(k+1, k)x_k + \Psi(k+1, k)u + \Gamma(k)w_{k+1}, \tag{8.51}$$

$$y_k = M(k)x_k + N(k)p + v_k. \tag{8.52}$$

The parameters in the vector u will be referred to as *dynamical param-*

eters, whereas those in the vector p will be called *measurement param-eters*. The *a priori* statistics of u and p will be taken as

$$\mathscr{E}\{u\} = 0, \qquad \mathscr{E}\{p\} = 0,$$
$$\mathscr{E}\{uu^T\} = U_0, \qquad \mathscr{E}\{pp^T\} = W_0. \tag{8.53}$$

We assume that u, p, x_0, $\{w_k\}$, and $\{v_k\}$ are uncorrelated.

Now these (constant) parameters may be regarded as outputs of the dynamical systems

$$u_{k+1} = u_k, \qquad u_0 = u,$$
$$p_{k+1} = p_k, \qquad p_0 = p. \tag{8.54}$$

Then, defining the *augmented* state vector,

$$X_k = \begin{bmatrix} x_k \\ u_k \\ p_k \end{bmatrix}, \tag{8.55}$$

we can combine the systems (8.51), (8.54) into the single *augmented* system

$$X_{k+1} = \begin{bmatrix} \Phi(k+1, k) & \Psi(k+1, k) & 0 \\ 0 & I & 0 \\ 0 & 0 & I \end{bmatrix} X_k + \begin{bmatrix} \Gamma(k) \\ 0 \\ 0 \end{bmatrix} w_{k+1} \tag{8.56}$$

with observations

$$y_k = [M(k) \quad 0 \quad N(k)]X_k + v_k. \tag{8.57}$$

Now the linear filter is clearly directly applicable to this augmented system, and produces estimates

$$\hat{X}(k \mid k) = \begin{bmatrix} \hat{x}(k \mid k) \\ \hat{u}(k \mid k) \\ \hat{p}(k \mid k) \end{bmatrix}.$$

Thus, we can estimate these uncertain parameters, together with the original state x, in a straightforward manner.

Although straightforward in theory, the estimation of parameters in addition to the state increases the computational load and therefore may not be desirable. On the other hand, uncertain parameters cannot be ignored, since they can produce divergence of the filter (see Sections 7.8 and 8). An alternative, developed by Schmidt [31], is to take into account the effect of the uncertain parameters in degrading the estimate of the state, without estimating the parameters themselves. Such a filter can be derived by first writing the standard linear filter for the augmented system (8.56), (8.57), and then throwing away the estimation equations for the parameters. That is, we shall insist that

$$\hat{u}(k \mid k) = \hat{p}(k \mid k) \equiv 0 = \mathscr{E}\{u\} = \mathscr{E}\{p\},$$

and

$$\mathscr{E}\{[u - \hat{u}(k \mid k)][u - \hat{u}(k \mid k)]^{\mathrm{T}}\} \equiv U_0,$$

$$\mathscr{E}\{[p - \hat{p}(k \mid k)][p - \hat{p}(k \mid k)]^{\mathrm{T}}\} \equiv W_0.$$

As a consequence, since u and p are initially uncorrelated, they will remain uncorrelated during the filtering process. The resulting filter will clearly be suboptimal, since the parameters are not being estimated. For if the parameters were estimated, then their uncertainty would decrease, permitting the uncertainty in the state estimate to decrease.

In view of the aforementioned *a posteriori* assumptions, the error covariance matrix $\mathscr{P}$ for the augmented system will be of the form

$$\mathscr{P} = \begin{bmatrix} P & C_u & C_p \\ C_u^{\mathrm{T}} & U_0 & 0 \\ C_p^{\mathrm{T}} & 0 & W_0 \end{bmatrix}, \tag{8.58}$$

where P is the usual error covariance matrix for the state x,

$$P(k \mid l) = \mathscr{E}\{[x_k - \hat{x}(k \mid l)][x_k - \hat{x}(k \mid l)]^{\mathrm{T}}\}, \tag{8.59}$$

and where

$$\begin{aligned} C_u(k \mid l) &= \mathscr{E}\{[x_k - \hat{x}(k \mid l)]u^{\mathrm{T}}\}, \\ C_p(k \mid l) &= \mathscr{E}\{[x_k - \hat{x}(k \mid l)]p^{\mathrm{T}}\}. \end{aligned} \tag{8.60}$$

Using (8.2) for the augmented system, we get that

$$\hat{x}(k + 1 \mid k) = \Phi(k + 1 \mid k)\,\hat{x}(k \mid k), \tag{8.61}$$

and $\hat{u}$, $\hat{p}$ are invariant in prediction. Then, from (8.3), for the augmented system

$$\mathscr{P}(k + 1 \mid k) = \begin{bmatrix} \Phi(k + 1, k) & \Psi(k + 1, k) & 0 \\ 0 & I & 0 \\ 0 & 0 & I \end{bmatrix} \begin{bmatrix} P(k \mid k) & C_u(k \mid k) & C_p(k \mid k) \\ C_u^{\mathrm{T}}(k \mid k) & U_0 & 0 \\ C_p^{\mathrm{T}}(k \mid k) & 0 & W_0 \end{bmatrix}$$

$$\times \begin{bmatrix} \Phi^{\mathrm{T}}(k + 1, k) & 0 & 0 \\ \Psi^{\mathrm{T}}(k + 1, k) & I & 0 \\ 0 & 0 & I \end{bmatrix} + \begin{bmatrix} \Gamma(k)\,Q(k + 1)\,\Gamma^{\mathrm{T}}(k) & 0 & 0 \\ 0 & 0 & 0 \\ 0 & 0 & 0 \end{bmatrix},$$

which is

$$P(k + 1 \mid k) = \Phi(k + 1, k) \, P(k \mid k) \, \Phi^{\mathrm{T}}(k + 1, k)$$
$$+ \, \Phi(k + 1, k) \, C_u(k \mid k) \, \Psi^{\mathrm{T}}(k + 1, k)$$
$$+ \, \Psi(k + 1, k) \, C_u{}^{\mathrm{T}}(k \mid k) \, \Phi^{\mathrm{T}}(k + 1, k)$$
$$+ \, \Psi(k + 1, k) \, U_0 \Psi^{\mathrm{T}}(k + 1, k)$$
$$+ \, \Gamma(k) \, Q(k + 1) \, \Gamma^{\mathrm{T}}(k), \tag{8.62}$$

$$C_u(k + 1 \mid k) = \Phi(k + 1, k) \, C_u(k \mid k) + \Psi(k + 1, k) U_0, \tag{8.63}$$

$$C_p(k + 1 \mid k) = \Phi(k + 1, k) \, C_p(k \mid k), \tag{8.64}$$

and U_0 and W_0 are invariant in prediction.

Let $\mathcal{M}(k)$ be the measurement matrix for the augmented system,

$$\mathcal{M}(k) = [M(k) \quad 0 \quad N(k)].$$

Then

$$\mathcal{P}(k + 1 \mid k) \, \mathcal{M}^{\mathrm{T}}(k + 1)$$
$$= \begin{bmatrix} P(k + 1 \mid k) \, M^{\mathrm{T}}(k + 1) + C_p(k + 1 \mid k) \, N^{\mathrm{T}}(k + 1) \\ C_u{}^{\mathrm{T}}(k + 1 \mid k) \, M^{\mathrm{T}}(k + 1) \\ C_p{}^{\mathrm{T}}(k + 1 \mid k) \, M^{\mathrm{T}}(k + 1) + W_0 N^{\mathrm{T}}(k + 1) \end{bmatrix},$$

$$\mathcal{Y}(k + 1 \mid k) \triangleq \mathcal{M}(k + 1) \, \mathcal{P}(k + 1 \mid k) \, \mathcal{M}^{\mathrm{T}}(k + 1) + R(k + 1)$$
$$= M(k + 1) \, P(k + 1 \mid k) \, M^{\mathrm{T}}(k + 1)$$
$$+ \, M(k + 1) \, C_p(k + 1 \mid k) \, N^{\mathrm{T}}(k + 1)$$
$$+ \, N(k + 1) \, C_p{}^{\mathrm{T}}(k + 1 \mid k) \, M^{\mathrm{T}}(k + 1)$$
$$+ \, N(k + 1) \, W_0 N^{\mathrm{T}}(k + 1) + R(k + 1). \tag{8.65}$$

Now, from (8.5) for the augmented system,

$$\hat{x}(k + 1 \mid k + 1) = \hat{x}(k + 1 \mid k) + \mathcal{K}(k + 1)[y_{k+1} - M(k + 1) \, \hat{x}(k + 1 \mid k)], \tag{8.66}$$

where

$$\mathcal{K}(k + 1) = [P(k + 1 \mid k) \, M^{\mathrm{T}}(k + 1)$$
$$+ \, C_p(k + 1 \mid k) \, N^{\mathrm{T}}(k + 1)] \, \mathcal{Y}^{-1}(k + 1 \mid k), \tag{8.67}$$

and we throw away the estimation equations for u and p. From (8.6) for the augmented system,

$$P(k + 1 \mid k + 1) = P(k + 1 \mid k) - \mathcal{K}(k + 1)[M(k + 1) P(k + 1 \mid k)$$
$$+ N(k + 1) C_p{}^{\mathrm{T}}(k + 1 \mid k)], \qquad (8.68)$$

$$C_u(k + 1 \mid k + 1) = C_u(k + 1 \mid k) - \mathcal{K}(k + 1) M(k + 1) C_u(k + 1 \mid k), \qquad (8.69)$$

$$C_p(k + 1 \mid k + 1) = C_p(k + 1 \mid k) - \mathcal{K}(k + 1)[M(k + 1) C_p(k + 1 \mid k)$$
$$+ N(k + 1) W_0], \qquad (8.70)$$

and we keep U_0 and W_0 invariant at an observation.

We have derived what we shall call the *Schmidt–Kalman* filter with uncertain parameters. It is summarized in the following theorem.

Theorem 8.4 (Schmidt–Kalman Filter with Uncertain Parameters). *The system is* (8.51), (8.52), *where the uncertain parameters u and p have the statistics given in* (8.53). *The filter is for the state x only; the uncertain parameters are allowed only to degrade the estimate of the state, and are not being estimated themselves. This filter consists of equations for the state estimate $\hat{x}(k \mid k)$, the state estimation error covariance matrix P defined in* (8.59), *and the correlation matrices C_u and C_p defined in* (8.60). *Between observations, these quantities satisfy* (8.61), (8.62), (8.63), *and* (8.64), *respectively. At an observation, they satisfy* (8.66), (8.68), (8.69), *and* (8.70), *respectively. $C_u(0 \mid 0) = C_p(0 \mid 0) = 0$.*

The Schmidt–Kalman filter involves recursions for the correlation matrices C_u and C_p, in addition to the usual equations for $\hat{x}$ and P. Thus, the computational requirements are increased. However, they are considerably lower than the computational requirements for the augmented state filter. Thus, if improved estimates of the uncertain parameters are not required, the Schmidt–Kalman filter is an attractive alternative.

Example 8.1. Show, using the method of Example 7.4, that, under the stated *a posteriori* assumptions, the gain $\mathcal{K}$ in (8.67) is optimal. ▲

Example 8.2. Show directly from the system equation (8.51), and from (8.61) and (8.66), that the matrix P defined by (8.62) and (8.68) is indeed the estimation error covariance matrix, as defined in (8.59). That is, the covariance matrix computed in the filter is the covariance of filter estimation errors. ▲

The Schmidt–Kalman filter for uncertain parameters can be readily applied, by linearization, to the *nonlinear* system

$$dx_t/dt = f(x_t, u, t) + G(t)w_t, \qquad t \geqslant t_0, \qquad x_{t_0} \sim N(\hat{x}_{t_0}, P_{t_0}),$$
$$y_{t_k} = h(x_{t_k}, p, t_k) + v_k. \tag{8.71}$$

Generate a reference trajectory with a given $\bar{x}(t_0)$, and, for a nominal value of u, call it $\bar{u}$, satisfying

$$d\bar{x}(t)/dt = f(\bar{x}(t), \bar{u}, t), \qquad t \geqslant t_0.$$

Then

$$d(\delta x_t)/dt = F(t; \bar{x}(t_0), \bar{u}) \, \delta x_t + L(t; \bar{x}(t_0), \bar{u}) \, \delta u + G(t)w_t, \tag{8.72}$$

where

$$F(t; \bar{x}(t_0), \bar{u}) \triangleq \left[\frac{\partial f_i(\bar{x}(t), \bar{u}, t)}{\partial x_j} \right], \qquad L(t; \bar{x}(t_0), \bar{u}) \triangleq \left[\frac{\partial f_i(\bar{x}(t), \bar{u}, t)}{\partial u_j} \right].$$

Equation (8.72) can be discretized

$$\delta x_{t_{k+1}} = \Phi(t_{k+1}, t_k; \bar{x}(t_k), \bar{u}) \, \delta x_{t_k} + \Psi(t_{k+1}, t_k; \bar{x}(t_k), \bar{u}) \, \delta u + w_{k+1},$$
$$\delta x_{t_0} \sim N(\hat{x}_{t_0} - \bar{x}(t_0), P_{t_0}), \tag{8.73}$$

where Φ is the usual state transition matrix and $\psi(t, \tau)$ satisfies

$$d\Psi(t, \tau)/dt = F\Psi(t, \tau) + L, \qquad \Psi(\tau, \tau) = 0. \tag{8.74}$$

The measurement equation is linearized about $\bar{x}(t_k)$ and a nominal value $\bar{p}$,

$$\delta y_{t_k} = M(t_k; \bar{x}(t_k), \bar{p}) \, \delta x_{t_k} + N(t_k; \bar{x}(t_k), \bar{p}) \, \delta p + v_k, \tag{8.75}$$

where

$$M(t_k; \bar{x}(t_k), \bar{p}) \triangleq \left[\frac{\partial h_i(\bar{x}(t_k), \bar{p}, t_k)}{\partial x_j} \right], \qquad N(t_k; \bar{x}(t_k), \bar{p}) \triangleq \left[\frac{\partial h_i(\bar{x}(t_k), \bar{p}, t_k)}{\partial p_j} \right].$$

The Schmidt–Kalman filter can now be directly applied to the system (8.73), (8.75). An extended Kalman–Schmidt filter (see Section 3, especially Theorem 8.1) can be readily written down. We leave these details for the reader.

Example 8.3.   Show that

$$\Phi(t, \tau) = \left[\frac{\partial x_i(t)}{\partial x_j(\tau)} \right], \qquad \Psi(t, \tau) = \left[\frac{\partial x_i(t)}{\partial u_j(\tau)} \right]. \qquad \blacktriangle$$

5. A SIMPLE EXAMPLE

A limited number of simple examples of linear filter applications are available in the literature. These include the analytical examples in Kalman [19], and a numerical example in Schmidt [31]. The scarcity of such examples is, of course, due to the nonlinearity of the covariance equations, which precludes closed-form solutions in all but the simplest cases. Since the detailed solution of a simple problem gives one a "feel" for the method, we think it useful to present some detailed computations and analyses of such a simple problem.

Our problem consists of a noise-free second-order system representing a falling body in a constant field

$$\ddot{z} = -g, \quad t \geqslant 0, \tag{8.76}$$

z a scalar. Let the position be $z = (x)_1$, and the velocity $\dot{z} = (x)_2$. Then, defining the vector x,

$$x^{\mathrm{T}} = [(x)_1, (x)_2],$$

(8.76) can be written in system form

$$\dot{x} = \begin{bmatrix} 0 & 1 \\ 0 & 0 \end{bmatrix} x + \begin{bmatrix} 0 \\ -g \end{bmatrix}, \quad F \triangleq \begin{bmatrix} 0 & 1 \\ 0 & 0 \end{bmatrix}.$$

It is easy to verify that the state transition matrix of this system is

$$\Phi(t, \tau) = \begin{bmatrix} 1 & t - \tau \\ 0 & 1 \end{bmatrix}, \tag{8.77}$$

so that

$$x_t = \Phi(t, \tau)x_\tau + \int_\tau^t \begin{bmatrix} 0 & t - s \\ 0 & 1 \end{bmatrix} \begin{bmatrix} 0 \\ -g \end{bmatrix} ds$$

$$= \Phi(t, \tau)x_\tau - g \begin{bmatrix} (t - \tau)^2/2 \\ (t - \tau) \end{bmatrix}.$$

Therefore,

$$x_{t+1} = \begin{bmatrix} 1 & 1 \\ 0 & 1 \end{bmatrix} x_t - g \begin{bmatrix} \frac{1}{2} \\ 1 \end{bmatrix}, \quad t = 0, 1, \dots. \tag{8.78}$$

We will take (scalar) observations of position

$$y_t = [1 \quad 0] x_t + v_t; \quad t = 1, \dots, 6; \quad M \triangleq [1 \quad 0], \tag{8.79}$$

with $R(t) = 1$. We assume that $x_0 \sim N(\hat{x}_0, P_0)$,

$$\hat{x}_0 = \begin{bmatrix} 95 \\ 1 \end{bmatrix}, \quad P_0 = \begin{bmatrix} 10 & 0 \\ 0 & 1 \end{bmatrix}.$$

Let us first determine whether our system (8.78), (8.79) is observable. The information matrix [from (7.151)],

$$\mathcal{I}(t, t-1) = \sum_{i=t-1}^{t} \Phi^T(i, t) \, M^T(i) \, R^{-1}(i) \, M(i) \, \Phi(i, t),$$

is computed to be

$$\mathcal{I}(t, t-1) = \begin{bmatrix} 2 & -1 \\ -1 & 1 \end{bmatrix}.$$

Therefore, our system is uniformly completely observable. [Take $\alpha = 0.1$, $\beta = 3$ in (7.153), for example.]

Take $g = 1$ in (8.76). Assume that the "true" initial condition is $(x_0)_1 = z(0) = 100$, $(x_0)_2 = \dot{z}(0) = 0$. Then the true trajectory (at $t = 1,..., 6$) and a realization of the observations are given in Table 8.2. We show the details of one cycle of the filter computations as outlined in Section 2. The results for $t = 1,..., 6$ are summarized in Table 8.2.

TABLE 8.2

FALLING BODY IN A CONSTANT FIELD

t	0	1	2	3	4	5	6
$(x_t)_1$	100.0	99.5	98.0	95.5	92.0	87.5	82.0
$(x_t)_2$	0	−1.0	−2.0	−3.0	−4.0	−5.0	−6.0
v_t		+0.5	−0.1	−1.1	+0.7	−0.2	+0.1
y_t		100.0	97.9	94.4	92.7	87.3	82.1
$\hat{x}_1(t \mid t)$	95.0	99.6	98.4	95.3	92.6	87.9	82.4
$\hat{x}_2(t \mid t)$	1.0	+0.36	−1.2	−2.3	−3.25	−4.6	−5.7
$P_{11}(t \mid t)$	10.0	0.88	0.66	0.66	0.62	0.56	0.50
$P_{22}(t \mid t)$	1.0	0.92	0.57	0.30	0.16	0.10	0.06

(Step 2)

$$\hat{x}(1 \mid 0) = \begin{bmatrix} 1 & 1 \\ 0 & 1 \end{bmatrix} \begin{bmatrix} 95 \\ 1 \end{bmatrix} - \begin{bmatrix} 0.5 \\ 1.0 \end{bmatrix} = \begin{bmatrix} 95.5 \\ 0 \end{bmatrix},$$

(Step 3)

$$P(1 \mid 0) = \begin{bmatrix} 1 & 1 \\ 0 & 1 \end{bmatrix} \begin{bmatrix} 10 & 0 \\ 0 & 1 \end{bmatrix} \begin{bmatrix} 1 & 0 \\ 1 & 1 \end{bmatrix} = \begin{bmatrix} 11 & 1 \\ 1 & 1 \end{bmatrix},$$

$$P(1 \mid 0)M^T = \begin{bmatrix} 11 & 1 \\ 1 & 1 \end{bmatrix} \begin{bmatrix} 1 \\ 0 \end{bmatrix} = \begin{bmatrix} 11 \\ 1 \end{bmatrix},$$

$$MP(1 \mid 0)M^T = \begin{bmatrix} 1 & 0 \end{bmatrix} \begin{bmatrix} 11 \\ 1 \end{bmatrix} = 11,$$

$$Y(1 \mid 0) = MP(1 \mid 0)M^T + R = 12,$$

(Step 4)

$$K(1) = P(1 \mid 0) M^T Y^{-1}(1 \mid 0) = \frac{1}{12}\begin{bmatrix} 11 \\ 1 \end{bmatrix} = \begin{bmatrix} 0.92 \\ 0.08 \end{bmatrix},$$

(Step 5)

$$\hat{x}(1 \mid 1) = \begin{bmatrix} 95.5 \\ 0 \end{bmatrix} + \begin{bmatrix} 0.92 \\ 0.08 \end{bmatrix}(100.0 - 95.5) = \begin{bmatrix} 99.64 \\ 0.36 \end{bmatrix},$$

$$K(1)M = \begin{bmatrix} 0.92 \\ 0.08 \end{bmatrix}[1 \quad 0] = \begin{bmatrix} 0.92 & 0 \\ 0.08 & 0 \end{bmatrix},$$

$$I - K(1)M = \begin{bmatrix} 0.08 & 0 \\ -0.08 & 1 \end{bmatrix},$$

(Step 6)

$$P(1 \mid 1) = \begin{bmatrix} 0.08 & 0 \\ -0.08 & 1 \end{bmatrix}\begin{bmatrix} 11 & 1 \\ 1 & 1 \end{bmatrix} = \begin{bmatrix} 0.88 & 0.10 \\ 0.10 & 0.92 \end{bmatrix}.$$

Finally,

$$\hat{x}(6 \mid 6) = \begin{bmatrix} 82.4 \\ -5.7 \end{bmatrix}, \qquad P(6 \mid 6) = \begin{bmatrix} 0.50 & 0.14 \\ 0.14 & 0.06 \end{bmatrix}.$$

The filtering results are also shown in Figs. 8.1 and 8.2. Plotted in

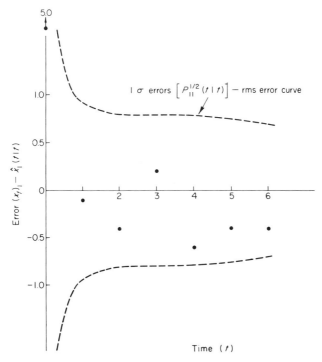

FIG. 8.1. Position estimation error.

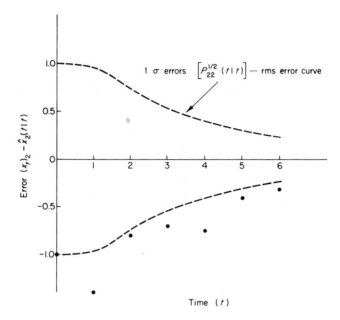

FIG. 8.2.   Velocity estimation error.

these figures are the position and velocity estimation errors. The dashed lines represent the error standard deviations, or root mean square (rms) errors. If the filter were simulated many times, each time with different samples of x_0 and $\{v_t, t = 1,..., 6\}$, and if the estimation errors (dots) were plotted for each simulation, then about 67% of the dots would plot within the dashed lines. The fact that, in our particular simulation, the velocity estimation errors fall somewhat outside the rms curves is due to our particular choice of x_0 and the measurement noise samples. This initial transient effect would eventually disappear, and the appropriate proportion (67%) of the dots would then fall within the rms error curves. Recall that in this problem we are observing the position. As a result, the rms position error drops dramatically as soon as the first observation is processed. On the other hand, the rms velocity error does not decrease substantially until the second observation is processed. This is because two position observations are required to determine both components of the state vector. The dynamics of our system are such that velocity affects the position, whereas position does not affect velocity. As a consequence, velocity must first be estimated rather accurately before good estimates of position can be had. This explains the fact that after the second observation the velocity rms errors steadily decrease, whereas position rms errors decrease much more slowly. Eventually, both rms errors will, of course, go to zero.

It is interesting to determine the effect of our initial uncertainty (P_0) on $P(6 \mid 6)$. This can be easily accomplished with the aid of the linear limited memory filter theory developed in Section 7.10. We first compute the predicted covariance matrix

$$P(6 \mid 0) = \Phi(6, 0)\, P_0 \Phi^T(6, 0) = \begin{bmatrix} 1 & 6 \\ 0 & 1 \end{bmatrix}\begin{bmatrix} 10 & 0 \\ 0 & 1 \end{bmatrix}\begin{bmatrix} 1 & 0 \\ 6 & 1 \end{bmatrix} = \begin{bmatrix} 46 & 6 \\ 6 & 1 \end{bmatrix},$$

and

$$P^{-1}(6 \mid 0) = \begin{bmatrix} 0.1 & -0.6 \\ -0.6 & 4.6 \end{bmatrix}.$$

Also,

$$P^{-1}(6 \mid 6) = \begin{bmatrix} 6.0 & -14 \\ -14 & 50 \end{bmatrix}.$$

Then, according to Theorem 7.10,

$$P^{-1}(6 \mid 6 \mid) = P^{-1}(6 \mid 6) - P^{-1}(6 \mid 0) = \begin{bmatrix} 5.9 & -13.4 \\ -13.4 & 45.4 \end{bmatrix},$$

and therefore

$$P(6 \mid 6 \mid) = \begin{bmatrix} 0.52 & 0.15 \\ 0.15 & 0.067 \end{bmatrix},$$

which is the covariance matrix of the error in the estimate conditioned on the observations *only*. Notice that $P(6 \mid 6 \mid)$ is only slightly "larger" than $P(6 \mid 6)$. That is, the prior data contain little information compared with the six observations.

What would the covariance matrix at $t = 6$ be had we used a different value of P_0, say P_0'? Take

$$P_0' = \begin{bmatrix} 5 & 0 \\ 0 & 2 \end{bmatrix}, \qquad P_0'^{-1} = \begin{bmatrix} 0.2 & 0 \\ 0 & 0.5 \end{bmatrix}.$$

Then

$$P'^{-1}(6 \mid 0) = \Phi^{-T}(6, 0)\, P_0'^{-1}\Phi^{-1}(6, 0) = \begin{bmatrix} 0.2 & -1.2 \\ -1.2 & 7.7 \end{bmatrix}.$$

Then, according to Theorem 7.10,

$$P'^{-1}(6 \mid 6) = P^{-1}(6 \mid 6 \mid) + P'^{-1}(6 \mid 0) = \begin{bmatrix} 6.1 & -14.6 \\ -14.6 & 53.1 \end{bmatrix},$$

and

$$P'(6 \mid 6) = \begin{bmatrix} 0.48 & 0.13 \\ 0.13 & 0.055 \end{bmatrix}.$$

Our difference in initial conditions is

$$\delta P_0 = P_0 - P_0' = \begin{bmatrix} 5 & 0 \\ 0 & -1 \end{bmatrix},$$

and

$$\delta P_6 = P(6 \mid 6) - P'(6 \mid 6) = \begin{bmatrix} 0.02 & 0.01 \\ 0.01 & 0.005 \end{bmatrix}.$$

If

$$\| A \| \triangleq \sum | a_{ij} |^2,$$

we see that

$$\| \delta P_0 \| = 26, \qquad \| \delta P_6 \| = 0.0006.$$

Thus we see that the effect of the initial condition on the estimation error is being washed out, as guaranteed by theory.

6. APPLICATIONS IN ORBIT DETERMINATION

We present in this section the results of some simulations of the extended Kalman filter (Theorem 8.1) in Earth satellite orbit determination. The reader unfamiliar with satellite orbit mechanics may refer to Appendix 8A. Simulated observations are used in these experiments. That is, a true orbit is assumed, observations are computed and contaminated with noise. As a consequence, the filter model is perfect, and none of the problems of divergence (Section 8) are encountered. Nevertheless, these simulations are indicative of results that can be obtained in real situations, and demonstrate the efficacy of the Kalman filter in determining the satellite orbit.[4]

Our example orbit determination problem takes place in two dimensions [(z, x) plane in Fig. 8A.2], and a nonrotating oblate Earth is used. The true orbit has the elements

$$a = 10{,}758.236 \quad \text{km (semimajor axis)},$$

$$e = 0.28427970 \quad \text{(eccentricity)},$$

$$\omega = -170.0 \quad \text{deg (argument of perigee)},$$

$$M = 5.0 \quad \text{deg (mean anomaly)}.$$

[4] The data presented in this section are due to Mr. R. K. Squires of the NASA Goddard Space Flight Center. We are indebted to Mr. Squires for making these data available.

Observations consist of range, range rate, and m direction cosines. The true orbit[5] and the observations are computed for 10.0 hours, or approximately three revolutions. The *a priori* estimates of the orbital elements were taken as

$$\hat{a} = 10{,}778.236 \quad \text{km},$$

$$\hat{e} = 0.28559740,$$

$$\hat{\omega} = -173.0 \quad \text{deg},$$

$$\hat{M} = 3.0 \quad \text{deg}.$$

The errors $(a - \hat{a}$, etc.) represent two standard deviations; that is, the initial covariance matrix P_0 is diagonal in the space of the orbital elements, with $(a - \hat{a})^2/4$, etc., along the diagonal. The noise in the observations has the following standard deviations:

$$\sigma \text{ (noise in range)} = 0.2 \quad \text{km},$$

$$\sigma \text{ (noise in range rate)} = 1.0 \quad \text{m/sec},$$

$$\sigma \text{ (noise in } m \text{ direction cosine)} = 0.005.$$

Three tracking stations, spaced 120 deg apart, observe the satellite. Station 1 sees the satellite for $\sim$0.6 hr, station 2 for $\sim$0.4 hr, and station 3 for $\sim$1.4 hr in each revolution. The standard data rate is one observation every $1/32$ hr, and the high data rate is one observation every $1/512$ hr ($\sim$7 sec). An observation may be a scalar (e.g., range) or a vector (e.g., range and range rate), as will be indicated.

Figure 8.3 shows the periodic variations in the semimajor axis caused by earth oblateness (see Appendix 8A). The oblateness causes a to depart from its constant (Keplerian) value; the element a is said to be *osculating*. This figure is presented here only for its general interest.

A comparison of the effectiveness of the various observation types in determining the orbit is given in Figs. 8.4 and 8.5. The magnitude of the actual estimation errors in the semimajor axis is plotted in Fig. 8.4. We see that range observations determine the orbit best. Note that, for range observations, the actual error in a is of the order of 10 m after one revolution, even though the noise in the range observations has a standard deviation of 200 m. We are indeed "filtering out" the noise. The rms errors in a are plotted in Fig. 8.5. These are the standard

---

[5] The filter computations were actually performed using the NASA variables [24] as state variables. These have certain advantages over the Cartesian position and velocity coordinates.

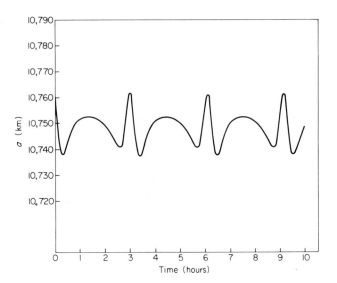

FIG. 8.3.   Osculating semimajor axis.

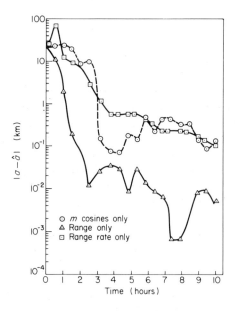

FIG. 8.4.   Estimation errors in a.

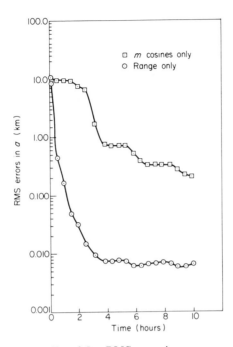

FIG. 8.5.   *RMS* errors in a.

deviations as predicted by the error covariance matrix. We see that the actual errors are consistent with their statistics. That is, the error covariance matrix faithfully represents the variance of the errors, as in theory it should.

The rate at which observations are taken has a profound effect on orbit estimation accuracy, especially if the observations are very noisy. This is illustrated in Fig. 8.6, where the m direction cosines were used in the estimation. Notice that, using a single station and the standard data rate, virtually no improvement in the relative position error is obtained. One can, however, determine the position quite accurately using all three stations and the high data rate. We have determined the position to about 10 m even with the very noisy m direction cosines.

In actual practice, a tracking station takes several observation types simultaneously. The effect of multiple data (at the same time instant) is shown in Fig. 8.7. We see that m direction cosines add little or nothing to the information contained in the range data. The addition of range rate observations improves the estimate somewhat, but not in any profound way. One should not attempt to generalize on the basis of Fig. 8.7, however, as station location, the number of stations, etc., have

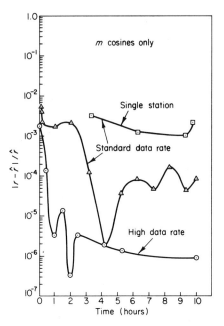

FIG. 8.6.   Relative errors in position magnitude.

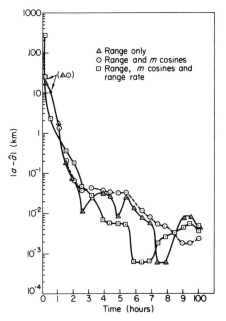

FIG. 8.7.   Effect of multiple data types.

some effect on the value of multiple data. Generally speaking, the
m cosines are so noisy that they do not enhance an estimate based on
range data alone. Range rate observations can, however, substantially
improve an estimate based on range alone.
The results of this section are indicative of what might be obtained in
real satellite orbit determination. The extended Kalman filter promises
to be a valuable tool in this field. As has been noted in the Introduction,
the Kalman filter is already in use in several aerospace applications. As
was also noted, the modeling problem still remains. We pursue that
problem in Section 8.

7. APPLICATIONS IN REENTRY

Reentry trajectory estimation differs from orbit determination in
several significant ways. For one thing, the dominant forces in reentry
are aerodynamic rather than gravitational. Because of our relatively
poor knowledge of the atmosphere (density, temperature, winds, etc.),
these forces are not well known. As a consequence, the modeling problem
is much more severe in reentry than it is in satellite orbit determination.
Furthermore, reentry dynamics are much more nonlinear than orbit
dynamics (see Appendix 8B). On the other hand, reentry trajectories
are of much shorter time duration than (several) satellite orbit revolutions, so that in reentry, model errors propagate over shorter time arcs.
We shall not attempt to determine which of these two estimation
problems is the more difficult. We can, however, expect that nonlinearities will have a more significant effect in reentry.

Through the courtesy of Mr. William E. Wagner of the Martin
Company, and by permission of the *Journal of Spacecraft and Rockets*
(*AIAA*), we reproduce here some reentry estimation experiments
performed by Mr. Wagner and reported by him [39]. These are simulated
experiments; that is, the data are generated, not real. Nevertheless,
these experiments are instructive, and indicate the kind of results which
can be expected in applying the Kalman filter to reentry trajectory
estimation. The reader is referred to Appendix 8B and to Wagner's
work [39] for a description of reentry mechanics.

The "true" reentry initial conditions (at time $t = 0$) in these experiments are

$$v = 25{,}093 \quad \text{fps}, \qquad h = 400{,}000 \quad \text{ft},$$

$$\gamma = -2.0422 \quad \text{deg}, \qquad \varphi = 24.8 \quad \text{deg},$$

$$\psi = -168.348 \quad \text{deg}, \qquad \theta = 215.25 \quad \text{deg}.$$

The prior estimates of these state variables are given in the second line of Table 8.3. Observations consist of ground-based radar tracking data (range, azimuth, and elevation) from three stations, and telemetered onboard acceleration measurements. Onboard accelerations are measured in two orthogonal directions. The three tracking zones and (line-of-sight) elevation angles are shown in Fig. 8.8. The (standard) noise in the observations has the following standard deviations:

$$3\sigma \quad (\text{range}) = 270 \quad \text{ft},$$

$$3\sigma \quad (\text{azimuth}) = 4.05 \quad \text{deg},$$

$$3\sigma \quad (\text{elevation}) = 4.23 \quad \text{deg},$$

$$3\sigma \quad (x\text{-acceleration}) = 3\%,$$

$$3\sigma \quad (z\text{-acceleration}) = 3\%.$$

The standard data rate is one set of these observations every 4 sec.

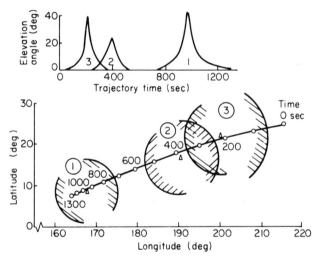

FIG. 8.8.   Tracking acquisition zones.

The linear filter is applied to this reentry estimation in two ways. In one case, a single reference trajectory is used as described in Section 3. In the second case, the extended Kalman filter (Theorem 8.1) is used. In each case, the first linearization is about the prior values (estimates). As is seen in Table 8.3, the prior estimation errors are rather large. The estimation results are displayed in Table 8.3 in terms of the estimates of (and estimation errors in) initial conditions (at time $t = 0$).

TABLE 8.3

INITIAL CONDITION ESTIMATION RESULTS

Iter.	v	$-\gamma$	$-\psi$	h	ϕ	θ	Δv	$\Delta\gamma$	$\Delta\psi$	Δh	$\Delta\phi$	$\Delta\theta$
true	25093.608	2.0422383	168.34792	400000.00	24.800000	215.25197	—	—	—	—	—	—
prior	24593.607	5.0422383	166.34791	420000.00	25.400000	215.85196	−500.	−3.00	2.00	20000.	0.600	0.600
Global iterations of extended Kalman filter												
1	25107.221	2.0996928	168.34696	402809.67	24.798588	215.24566	13.6	−0.057	0.001	2809.	−0.001	−0.006
2	25098.217	2.0537964	168.34372	400058.93	24.801576	215.24589	4.61	−0.012	0.004	58.9	0.002	−0.006
3	25098.097	2.0533670	168.34369	400039.77	24.801577	215.24579	4.49	−0.011	0.004	39.8	0.002	−0.006
Global iterations using single reference each iteration												
2	25098.217	2.0537964	168.34372	400058.93	24.801576	215.24589	4.61	−0.012	0.004	58.93	0.002	−0.006
3	25080.183	2.0014297	168.34544	398775.88	24.801116	215.24873	−13.43	0.0408	0.002	−1224.1	0.001	−0.003
4	25088.111	2.0266663	168.34500	399610.04	24.801433	215.24854	−5.50	0.0156	0.003	−390.0	0.001	−0.003
5	25089.525	2.0308211	168.34487	399713.25	24.801435	215.24817	−4.08	0.0114	0.003	−286.8	0.001	−0.004

That is, after all the data are processed, the estimate is smoothed back to time zero.

Wagner found that the filter that used a single reference trajectory diverged and was unusable. This is due to the large initial estimation errors and the nonlinearities in the problem. The extended Kalman filter performed relatively well. Its results are summarized in line 1 of Table 8.3. The errors in all the state components, except altitude, are relatively small. The altitude error is, however, unsatisfactory.

Assuming that the unsatisfactory altitude error is caused by significant nonlinearities neglected in the filter, the extended Kalman filter was iterated globally, as described in Section 3. Three iterations are summarized in Table 8.3. It is seen that these iterations lead to very satisfactory results, inasmuch as the estimation errors are rather small.

Using the results of the extended Kalman filter as the first iteration, the single reference filter was iterated several times. The results of these iterations are also given in Table 8.3. It is seen that altitude errors are quite large even after five iterations. This demonstrates the advantage of the relinearization procedure used in the extended Kalman filter. These results also show that nonlinear effects are quite significant in reentry problems.

Timewise convergence of the estimates is shown (for altitude) in Fig. 8.9. We see that the true altitude is acquired very rapidly, and that the estimate subsequently oscillates about the true value with ever-decreasing error amplitude.

An error analysis for this reentry problem is given in Fig. 8.10.

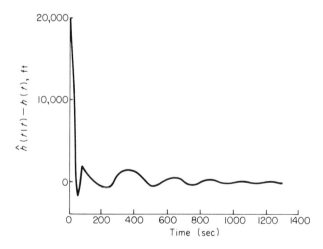

FIG. 8.9.   Timewise convergence.

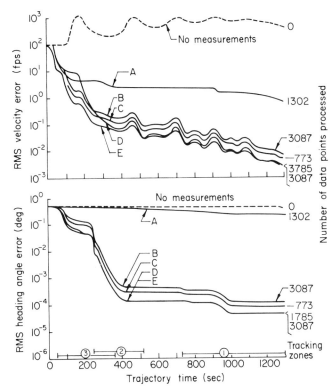

FIG. 8.10.   Standard deviations of Errors in v and ψ. A: telemetry, 2-sec data rate; B: tracking and telemetry, 2-sec data rate; C: tracking and telemetry, 8-sec data rate; D: tracking, 2-sec data rate; E: tracking and telemetry, 2-sec data rate (A–E, all standard measurement noise).

Plotted are the rms errors in velocity and heading angle (square roots of the appropriate elements of the error covariance matrix P). Figure 8.10 shows the effect of using tracking and/or telemetered acceleration data, measurement noise, and data rate. We see that tracking data are of greater value in estimating the trajectory than are the acceleration data. The fluctuations in the rms velocity error are caused by the skipping motion of the reentry vehicle.

8. FILTER DIVERGENCE AND ERROR COMPENSATION TECHNIQUES

The problem of filter divergence was already noted in Sections 6.11, 7.8, and elsewhere. In essence, the problem is this. If the filter is constructed on the basis of an erroneous model, it "learns the wrong state too well" when it operates over many data. The problem is particularly acute when the noise inputs to the system are small and when measurement noise is small. For then the filter is capable of learning the state very well. Eventually, the error covariance matrix becomes very small, the filter gain is therefore small, and subsequent observations have little effect on the estimate. But the dynamical system model in the filter is different from the actual system model, so that the estimate and the state can diverge. In actual applications, the onset of divergence manifests itself by the inconsistency of residuals [e.g., Eq. (8.8)] with their predicted statistics. Residuals become biased [nonzero mean (8.9)] and larger in magnitude than their rms values, as predicted by (8.10).

Divergence problems have been particularly acute in orbit determination [41]. This is because measurement devices (radars) are very accurate, many data are available, and the filter is required to operate over many revolutions of the satellite orbit. As we noted in Section 1, the modeling or divergence problem remains a serious problem in orbit determination.

To give the reader a "feel" for the divergence problem, we present a simple analytical example[6] of divergence borrowed from Schlee *et al.* [29]. Consider the problem of estimating the altitude h of a space vehicle from altimeter data. Suppose the filter is designed assuming that the altitude is a constant, whereas in actual fact the vehicle is climbing at a constant rate. The observations are

$$y_k = h_k + v_k,$$ (8.80)

and, in the notation of Section 7.8,

$$\bar{h}_{k+1} = \bar{h}_k.$$ (8.81)

Our filter design is [see (7.206)–(7.208)]

$$\hat{h}_k^k = \hat{h}_{k-1}^{k-1} + \frac{P_c(k-1 \mid k-1)}{P_c(k-1 \mid k-1) + R}(y_k - \hat{h}_{k-1}^{k-1}),$$ (8.82)

where the recursion for the computed error variance is

$$P_c(k \mid k) = P_c(k-1 \mid k-1) - \frac{P_c^2(k-1 \mid k-1)}{P_c(k-1 \mid k-1) + R}.$$ (8.83)

---

[6] The paper by Schlee *et al.* [29] contains other examples of filter divergence in orbit determination. Additional examples may also be found in Fitzgerald [13] and in Sections 11 and 12.

Equation (8.83) has the solution

$$P_c(k \mid k) = \frac{P_0 R}{k P_0 + R},$$ (8.84)

so that (8.82) becomes

$$\hat{h}_{k+1}^{k+1} = \hat{h}_k^{\ k} + \frac{P_0}{(k+1)P_0 + R}(y_{k+1} - \hat{h}_k^{\ k}).$$ (8.85)

Since, in actuality,

$$h_k = h_0 + ks \qquad (h_{k+1} = h_k + s)$$ (8.86)

(the vehicle is climbing at a constant speed), the recursion for the actual estimation error

$$\tilde{h}_k^{\ k} \triangleq h_k - \hat{h}_k^{\ k}$$

is

$$\tilde{h}_{k+1}^{k+1} = [\tilde{h}_k^{\ k} + s]\left[\frac{k P_0 + R}{(k+1)P_0 + R}\right] - \frac{P_0 v_{k+1}}{(k+1)P_0 + R},$$ (8.87)

which has the solution

$$\tilde{h}_k^{\ k} = \frac{R}{k P_0 + R}\tilde{h}_0^{\ 0} + \frac{[(k-1)k/2]P_0 + kR}{k P_0 + R}s - \frac{P_0}{k P_0 + R}\sum_{i=1}^{k} v_i.$$ (8.88)

Now the first term in (8.88) goes to zero as $k \to \infty$. The last term also goes to zero in the mean square sense, say (the familiar "square root of k" law). However, the middle term involving the speed s is unbounded in k and goes to infinity as $(k-1)/2$. As we see, the estimation errors are unbounded; the filter is diverging. Note, however, that, if s were zero, we could learn the altitude arbitrarily well given sufficient observations.

Example 8.4. Use the techniques of Section 7.8 to analyze the effect of the model error in this example. The actual system can be considered to be the second-order system

$$h_{k+1} = h_k + s_k, \qquad s_{k+1} = s_k,$$

with the state transition matrix

$$\Phi(k+1, k) = \begin{bmatrix} 1 & 1 \\ 0 & 1 \end{bmatrix},$$

whereas the erroneous model has the transition matrix

$$\Phi_c(k+1, k) = \begin{bmatrix} 1 & 0 \\ 0 & 1 \end{bmatrix}.$$

The model error can thus be viewed as an error in Φ. Alternatively, the model error can be viewed as the error $\Delta\varphi = s$ in the first-order system (8.86). ▲

Schlee *et al.* [29] show that actual estimation errors can be bounded (in this example) by adding noise to the system (8.81):

$$\bar{h}_{k+1} = \bar{h}_k + \bar{w}_k \,.$$

This is, in fact, a special case of the corollary to Theorem 7.7. Identify s in (8.86) with $\Delta\varphi(k)$ in Section 7.8. We see that the addition of system noise gives us uniform complete controllability. The other hypotheses of the corollary are clearly satisfied and, as a consequence, estimation errors are bounded. The important question is, of course, how much noise to add. Furthermore, it is one thing to bound the errors and quite another to assure that the resulting filter performance is satisfactory. The "bounded" errors may be too large for the results to be usable.

The error analyses of Section 7.8 provide us with numerical tools for evaluating the effect of model errors on filter performance. As was mentioned there, such errors may be intentional. That is, certain approximations might be made to simplify the filter computations. On the other hand, model errors may be present because of our limited knowledge of the physical system. Furthermore, computational errors resulting from the use of finite precision arithmetic effectively produce errors in the model. In any case, the tools and qualitative results developed in Section 7.8 are not enough. Faced with model errors, we require techniques for synthesizing a filter that works. That is, a filter that not only produces bounded errors, but also one that gives satisfactory performance. The actual error variance should be sufficiently small, and the estimation errors themselves should be consistent with their predicted statistics.

As was already noted, divergence takes place when the covariance matrix becomes too small or optimistic. For then the filter gain K (8.4) becomes small, and subsequent observations are ignored. In the example of divergence discussed previously, we see from (8.84) that

$$P_c(k \mid k) \to 0 \qquad (k \to \infty),$$

and consequently

$$K_c(k) = \frac{P_c(k-1 \mid k-1)}{P_c(k-1 \mid k-1) + R} \to 0 \qquad (k \to \infty).$$

Then the estimate evolves according to (8.81), whereas the actual state evolves according to (8.86), and divergence takes place. The basic idea, then, in order to prevent divergence, is to increase the covariance matrix P. This is justified, since such an increase in the covariance matrix *compensates* for the model errors. Model errors in some sense add uncertainty to the system which should reflect itself in some degradation in certainty (increase in P).

Two somewhat different points of view can be taken with respect to model error compensation. One is to provide a (fictitious) noise input to the system or to attribute the errors to inaccuracies (uncertainties) in some of the parameters of the system, thus effecting an increase in the covariance matrix directly. The other point of view is to overweight the most recent data in some way. This causes the filter to forget old information, and thus indirectly increases the covariance matrix. The physical justification for forgetting old observations is that old observations, when predicted over long time arcs through an erroneous system, can be valueless. We saw that the first approach of adding a fictitious noise input is effective in the example described in this section. Recalling, however, the discussion of Theorem 7.7, we note that fictitious noise may not always be effective in preventing divergence.

The next four sections are devoted to the development of these two points of view in model error compensation. Several error compensation techniques are described and discussed. Sections 11 and 12 describe more sophisticated approaches to error compensation, together with some interesting applications in orbit determination.

9. FICTITIOUS NOISE INPUTS AND
PARAMETER UNCERTAINTIES

It is possible to parametrize those portions of the dynamics (and/or measurement functions) which are poorly known, and consider the parameters in such a representation as unknown or poorly known quantities. Any well-behaved function can be approximated arbitrarily well by polynomials or by other series. One could then augment the state vector with these parameters and estimate them together with the state, as described in Section 4. This approach may, of course, lead to a high-dimensional state vector which, in turn, greatly increases the filter computational load. Furthermore, unless a sufficient number of parameters are provided, we may be simply postponing the inevitable. Once we learn the wrong parametric representation, the filter can again diverge. On the other hand, if too many parameters are used, there may be insufficient data for filtering. As a rather absurd extreme, we could imagine fitting an nth-order polynomial to $n + 1$ data points. In this connection, the reader is referred to an interesting nonparametric approach of Dennis [9].

Instead of estimating these uncertain parameters, we could, via the Schmidt–Kalman filter (Theorem 8.4), include their uncertainty in the state filter. The parameter uncertainty will always degrade the state estimate, since we are not improving our estimate of these parameters themselves. This is a desirable feature that will tend to keep the covariance matrix open. The Schmidt–Kalman filter has, for example, been successfully used by Wolf [41] for drag and earth-mass uncertainties in earth satellite orbit determination. It was found, however, that uncertain parameters, which lead to successful orbit determination for one satellite, do not produce good results in the case of another satellite. Obviously, different parameters were dominant in each case. Which parameters to use in a given problem has to be determined by experimentation.

As we saw in the preceding section, fictitious noise inputs, which essentially cover (deterministic) model errors, can be effective in preventing divergence. In taking this approach to divergence, one has to determine what noise to add. This amounts to specifying $\Gamma Q \Gamma^T$ in (8.3). An effective $\Gamma Q \Gamma^T$ can be determined by trial and error, and this has been successfully done by Fitzgerald [13], for example. One simulates the filter for various values of $\Gamma Q \Gamma^T$ until one gets satisfactory filter performance.

Wolf [41] has used more systematic approaches to determine the noise input. One such approach is to determine the major "direction(s)" in state space which are affected by model errors. For example, errors in drag affect in-track satellite motion. This essentially defines the noise coefficient matrix Γ, and one can set

$$\Gamma Q \Gamma^T = k^2 \, \Delta t^2 \, \Gamma \Gamma^T, \tag{8.89}$$

where Δt is the time interval between observations and the parameter k is to be determined by experimentation. Another approach is to pick (perhaps by experimentation) a likely parameter error Δa, say, and, using the dynamical equations, actually compute $\Delta x_{t_{k+1}}$ due to Δa at t_k. Then set

$$\Gamma(t_k) Q(t_{k+1}) \Gamma^T(t_k) = \Delta x_{t_{k+1}} \Delta x_{t_{k+1}}^T. \tag{8.90}$$

Yet another approach due to Wolf [41] is the following. Assume a likely noise level due to errors in the system. This would be noise in the accelerations in satellite orbit problems. Generate a random sample of this noise and pass it through the dynamical system (from t_k to t_{k+1}). The noise sample produces a $\Delta x_{t_{k+1}}$, which is used in (8.90). These techniques take into account the effect that model errors have on the system, and are aids in determining appropriate fictitious noise inputs.

Pines *et al.* [24] have developed a computational (machine) error noise model that is useful when the computations are performed with limited (e.g., single) precision. Suppose the computer has a p-digit word length. Then a number x in the computer is of the form

$$x = . x_1 x_2 \cdots x_p 10^n.$$

Assume that the computer simply drops the number beyond the least significant digit in addition. Then the addition round-off error is of the same sign as x, and has magnitude 10^{n-p},

$$v_x \sim \frac{x}{|x|} 10^{n-p}.$$

Taking $|x| \sim 10^n$, we have

$$v_x \sim x 10^{-p}.$$

Assuming that this round-off takes place in the prediction (8.2), we use the noise covariance matrix

$$\Gamma(t_k) Q(t_{k+1}) \Gamma^T(t_k) = 10^{-2p} \begin{bmatrix} \hat{x}_1^2(t_{k+1} \mid t_k) & & 0 \\ & \ddots & \\ 0 & & \hat{x}_n^2(t_{k+1} \mid t_k) \end{bmatrix}, \quad (8.91)$$

and assume that round-off errors are uncorrelated.

This computational error model is useful when model errors are in the machine noise level. Schlee *et al.* [29] have used it successfully in cases where model errors are larger than computational errors, by simply making p smaller. They used single precision arithmetic ($p = 8$) and engineered a $p = 6.5$.

The proliferation of so many techniques of adding uncertainty to the system to account for model errors is evidence of the fact that none of them work very well. All these techniques have their "fiddle parameters," which have to be adjusted in each application by experimentation. As a consequence, these techniques are not applicable in real-time filtering. In Section 11, we describe an *adaptive* technique of estimating noise input variances in real time. We let the actual filter measurement residuals determine the noise inputs for us.

10. OVERWEIGHTING THE MOST RECENT DATA

As we noted in Section 8, an indirect approach to increasing the covariance matrix, and thus increasing the filter gain, is to degrade old data (observations) or, equivalently, to overweight more recent observations. This has the effect of decreasing the information (matrix) on which the estimate is based. (Recall that error variance is inversely proportional to information.) The reasoning behind this approach is that model errors themselves degrade the value of information in the distant past. We now describe several techniques for this type of model error compensation.

Fagin [10] describes the following technique of *exponentially age-weighting* old data. Consider the recursive least squares derivation of the linear filter in Example 7.2. In the minimization of J_k (7.37), replace

$$R_1, R_2, ..., R_k$$

by

$$e^{(t_k-t_1)/\tau}R_1, \, e^{(t_k-t_2)/\tau}R_2, ..., \, e^{(t_k-t_k)/\tau}R_k, \quad \tau > 0.$$

In the minimization of J_{k+1}, replace

$$R_i, \quad i = 1, ..., k$$

by

$$e^{(t_{k+1}-t_k)/\tau}R_i, \quad i = 1, ..., k.$$

This leads to a filter with gain matrix

$$K_e(k+1) = P(k+1 \mid k) M^T(k+1)$$
$$\times [M(k+1) P(k+1 \mid k) M^T(k+1) + e^{-(t_{k+1}-t_k)/\tau}R(k+1)]^{-1}, \quad (8.92)$$

and the recursion for P (at an observation)

$$P(k+1 \mid k+1) = e^{(t_{k+1}-t_k)/\tau}[I - K_e(k+1) M(k+1)] P(k+1 \mid k). \quad (8.93)$$

The exponentially age-weighted filter then consists of the filter described in Section 2 [(8.2)–(8.6)] with (8.4) replaced by (8.92), and (8.6) replaced by (8.93). Note that, the smaller τ is, the faster old observations are "forgotten."

A simple means of keeping the filter gain from becoming small is to fix the covariance matrix P at some value, say $\bar{P}$. $\bar{P}$ could be engineered *a priori*, or could be the result of filtering a given amount of data. The filter gain would then be

$$K_f(k+1) = \bar{P}M^T(k+1)[M(k+1) \bar{P}M^T(k+1) + R(k+1)]^{-1}, \quad (8.94)$$

and

$$\hat{x}(k+1 \mid k+1) = \hat{x}(k+1 \mid k) + K_f(k+1)[y_{k+1} - M(k+1) \hat{x}(k+1 \mid k)]. \quad (8.95)$$

The recursion for the estimation error covariance matrix for this filter is easily computed to be

$$P(k + 1 \mid k + 1) = [I - K_f(k + 1) M(k + 1)]$$
$$\times P(k + 1 \mid k)[I - K_f(k + 1) M(k + 1)]^T$$
$$+ K_f(k + 1) R(k + 1) K_f^T(k + 1). \quad (8.96)$$

This technique was successfully applied by Fitzgerald [13]. It appears quite attractive for a real-time filtering situation that can be extensively simulated to arrive at a design for $\bar{P}$.

Schmidt [32, 33] has proposed several techniques of overweighting current observations. They are applicable to *scalar* observations; vector observations can usually be filtered by processing their components one at a time. One of these techniques, which appears to be the most interesting, is described next.

Schmidt's technique consists of computing an estimate that is a linear combination of the estimate based on the current observation and past data, and the estimate based on the current observation alone. Let

$$\Delta x \triangleq x(k + 1) - \hat{x}(k + 1 \mid k), \quad (8.97)$$

$$\Delta \hat{x} \triangleq \hat{x}(k + 1 \mid k + 1) - \hat{x}(k + 1 \mid k), \quad (8.98)$$

$$\Delta y \triangleq y_{k+1} - M(k + 1) \hat{x}(k + 1 \mid k) = M(k + 1) \Delta x + v_{k+1}. \quad (8.99)$$

The estimate based on the current observation and past data is the usual Kalman estimate

$$\Delta \hat{x} = P(k + 1 \mid k) M^T(k + 1)$$
$$\times [M(k + 1) P(k + 1 \mid k) M^T(k + 1) + R(k + 1)]^{-1} \Delta y. \quad (8.100)$$

Note that the quantity in brackets is a scalar, since observations are scalar. Let us determine the estimate of Δx based on y_{k+1} only. From (8.99), the least squares estimate is

$$\Delta \hat{x} = [M^T(k + 1) R^{-1}(k + 1) M(k + 1)]^\# M^T(k + 1) R^{-1}(k + 1) \Delta y, \quad (8.101)$$

where $A^\#$ is the pseudo-inverse of A. But

$$M^\# = M^T / MM^T,$$

since M is a row matrix, so that (8.101) becomes

$$\Delta \hat{x} = \frac{M^T(k + 1)}{M(k + 1) M^T(k + 1)} \Delta y,$$

or

$$\Delta \hat{x} = cM^{\mathrm{T}}(k+1)\,R(k+1)[M(k+1)\,M^{\mathrm{T}}(k+1)]^{-1}$$
$$\times [M(k+1)\,P(k+1\mid k)\,M^{\mathrm{T}}(k+1) + R(k+1)]^{-1}\,\Delta y, \quad (8.102)$$

with the appropriate constant c. Taking our estimate to be a linear combination of (8.100) and (8.102), we have the estimator

$$\hat{x}(k+1\mid k+1) = \hat{x}(k+1\mid k) + [P(k+1\mid k)\,M^{\mathrm{T}}(k+1)$$
$$+ cM^{\mathrm{T}}(k+1)\,R(k+1)[M(k+1)\,M^{\mathrm{T}}(k+1)]^{-1}]$$
$$\times [M(k+1)\,P(k+1\mid k)\,M^{\mathrm{T}}(k+1) + R(k+1)]^{-1}$$
$$\times [y_{k+1} - M(k+1)\,\hat{x}(k+1\mid k)]. \quad (8.103)$$

Multiplying (8.103) on the left by M, we have

$$M(k+1)\,\hat{x}(k+1\mid k+1)$$
$$= M(k+1)\,\hat{x}(k+1\mid k)$$
$$+ \frac{[M(k+1)\,P(k+1\mid k)\,M^{\mathrm{T}}(k+1) + cR(k+1)]}{[M(k+1)\,P(k+1\mid k)\,M^{\mathrm{T}}(k+1) + R(k+1)]}\,\Delta y.$$

Thus, with $c = 1$,

$$M(k+1)\,\hat{x}(k+1\mid k+1) = M(k+1)\,\hat{x}(k+1\mid k) + \Delta y = y_{k+1}\,;$$

the estimate of the observation is the observation itself. With $c = 0$, (8.103) reduces to the Kalman filter.

We easily compute the recursion for the error covariance matrix associated with the estimator (8.103):

$$P(k+1\mid k+1)$$
$$= P(k+1\mid k) - P(k+1\mid k)\,M^{\mathrm{T}}(k+1)$$
$$\times [M(k+1)\,P(k+1\mid k)\,M^{\mathrm{T}}(k+1) + R(k+1)]^{-1}$$
$$\times M(k+1)\,P(k+1\mid k)$$
$$+ \frac{c^2 R^2(k+1)[M^{\mathrm{T}}(k+1)\,M(k+1)]}{[M(k+1)\,P(k+1\mid k)\,M^{\mathrm{T}}(k+1) + R(k+1)][M(k+1)\,M^{\mathrm{T}}(k+1)]^2}.$$
$$(8.104)$$

We note that the new term involving c^2 produces a desirable increase in P. Clearly, the appropriate value of c is not known *a priori*, and must be determined by simulations. In general, different values of c will be required for each observation type. Schmidt, in the references cited, has successfully applied this filter in some interesting example problems. The limited memory filter developed in Section7.10 falls in the present category of model error compensation techniques. Some simulations of the limited memory filter are presented in Section 12.

11. ADAPTIVE NOISE ESTIMATION

In Section 9, we described several systematic methods of determining the noise input levels (variances) as a means of compensating for model errors. All the methods described depend on simulation to establish appropriate values of certain parameters [e.g., k in (8.89), p in (8.91)]. In a simulation using real data, one looks at measurement residuals, such as (8.8), to evaluate filter performance. Residuals should be small, random, and should possess statistical properties consistent with (8.9) and (8.10). Clearly, we cannot be thinking here of real-time filtering. Such simulations can be performed only after the fact.

Instead of looking at residuals after the fact to evaluate filter performance, we propose to look at them in real time and to let the residuals themselves determine appropriate noise input levels. We let the noise input levels *adapt* to the residuals. Our development here is based on the work of Jazwinski [15, 17, 18].

Consider first the special case of *uncorrelated* and *identically distributed* noise inputs, that is, $Q(k) = qI$ in (8.1). We suppose that the noise coefficient matrix $\Gamma(k)$ has been engineered for us, perhaps by one of the methods described in Section 9. In this case, the noise variance term in the error covariance matrix prediction equation (8.3) is simply $q\Gamma(k)\,\Gamma^{\mathrm{T}}(k)$. We assume henceforth, and throughout this section, that measurements are scalar ($m = 1$). This is usually no loss of generality, since vector measurements can be filtered by processing their components one at a time.

Suppose that we are at time t_k, we have just processed observation y_k, and we are about to predict to t_{k+1} via (8.2) and (8.3). Let us use that value of the noise variance q which produces the *most probable* predicted residual $r(k + 1 \mid k)$, as defined in (8.8), that is, determine the q value (call it $\hat{q}_{k,1}$) by the operation

$$\max_{q \geqslant 0} p[r(k + 1 \mid k)], \qquad (8.105)$$

where $p(\cdot)$ is the probability density function of the residual. The restriction $q \geqslant 0$ is consistent with the property of a variance.

Now the density in (8.105) is zero-mean, Gaussian, with the variance given in (8.10). It is easy to show [15] that the maximizing q is determined by

$$r^2(k+1 \mid k) = \mathscr{E}\{r^2(k+1 \mid k)\} \qquad (8.106)$$

if

$$r^2(k+1 \mid k) > M(k+1)\, \Phi(k+1, k)\, P(k \mid k)\, \Phi^T(k+1, k)\, M^T(k+1)$$
$$+ R(k+1), \qquad (8.107)$$

and is zero otherwise. Since

$$\mathscr{E}\{r^2(k+1 \mid k)\} = M(k+1)\, \Phi(k+1, k)\, P(k \mid k)\, \Phi^T(k+1, k)\, M^T(k+1)$$
$$+ qM(k+1)\, \Gamma(k)\, \Gamma^T(k)\, M^T(k+1) + R(k+1), \qquad (8.108)$$

and if we use the slightly abused notation

$$\mathscr{E}\{r^2(k+1 \mid k) \mid q \equiv 0\} = M(k+1)\, \Phi(k+1, k)\, P(k \mid k)$$
$$\times\, \Phi^T(k+1, k)\, M^T(k+1) + R(k+1), \qquad (8.109)$$

our estimate of q is

$$\hat{q}_{k,1} = \begin{cases} \dfrac{r^2(k+1 \mid k) - \mathscr{E}\{r^2(k+1 \mid k) \mid q \equiv 0\}}{M(k+1)\, \Gamma(k)\, \Gamma^T(k)\, M^T(k+1)}, & \text{if positive} \\ 0 & \text{otherwise.}^7 \end{cases} \qquad (8.110)$$

The usual linear filter [(8.2)–(8.7)], with the estimator (8.110) used for q in (8.3), is our basic *adaptive filter* for uncorrelated and identically distributed noise inputs. Let us analyze the implications of the noise variance estimator (8.110). Except for an appropriate scaling factor, $\hat{q}_{k,1}$ is the excess, if any, of the residual squared over the expected value of the residual squared, under the assumption of no input noise. If the estimate $\hat{q}_{k,1}$ is positive, as determined by (8.110), then it produces the equality in (8.106). In any case, it produces *consistency* between residuals and their statistics.

Our adaptive filter is adaptive in the following sense. As long as residuals remain within their 1σ limits, the noise input level in the filter is zero. This is as it should be, since residuals are small and consistent with their statistics; the filter is operating satisfactorily. When residuals become large relative to their *predicted* 1σ values, the filter is diverging. $\hat{q}_{k,1}$ increases $P(k+1 \mid k)$, which increases the filter gain and "opens"

7 We of course assume that $M\Gamma\Gamma^T M^T > 0$. That is, M is not orthogonal to Γ, which is an observability condition.

the filter to incoming observations. Subsequent filtering reduces $P(k + 1 \mid k)$ again.

The adaptive filter we have described has a rather serious defect. The estimate of q is based on only one residual and therefore has little statistical significance. The estimator (8.110) will respond to (large) measurement noise samples as well as large residuals caused by model errors. This can be remedied in several ways. The simplest remedy is to smooth the estimates of q. Another remedy, which is described in detail by Jazwinski [17], is to replace the one predicted residual by the *sample mean* of N predicted residuals:

$$m_r \triangleq N^{-1} \sum_{l=1}^{N} r(k + l \mid k)/R^{1/2}(k + l).$$

The resulting estimator of q has the same structure as (8.110). This has been found very effective [17], but we shall not pursue the matter here. We note in passing that the present theory can also be applied to the estimation of the measurement noise variance R [15].

We next consider the more general case of n *independent* noise inputs in (8.1), that is, $\Gamma(k) = I$, Q $n \times n$ and diagonal. This structure obviates the requirement of engineering a noise coefficient matrix, which, as we noted in Section 9, is not a simple matter. Let us define the *predicted residuals*

$$r(k + l \mid k) = y_{k+l} - \mathscr{E}\{y_{k+l} \mid Y_k\}, \qquad l > 0. \tag{8.111}$$

Note that $r(k + 1 \mid k)$ has already been defined in (8.8). These residuals are zero-mean, Gaussian, and it is straightforward to compute

$$r(k + l \mid k) = M(k + l)\,\Phi(k + l, k)[x_k - \hat{x}(k \mid k)]$$
$$+ M(k + l) \sum_{i=1}^{l} \Phi(k + l, k + i)\,\Gamma(k + i - 1)\,w_{k+i} + v_{k+l} \tag{8.112}$$

and

$$\mathscr{E}\{r(k + l \mid k)\,r(k + m \mid k)\}$$

$$= M(k + l)\,\Phi(k + l, k)\,P(k \mid k)\,\Phi^{\mathrm{T}}(k + m, k)\,M^{\mathrm{T}}(k + m)$$

$$+ M(k + l) \sum_{i=1}^{l} [\Phi(k + l, k + i)\,\Gamma(k + i)\,Q\Gamma^{\mathrm{T}}(k + i)$$

$$\times \Phi^{\mathrm{T}}(k + m, k + i)\,M^{\mathrm{T}}(k + m)]$$

$$+ R(k + l)\,\delta_{lm}, \qquad m \geqslant l. \tag{8.113}$$

By analogy with the special case developed previously ($Q = qI$), and in particular, by analogy with (8.106), we determine a diagonal Q from the *consistency requirement*

$$r^2(k + l \mid k) = \mathscr{E}\{r^2(k + l \mid k)\}, \qquad l = 1,..., N. \tag{8.114}$$

If (8.114) is satisfied, residuals will be consistent with their statistics. Equation (8.114) is related to the maximization of the joint density

$$p[r(k + 1 \mid k),..., r(k + N \mid k)],$$

but we do not pursue that relationship here. In our present case $[\Gamma(k) = I, Q$ diagonal$]$, we have, from (8.113),

$$\mathscr{E}\{r^2(k + l \mid k)\} = \mathscr{E}\{r^2(k + l \mid k) \mid Q \equiv 0\} + \sum_{j=1}^{n} a_{lj}q_{jj}, \tag{8.115}$$

where

$$\mathscr{E}\{r^2(k + l \mid k) \mid Q \equiv 0\} = M(k + l)\, \Phi(k + l, k)\, P(k \mid k)$$
$$\times\, \Phi^{\mathrm{T}}(k + l, k)\, M^{\mathrm{T}}(k + l) + R(k + l), \tag{8.116}$$

$$a_{lj} = \sum_{i=1}^{l} [M(k + l)\, \Phi(k + l, k + i)]_j^2, \tag{8.117}$$

and q_{jj} are the diagonal elements of Q. Introducing the vectors and matrix

$$\epsilon^{\mathrm{T}} = [r^2(k + 1 \mid k) - \mathscr{E}\{r^2(k + 1 \mid k) \mid Q \equiv 0\},..., r^2(k + N \mid k)$$
$$- \mathscr{E}\{r^2(k + N \mid k) \mid Q \equiv 0\}],$$

$$q^{\mathrm{T}} = [q_{11},..., q_{nn}], \qquad A = [a_{lj}] \quad (N \times n),$$

the consistency requirement (8.114) becomes

$$Aq = \epsilon. \tag{8.118}$$

The solution of (8.118) can be written in general as

$$\bar{q}_{k,N} = (A^{\mathrm{T}}A)^{\#} A^{\mathrm{T}}\epsilon, \tag{8.119}$$

where $B^{\#}$ is the pseudo-inverse of B. Various special cases of (8.119), depending on N, are discussed by Jazwinski [15, 17]. Now, in case N is taken small, an estimate of q based on (8.119) may have little statistical significance. Various devices that give such an estimate statistical significance are discussed by Jazwinski [17]. Here, motivated by least squares estimation theory, we make the *ad hoc* modification to (8.119):

$$\bar{q}_{k,N} = (C^{-1} + A^{T}A)^{-1} (A^{T}\epsilon + C^{-1}q^{0}).$$ (8.120)

q^{0} is an *a priori* estimate of the vector q, and C^{-1} (diagonal) is a measure of belief in the *a priori* estimate. Based on (8.120), our estimator of q is

$$\hat{q}_{k,N} = \bar{q}_{k,N} ,$$ (8.121)

subject to the rule that, if $(\hat{q}_{jj})_{k,N} < 0$, we set $(\hat{q}_{jj})_{k,N} = 0$. The extended Kalman filter (Theorem 8.1), with noise variance estimator (8.121), is the adaptive filter used in the simulations described below.

We present simulations of this adaptive filter in IMP[8]-type orbit estimation. The IMP satellite has a highly eccentric orbit that is extremely sensitive to model errors (see Wolf [41]). As a result, it represents an excellent test for the adaptive filter. The material presented below is taken from Jazwinski [18].

For the purpose of our adaptive filter simulation study, the following (true) Keplerian osculating orbit elements were assumed (at time zero):

$a = 16.37317$ earth radii (semimajor axis),

$e = 0.9370696$ (eccentricity),

$i = 33.33961$ deg (inclination),

$\omega = 133.1376$ deg (argument of perigee),

$\Omega = -101.6029$ deg (right ascension of ascending node),

$P = 93.29312$ hr (period),

$\nu = 1.492517$ rad (true anomaly),

(earth radius = 6378.165   km).

The dynamics used in the filter are two-body (Keplerian) dynamics, so that, as far as the filter dynamics are concerned, the orbital elements are constants of the motion. On the other hand, the true motion or satellite trajectory is generated via the theory given by Pines [25], which includes the secular effect of the second zonal harmonic in the earth's gravitational potential. The effect of this model error is a rigid rotation of the true orbit which causes a change in the argument of perigee (ω)

---

[8] Interplanetary Monitoring Probe, although IMP is an appropriate name for this satellite from the point of view of its orbit determination.

and the right ascension of the ascending node (Ω). In the case to be reported here, the stated model error produces $|\,\Delta\omega\,| = 0.0014$ and $|\,\Delta\Omega\,| = 0.0010$ at 747 hr. The other orbital elements are unaffected by this rotation. To summarize, the filter dynamics are two-body, whereas the real dynamics are not. Observations are generated on the basis of the real dynamics. The model error chosen is a realistic representation of the type of errors which are present in orbit estimation models. The initial estimation errors are

$$a - \hat{a} = -0.1336430; \qquad \omega - \hat{\omega} = -0.1266091;$$

$$e - \hat{e} = -0.0004206; \qquad \Omega - \hat{\Omega} = 0.0912469;$$

$$i - \hat{i} = 0.0004442; \qquad \nu - \hat{\nu} = 0.0013183;$$

$$P - \hat{P} = -1.144560;$$

and the initial estimation error covariance matrix, P_0 , is consistent with these errors.

Observation types and times used in this study correspond to those of a real tracking schedule. Range, range rate, and l and m direction cosine observations were generated and contaminated with white Gaussian noise, with the respective variances: 0.676×10^{-7} earth radii, 0.529×10^{-7} earth radii/hr, 0.400×10^{-7}, 0.179×10^{-6}. The temporal distribution of observations is indicated in Fig. 8.11. A total of 1140

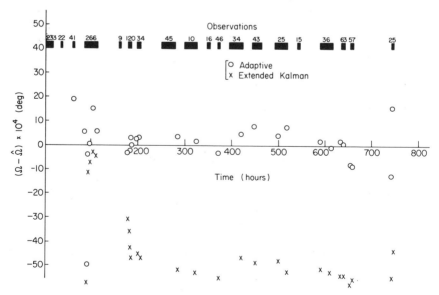

FIG. 8.11. Estimation errors in Ω.

observations were available, covering about eight revolutions of the satellite. Note that most of the observations were taken during the first two revolutions, and subsequent data are rather sparse.

We compare the performances of the extended Kalman filter and the adaptive filter. Both filters use two-body dynamics, but the adaptive filter estimates a (fictitious) noise covariance matrix, whereas the extended Kalman filter contains no system noise. We use six residuals in the adaptive filter ($N = 6$), $q^0 = 0$, and $C^{-1} = 10^{22}$ on the diagonal. The latter number is relatively easily arrived at after processing six observations. Fresh estimates of q are only computed every sixth observation.

Recall that the true orbit is rotated as described, producing a secular change in ω and Ω. The estimation errors[9] in ω and Ω, for the two filters, are shown in Figs. 8.11 and 8.12. Not shown are the initial transient

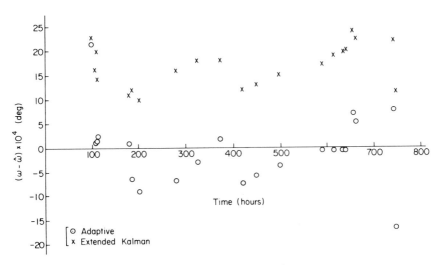

Fig. 8.12. Estimation errors in ω.

errors, similar for both filters, as they are off-scale. We see that, shortly after one revolution, the extended Kalman filter acquires the orbit quite well. Subsequently, however, the extended Kalman estimation errors drift off and are biased. The sense of the errors is precisely such that the extended Kalman filter is tracking the nonrotating (erroneous) orbit. The rms Kalman errors (square roots of the diagonal elements of the

---

[9] The errors plotted are representative; actual errors at each of the 1140 observations are not plotted.

P matrix) are small, the filter gain is small, and the extended Kalman filter is no longer correcting the orbit. In fact, the estimation errors range up to 10 standard deviations. The residuals are inconsistent with their predicted statistics. This is a typical example of filter divergence.

As is seen in Figs. 8.11 and 8.12, the adaptive filter acquires the orbit and tracks the (true) rotating orbit. Estimation errors are small and relatively random. The adaptive filter residuals are always consistent with their predicted statistics. That is, the estimation errors are generally less than one standard deviation, and almost always less than two standard deviations.

The approximate magnitudes of the errors in the other orbital elements at 747 hr are listed in the following tabulation.

	Extended Kalman	Adaptive
$\lvert a - \hat{a} \rvert$	1×10^{-7}	4×10^{-6}
$\lvert e - \hat{e} \rvert$	2×10^{-6}	5×10^{-7}
$\lvert i - \hat{i} \rvert$	1×10^{-3}	4×10^{-4}
$\lvert P - \hat{P} \rvert$	1×10^{-6}	4×10^{-5}
$\lvert \nu - \hat{\nu} \rvert$	2×10^{-5}	5×10^{-6}

We see that, in all the orbital elements but the period (or, equivalently, the semimajor axis), the adaptive filter errors are smaller than the extended Kalman filter errors. Now in the IMP orbit, the period is particularly sensitive to errors. Since many observations are taken in the first 1.3 revolutions, before the model errors cause the filter to diverge, the extended Kalman filter is able to obtain a very good estimate of the period. Subsequent model errors do not affect the period. On the other hand, because of the sensitivity, the uncertainties generated by the adaptive filter somewhat degrade the estimate of the period. This is probably legitimate, since uncertainties in the argument of perigee should degrade the knowledge of the period.

These and other simulations indicate that the adaptive filter is a promising model error compensation technique.

12. LIMITED MEMORY FILTERING

In this section, we present an application of the limited memory filter developed in Section 7.10 to orbit determination in the presence of model errors. We use the limited memory procedure described in the discussion of Theorem 7.10. That is, conditioning of the estimate on old data is discarded in batches of N; the filter memory varies from N to $2N$. The data presented below are taken from Jazwinski [16].

Our dynamical system for the limited memory filtering study is the "rectilinear" orbit problem with dynamics

$$\ddot{x} = -(\mu/x^2) \qquad (x \text{ scalar}).$$

μ is the gravitational constant times earth mass, 19.9094165 earth radii^3/hr^2. In first-order form,

$$\dot{x}_1 = x_2, \qquad \dot{x}_2 = -(\mu/x_1^2),$$

where x_1 is position and x_2 velocity. Our observations consist of position (range) data

$$y_k = (x_1)_k + v_k,$$

with

$$\mathscr{E}\{v_k\} = 0, \qquad \mathscr{E}\{v_k v_l\} = r\delta_{kl}, \qquad r = 1.0 \times 10^{-7} \quad \text{earth radii}^2.$$

The model error is a bias in μ; observations are generated on the basis of the value of μ given in the foregoing, whereas the filter uses the value 19.9244165.

The true trajectory is a "rectilinear ellipse" starting at approximately 8 earth radii at zero time, going to apogee of 34 earth radii at 42 hr, and terminating at 26 earth radii at 70 hr. The range data were generated at the rate of 1 observation every 0.1 hour, but a gap in the data was left between 40 and 60 hours. The data gap will demonstrate how the limited memory filter performs when faced with prediction errors caused by system nonlinearity.

The performances of the extended Kalman filter and the limited memory filter will be compared on the basis of actual estimation errors and the RSS/RMS ratios. $RSS(k)$ is the square root of the time average of the squares of the estimation errors, whereas $RMS(k)$ is the square root of the appropriate diagonal element of the error covariance matrix. That is,

$$RSS(k) = \left\{ k^{-1} \sum_{i=1}^{k} [x_l(i) - \hat{x}_l(i \mid i)]^2 \right\}^{1/2},$$

$$RMS(k) = [P_{ll}(k \mid k)]^{1/2},$$

where $l = 1, 2$ for position and velocity, respectively. Clearly, $RSS(k)/RMS(k)$ should approach one for large k.

Figures 8.13 and 8.14 show how the extended Kalman filter behaves

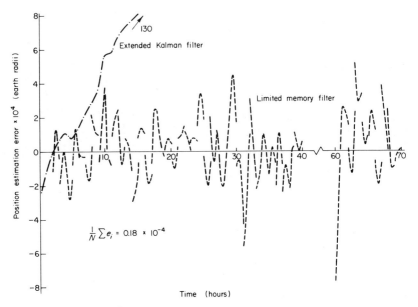

FIG. 8.13.   Position estimation errors.

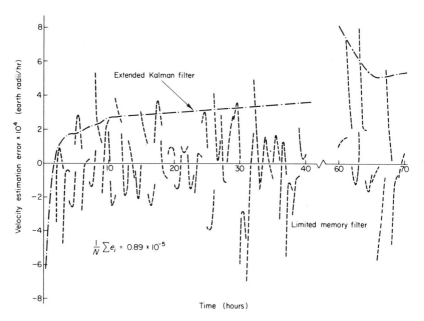

FIG. 8.14.   Velocity estimation errors.

in terms of actual estimation error. The position estimation error, in particular, is seen to grow very rapidly. The velocity estimation error does not grow as much, but is definitely not random. Figures 8.15 and 8.16 show the histories of the RSS/RMS ratios for position and velocity. These are seen to grow almost exponentially in the extended Kalman

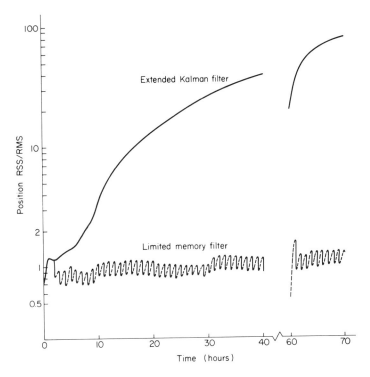

FIG. 8.15.   Position RSS/RMS history

filter. Thus, in the presence of the bias, the Kalman error uncertainties quickly become small since many accurate data have been processed, and then the filter fails to track the orbit.

The limited memory filter performance is shown in the same figures. In these simulations, old data are discarded in batches of 10, and the filter memory varies between 10 and 20. Reference to Figs. 8.13 and 8.14 shows that the estimation errors are rather random, with appropriately small averages. The estimation errors remain within approximately 2 sigma measurement noise level, and the filter is tracking the orbit. More accurate determination of the orbit is not possible in view of the

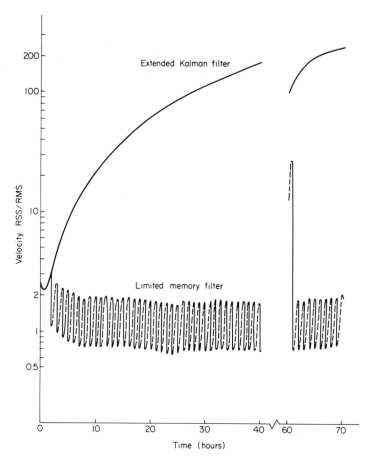

Fig. 8.16. Velocity *RSS/RMS* history.

μ bias. More accurate determination would require the estimation of the bias itself.

Figures 8.15 and 8.16 contain several interesting features. First, it is seen that the limited memory filter covariance matrix is consistent with the estimation errors ($RSS/RMS < 3$), except at 60 hours in Fig. 8.16, which will be discussed later. Note that the RSS/RMS ratios always grow and then decrease the instant old data are discarded. Apparently the filter begins to diverge; the model is not quite accurate enough over time-arcs of 2 hr. Note the unusual RSS/RMS ratios at 60 hr. Apparently, the computation of the covariance matrix over the data gap was inaccurate because of system nonlinearity. Thus at 60 hours the covariance matrix is in error. This error is eliminated, together with old data, at 61 hours. The erroneous covariance matrix is discarded in the same way as is an initial condition (see discussion following Theorem 7.10). The limited memory filter is clearly computationally stable.

These simulations indicate that the limited memory filter is effective in model error compensation. No extensive study of this filter has yet been made, however. In any given estimation problem, it is necessary to determine the best memory length of the filter. The memory length must be, of course, related to the effective errors in the model, as well as the amount of data per unit time which is available. The idea of limited memory filtering actually goes to the heart of the real problem in estimation. For a given required estimation accuracy, our dynamical model must be sufficiently accurate, say over $t_1 \leqslant t \leqslant t_2$, so that the data in that interval can give us the required estimation accuracy. Conversely, given the best model available, adequate over a given time interval, there must be enough data available in that interval to produce the desired estimation accuracy.

13. MISCELLANEOUS TOPICS

There are a number of interesting topics in linear filtering theory which we do not cover in this book. These topics are peripheral to the basic theory but are, nevertheless, important in certain applications. We briefly outline some of these topics here.

Throughout this book, we suppose that a measurement schedule is prespecified for us, and all that remains is to filter the data. In designing a system, however, part of the design is to engineer a measurement subsystem, and to decide what kind of data and how many data will be made available to the filter. Here we run into various tradeoffs, since a measurement system costs money. Various optimization problems thus arise. One such problem is the following. Given a fixed total number of observation times, what is the optimum observation schedule that minimizes *rms* estimation errors? Conversely, given a desired *rms* estimation error, what is the minimum number and schedule of observations? These and related problems have been treated by Denham and Speyer [8] and by Potter and Fagan [26], for example. These authors give additional references on the subject.

A review of the linear filter computational procedure (Section 2) quickly reveals that the necessary filter computations rapidly increase with the dimension (n) of the dynamical system. The larger n is, the larger the computer required for the filter. This has motivated some research on the design of suboptimal filters of dimension less than n. In this category is the work of Johansen on the minimal-order observer, and the partitioning approach of Joseph. This and related work are reviewed by Aoki [2].

The computation of the estimation error covariance matrix P is often plagued with computational error (round-off) problems. These problems are particularly acute when much information is available. For then the covariance matrix may decrease by many orders of magnitude. The typical numerical problem of subtracting small and almost equal numbers arises. This problem can clearly be solved by using higher precision arithmetic. An interesting alternative is to compute the *square root* of P rather than P itself. Recursions for $P^{1/2}$ have been developed by Bellantoni and Dodge [6], and also by Andrews [1]. In essence, one can maintain the same precision of $P^{1/2}$ in single-precision arithmetic as one can maintain using P in double-precision arithmetic. We also note the related work of Potter and Fraser [27] which develops recursions for the determinant of P, which is useful in some error analysis applications.

APPENDIX 8A

ORBIT MECHANICS

The purpose of this appendix is to summarize briefly the mechanics of motion of an orbiting (earth) satellite. An extensive treatment of this subject may be found in [3].

The motion of a satellite of negligible mass in a point-mass gravitational field is described by the second-order differential equations

$$\ddot{R} = -(\mu/r^3)R, \tag{8A.1}$$

where

$$R^T = [x\ y\ z], \qquad r = \|R\| = [R^T R]^{1/2}$$

x, y, and z are the position coordinates of the satellite in an inertial coordinate system centered at the point-mass (center of the earth). μ is the (earth's) gravitational constant. Thus $\dot{R}$ is the velocity vector, and $\ddot{R}$ is the acceleration vector ($v \triangleq \|\dot{R}\|$).

Equations (8A.1) describe the so-called *two-body* motion, and we sometimes refer to the dynamics (8A.1) as two-body dynamics. As is well known, the satellite orbit is an ellipse, as shown in Fig. 8A.1. The point F is the focus of the ellipse containing the point-mass; C is the center of the ellipse or concentric circle of radius a. The ellipse (orbit)

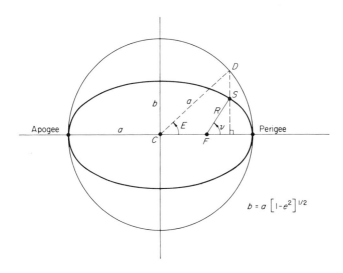

FIG. 8A.1. Satellite orbit plane.

is defined by its *semimajor axis a* and *eccentricity e*. The position of the satellite *S* in the orbit can be defined in terms of one of three angles. The *true anomaly v* is the angle subtended at the focus by the line of apsides (perigee-apogee) and the radius vector *R*. The *eccentric anomaly E* is the angle subtended at the center of the ellipse by the line of apsides and the line *CD*. The *mean anomaly M* is defined by *Kepler's equation*

$$M = E - e \sin E. \tag{8A.2}$$

These angles are measured positive from perigee in the direction of the satellite's motion. The (perigee-to-perigee) *period p* and the *mean motion n* of the satellite are functions of the semimajor axis (and μ) given by

$$p = 2\pi a^{3/2}\mu^{-1/2}, \tag{8A.3}$$

$$n = \mu^{1/2}a^{-3/2}. \tag{8A.4}$$

It remains to describe the orientation of the satellite orbit in inertial space. This is done via three angles as shown in Fig. 8A.2. First, the inertial coordinate system, centered at the earth's center (point-mass), is defined as follows. The *x* axis passes through the vernal equinox, and the (*x, y*) plane is the earth's equatorial plane. The *z* axis passes through the earth's north pole. The *y* axis completes a right-handed coordinate system. The intersection of the orbit plane with earth's equatorial plane is called the *line of nodes* (*FN*). Then, the *inclination i*

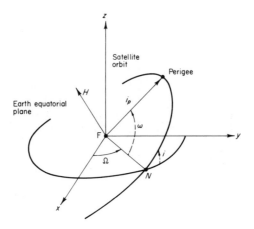

FIG. 8A.2.   Inertial orientation of satellite orbit plane.

is the angle between the orbit plane and earth's equatorial plane. The *argument of perigee* ω, measured in the orbit plane, is the angle between the line of nodes and the line of apsides. Finally, *the right ascension of the ascending node Ω*, measured in the plane of the equator, is the angle between the x-axis and the line of nodes (positive east to the ascending node).

To summarize, the six (independent) orbital elements of the satellite's motion are

$$a \text{ (or } p, \text{ or } n), \qquad i,$$

$$e, \qquad\qquad \omega, \qquad\qquad (8A.5)$$

$$\nu \text{ (or } E, \text{ or } M), \qquad \Omega.$$

They are called the *Keplerian orbital elements* and, at any time t, uniquely define the position (R) and velocity $(\dot{R})$ of the satellite. If we define the angular momentum vector

$$H = R \times \dot{R} \quad \text{(vector cross-product)}, \qquad h = \| H \|, \qquad (8A.6)$$

with components (H_x, H_y, H_z), and the vector i_p from the focus to perigee (see Fig. 8A.2),

$$e i_p = \left\{ \frac{1}{r} - \frac{1}{a} \right\} R - \frac{d}{\mu} \dot{R}, \qquad (d = R^\mathsf{T} \dot{R}), \qquad (8A.7)$$

then the orbital elements are given in terms of position and velocity by

$$a = \left\{ \frac{2}{r} - \frac{v^2}{\mu} \right\}^{-1},$$

$$e \cos E = 1 - r/a \quad [M = E - e \sin E],$$

$$e \sin E = d/(\mu a)^{1/2} \quad [r = a(1 - e^2)/(1 + e \cos \gamma)], \qquad (8A.8)$$

$$\cos i = H_z/h \quad \text{(first, fourth quadrant)},$$

$$\sin \Omega = H_x/h \sin i, \qquad \cos \Omega = -H_y/h \sin i,$$

$$\cos \omega = (i_p)_x \cos \Omega + (i_p)_y \sin \Omega, \qquad \sin \omega = (i_p)_z/\sin i.$$

Now the earth is not a point-mass and, as a result, (8A.1) is an incomplete description of earth satellite motion. Accelerations a_j, caused by the earth's gravitational harmonics, act on the satellite, and therefore

$$\sum a_j$$

must be added to the right-hand side of (8A.1). These accelerations cause both periodic and secular variations in the elements of the satellite's orbit. Earth oblateness, for example, causes a secular variation in the argument of perigee (ω) and in the right ascension of the ascending node (Ω), in addition to producing periodic variations in the other elements. The secular variations in ω and Ω, produced by the earth's second zonal harmonic, are used in [25] to simulate dynamical model errors in orbit determination. We use this device to demonstrate the performance of the adaptive filter in Section 11.

Additional innumerable accelerations act on the earth satellite. These include the gravitational accelerations of the sun, moon, and planets, solar radiation pressure, and drag caused by the earth's atmosphere. It is no wonder that the dynamical modeling problem is an extremely difficult one in earth satellite orbit determination, and that filter divergence is a real problem in this field (see Section 8).

We next summarize the radar observations most commonly used in orbit determination. These are *range* ρ,

$$\rho = [(x - x_s)^2 + (y - y_s)^2 + (z - z_s)^2]^{1/2}, \qquad (8A.9)$$

where (x, y, z) is the satellite position and (x_s, y_s, z_s) is the position of the radar station; *range rate* $\dot{\rho}$,

$$\dot{\rho} = \rho^{-1}[(x - x_s)(\dot{x} + \Omega_E y_s) + (y - y_s)(\dot{y} - \Omega_E x_s) + (z - z_s)\dot{z}], \qquad (8A.10)$$

where Ω_E is the earth's rotation rate; and the l and m *direction cosines.* The m direction cosine is the cosine of the angle subtended at the station between the horizon looking north and the satellite as it pierces the meridian plane containing the station. The l direction cosine is the cosine of the angle subtended at the station between the horizon looking east and the satellite as it pierces the plane of the parallel of latitude containing the station. There are many errors involved in the mathematical models for observations. These include errors in station location, atmospheric refraction that affects the radar signals, and so on. These, of course, compound the modeling problem mentioned earlier.

APPENDIX 8B

REENTRY MECHANICS

The coordinate system commonly used in reentry trajectory analysis is shown in Fig. 8B.1. The reentry vehicle state in three-dimensional reentry can be taken to be

$$x^{\mathrm{T}} = [v, \gamma, \psi, h, \varphi, \theta],$$

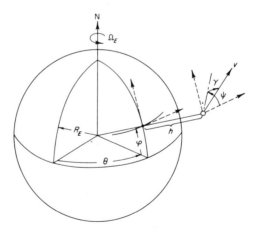

FIG. 8B.1. Reentry coordinate system.

where v is velocity (magnitude), γ is the flight path angle measured from the local horizontal, ψ is the heading angle measured from east toward north in the horizon plane, h is the altitude ($r \triangleq h + R_E$), φ the latitude, and θ the longitude.

In this coordinate system, the motion of the vehicle is described by the differential equations

$$\dot{v} = \frac{F_{xv}}{mv} - \frac{\mu}{r^2} \sin \gamma + \Delta \dot{v}_{\text{rot}},$$

$$\dot{\gamma} = \frac{F_{zv}}{mv} + \frac{v \cos \gamma}{r} - \frac{\mu}{vr^2} \cos \gamma + \Delta \dot{\gamma}_{\text{rot}},$$

$$\dot{\psi} = \frac{F_{yv}}{mv} - \frac{v}{r} \cos \gamma \cos \psi \tan \varphi + \Delta \dot{\psi}_{\text{rot}}, \qquad (8\text{B}.1)$$

$$\dot{h} = v \sin \gamma,$$

$$\dot{\varphi} = \frac{v}{r} \cos \gamma \sin \psi,$$

$$\dot{\theta} = \frac{v}{r \cos \varphi} \cos \gamma \cos \psi.$$

In these equations, μ is the earth's gravitational constant and m the vehicle mass (constant). F_{xv}, F_{yv}, and F_{zv} are the aerodynamic forces; and $\Delta \dot{v}_{\text{rot}}$, $\Delta \dot{\gamma}_{\text{rot}}$, and $\Delta \dot{\psi}_{\text{rot}}$ are due to earth's rotation. Earth oblateness has been neglected in (8B.1). Thus our reentry model corresponds to a *three-dimensional, constant-mass* reentry about a *spherical, rotating* earth. The physical and mathematical details of this reentry model may be found in [39]. We mention here that the aerodynamic forces depend, among other things, on the orientation of the reentry vehicle relative to the wind. In case of a maneuverable lifting reentry vehicle, this orientation can be controlled. In the numerical experiments of Section 7, these controls are specified.

REFERENCES

1. A. Andrews, A Square Root Formulation of the Kalman Covariance Equations, *AIAA J.* **6**, 1165–1166 (1968).
2. M. Aoki, "Optimization of Stochastic Systems." Academic Press, New York, 1967.
3. R. M. L. Baker and M. W. Makemson, "An Introduction to Astrodynamics." Academic Press, New York, 1967.
4. R. H. Battin, A Statistical Optimizing Navigation Procedure for Space Flight, *ARS J.* **32**, 1681–1696 (1962).
5. R. H. Battin *et al.*, Guidance System Operations Plan for Manned CM Earth Orbital Missions. M.I.T., Cambridge, Massachusetts, Report R-547, December 1967.
6. J. F. Bellantoni and K. W. Dodge, A Square Root Formulation of the Kalman-Schmidt Filter, *AIAA J.* **5**, 1309–1314 (1967).
7. W. F. Denham and S. Pines, Sequential Estimation when Measurement Function Nonlinearity Is Comparable to Measurement Error, *AIAA J.* **4**, 1071–1076 (1966).

8. W. F. Denham and J. L. Speyer, Optimal Measurement and Velocity Correction Programs for Midcourse Guidance, *AIAA J.* **2**, 896–907 (1964).

9. A. R. Dennis, Functional Updating and Adaptive Noise Variance Determination in Recursive-type Trajectory Estimators, *Special Projects Branch Astrodynamics Conf. NASA-GSFC, Greenbelt, Maryland,* May 1967.

10. S. L. Fagin, Recursive Linear Regression Theory, Optimal Filter Theory, and Error Analysis of Optimal Systems, *IEEE Intern. Conv. Record* **12**, 216–240 (1964).

11. J. L. Farrell, Simulation of a Minimum Variance Orbital Navigation System, *J. Spacecraft Rockets* **3**, 91–98 (1966).

12. R. J. Fitzgerald, Filtering Horizon-Sensor Measurements for Orbital Navigation, *J. Spacecraft Rockets* **4**, 428–435 (1967).

13. R. J. Fitzgerald, Error Divergence in Optimal Filtering Problems, *Second IFAC Symp. Automatic Control Space, Vienna, Austria,* 1967.

14. T. L. Gunckel, Orbit Determination Using Kalman's Method, *J. Inst. Navigation* **10**, 273–291 (1963).

15. A. H. Jazwinski and A. E. Bailie, Adaptive Filtering, Analytical Mechanics Associates, Inc., Seabrook, Maryland, Interim Report 67-6, 1967.

16. A. H. Jazwinski, Limited Memory Optimal Filtering, *Proc. 1968 Joint Automatic Control Conf., Ann Arbor, Michigan,* June 1968. *IEEE Trans. Automatic Control* **13**, 558–563 (1968).

17. A. H. Jazwinski, Adaptive Filtering, *Proc. IFAC Symp. Multivariable Control Systems, Dusseldorf, Germany,* October 1968, **2**, 1–15.

18. A. H. Jazwinski, Adaptive Filtering in Satellite Orbit Estimation, *Second Asilomar Conference on Circuits and Systems, Pacific Grove, California,* October 1968 (presented paper).

19. R. E. Kalman, New Methods in Wiener Filtering Theory, *Proc. 1st Symp. Eng. Appl. Random Function Theory and Probability* (J. L. Bogdanoff and F. Kozin, eds.), Wiley, New York, 1963.

20. H. J. Kushner, "Stochastic Stability and Control." Academic Press, New York, 1967.

21. G. Lee, R. Falce, and F. Hopper, Interplanetary Trajectory Error Analysis: An Introduction and Survey, Martin-Marietta Corp., Denver Division, Rept. MCR-67-441, 1967.

22. J. D. McLean, S. F. Schmidt, and L. A. McGee, Optimal Filtering and Linear Prediction Applied to a Midcourse Navigation System for the Circumlunar Mission, NASA TN D-1208, 1962.

23. S. Pines, H. Wolf, D. Woolston, and R. Squires, Goddard Minimum Variance Orbit Determination Program, NASA-GSFC Rept. X-640-62-191, October 1962.

24. S. Pines, H. Wolf, A. Bailie, and J. Mohan, Modifications of the Goddard Minimum Variance Program for the Processing of Real Data, Analytical Mechanics Associates, Inc., Westbury, New York, Rept. Contract NAS 5-2535, October 1964.

25. S. Pines, On the Modeling of Real-World Error Sources for Near-Earth Orbits, Analytical Mechanics Associates, Inc., Westbury, New York, Rept. presented at the *7th Semi-Annual Astrodynamics Conference, NASA-GSFC, Greenbelt, Maryland,* April 1968.

26. J. E. Potter and J. H. Fagan, Optimization of Navigation Measurements, unpublished MIT Rept., October 1966.

27. J. E. Potter and D. C. Fraser, A Formula for Updating the Determinant of the Covariance Matrix, *AIAA J.* **5**, 1352–1353 (1967), see also *AIAA J.* **6**, 1439 (1968).

28. H. E. Rauch, Optimum Estimation of Satellite Trajectories Including Random Fluctuations in Drag, *AIAA J.* **3**, 717–722 (1965).

29. F. H. Schlee, C. J. Standish, and N. F. Toda, Divergence in the Kalman Filter, *AIAA J.* **5**, 1114–1120 (1967).

30. S. F. Schmidt, State Space Techniques Applied to the Design of a Space Navigation System, *1962 Joint Automatic Control Conf.*

31. S. F. Schmidt, Application of State-Space Methods to Navigation Problems, *Advan. Control Systems* **3**, 293–340 (1966).

32. S. F. Schmidt, Estimation of State with Acceptable Accuracy Constraints, Analytical Mechanics Associates, Inc., Westbury, New York, Interim Rept. 67-4, 1967.

33. S. F. Schmidt, Compensation for Modeling Errors in Orbit Determination Problems, Analytical Mechanics Associates, Inc., Westbury, New York, Interim Rept. 67-16, 1967.

34. S. F. Schmidt, J. D. Weinburg, and J. S. Lukesh, Case Study of Kalman Filtering in the C-5 Aircraft Navigation Systems, Case Studies in System Control, Workshop Preprints, IEEE-GAC and Univ. of Michigan, Ann Arbor, Michigan, June 1968.

35. G. L. Smith, S. F. Schmidt, and L. A. McGee, Application of Statistical Filter Theory to the Optimal Estimation of Position and Velocity On Board a Circumlunar Vehicle, NASA TR R-135, 1962.

36. G. L. Smith and L. A. McGee, Midcourse Navigation Scheme, *NASA Tech. Conf. on Control, Guidance and Navigation Res. Manned Lunar Missions*, July 1962.

37. G. L. Smith, Multivariable Linear Filter Theory Applied to Space Vehicle Guidance, *SIAM J. Control* **2**, 19–32 (1964).

38. H. W. Sorenson, Kalman Filtering Techniques, *Advan. Control Systems* **3**, 219–300 (1966).

39. W. E. Wagner, Re-Entry Filtering, Prediction and Smoothing, *J. Spacecraft Rockets* **3**, 1321–1327 (1966).

40. R. P. Wishner, J. A. Tabaczynski, and M. Athans, On the Estimation of the State of Noisy Nonlinear Multivariable Systems, *IFAC Symp. Multivariable Control Systems, Dusseldorf, West Germany*, October 1968.

41. H. Wolf, An Assessment of the Work in Minimum Variance Orbit Determination, Analytical Mechanics Associates, Inc., Westbury, New York, Report No. 68-3, January 1968.

9

Approximate Nonlinear Filters

1. INTRODUCTION

In nonlinear filtering problems as well as in the linear ones, we are interested in computing the conditional mean and covariance matrix. The conditional mean is our minimum variance estimate, and the covariance matrix measures the uncertainty in the estimate. Throughout our development of nonlinear filter theory in Chapter 6 (see, for example, the discussion following Theorem 6.2), we observed that, in general, the equations of evolution for the conditional mean and covariance matrix depend on all the moments of the conditional density function. In general, the conditional density cannot be characterized by a finite set of parameters (e.g., moments). A notable exception to this, of course, is the linear filtering problem. In that situation, the conditional density is Gaussian and is, therefore, completely determined by its mean vector and covariance matrix. The mean and covariance matrix may be thought of as the *state* of the linear filter. In the nonlinear situation, the state of the filter is essentially infinite, consisting of the whole conditional density function. If we had the conditional density in hand, we could, of course, compute any moment.

We saw in Chapter 6 that numerical problems in computing the whole conditional density function are, in general, intractable, since they involve solution of partial integro-differential equations (Theorem 6.5) or functional integral difference equations (Theorem 6.1). Thus, we are forced to consider approximations to the conditional density function. We should like to *parametrize* the conditional density via a finite and small set of parameters. Our nonlinear filter would then consist of equations of evolution for these parameters, which would comprise the *state* of our filter. If we could find a set of parameters that completely determine the conditional density, these parameters would be *sufficient statistics*. Unfortunately, it is virtually impossible to find sufficient statistics for nonlinear problems. The linear case is a unique exception.

This chapter is devoted to a study of several approximations for the conditional density function. A number of approximate nonlinear filters are developed. The reader will recall that one such approximation was already used in Section 3.9 and Example 4.21 in obtaining approximate equations of evolution for unconditioned moments. We critically discuss these approximations, noting the effects of nonlinearities in nonlinear systems, and describe results of numerical simulations. The approximate nonlinear filters are also compared with the extended Kalman filter and the local iteration algorithms developed in Chapter 8. Finally, we present simulation results in a simple, but nonlinear, problem for which we can actually compute the conditional density function, and describe certain applications in orbit determination and reentry. The chapter closes with a summary that attempts to evaluate nonlinear filtering work, its value and prospects.

2. APPROXIMATION TECHNIQUES:

PARAMETRIZATION OF DENSITY FUNCTIONS

A parametrization of the density function which immediately suggests itself is an *orthogonal series* expansion. Restricting ourselves, for simplicity, to scalar systems, we might approximate the conditional density by a (finite) series expansion

$$p(x_t \mid Y_t) \approx \sum_{i=1}^{N} c_i \varphi_i(x), \qquad (9.1)$$

where

$$\int \varphi_i(x) \, \varphi_j(x) \, dx = \delta_{ij},$$

so that the coefficients c_i are given by

$$c_i = \int \varphi_i(x) \, p(x \mid Y_t) \, dx.$$

Equations of evolution for the coefficients c_i can be determined from the law of evolution of the conditional density function. The filter then consists of the equations for these coefficients. In view of (9.1), the conditional density can always be reconstructed, and its moments can be computed.

The problem, of course, is to find an appropriate approximating series. An apparently useful series is the *Edgeworth* series (see Cramer [8]). This series, consisting of Hermite polynomials, is useful for nearly Gaussian densities. Developed originally for scalar random variables, it has been extended to the vector case by Kuznetsov *et al.* [20]. The coefficients c_i in this series are called *quasi-moments* by Stratonovich. The Edgeworth series expansion has been used by Sorenson and Stubberud [25] in developing approximate filters for the scalar, discrete system. They had difficulties in developing difference equations for the quasi-moments across an observation (through Bayes' rule) and had to rely on an additional expansion to which the filter was very sensitive. A method of propagating moments through Bayes' rule is described in Section 4, and discussion of Sorenson and Stubberud's results is deferred to Section 6. Quasi-moments have also been used by Fisher [11] in developing an approximate filter for continuous nonlinear systems. His filter is equivalent to one which we develop in Section 3.

A related method of parametrizing the conditional density is *via moments*. Now clearly it is not feasible to develop many moments, as it is not feasible to develop many terms in the expansion (9.1). Thus, it is necessary to truncate the moments or to make some other approximation. In Section 3.9 and Example 4.21, we used the approximation of neglecting third- and higher-order central moments. This sort of approximation is appropriate if the density is (almost) symmetric and concentrated close to the mean (small variance). Another approximation is to suppose that the density is nearly Gaussian. In that case, all the odd central moments can be neglected, and higher-order even moments can be written in terms of the variance. We use these moment approximations to develop nonlinear filters in Sections 3 and 4, and discussion of contributions in this area is deferred to those sections.

Now quasi-moments are closely related to moments. In fact, the first two quasi-moments are nothing more than the mean and variance, and the third quasi-moment is proportional to the third central moment. The fourth quasi-moment q_4 is given by

$$q_4 = (m_4/\sigma^4) - 3, \qquad (9.2)$$

where m_4 is the fourth central moment and σ^2 is the variance. In general, if $q_i = 0$, $i \geqslant 3$, the density is Gaussian. As a consequence, an expansion using Edgeworth-type series and quasi-moments, which is truncated at q_2, is equivalent to our parametrization via moments if it is assumed that the density is nearly Gaussian. Now in problems of any sizable dimension, it is not feasible to carry third and higher order moments or quasi-moments. Therefore, in practice, there is little to be gained in using quasi-moments. For this reason, we will not consider quasi-moments further, but will deal with moment approximation.

An objection to the approximations we have described is that, by neglecting high-order moments or quasi-moments, the approximate density function which results is not really a density function. It can, for instance, assume negative values. This point was raised by Kushner [19], and he suggested computing the expectations in the equations for the conditional mean and covariance (Theorems 6.2 and 6.6) on the basis of some assumed density. We in fact do this in Section 8 for a simple scalar problem. In order to do this in a computationally practical way, we have to find density functions that are characterized by a finite and small set of parameters. Unfortunately, this is not a simple matter. An approach suggested by Kushner is via *moment sequences*. Given two moments, m_1 and m_2 say, find m_3 ,..., m_N so that $\{m_1 ,..., m_N\}$ define a density function. Kushner shows how moment sequences can be found in the scalar case, but there does not appear to be a vector generalization. In the final analysis, Kushner assumes that the density is Gaussian.

Reference to Theorems 6.2 and 6.6 shows that the right-hand sides of the equations for the conditional mean and covariance matrix involve expectations of nonlinear functions [involving the system function $f(x, t)$ and measurement function $h(x, t)$] with respect to the conditional density. Even if a particular form of the conditional density is assumed, these expectations are difficult to evaluate because of the nonlinearity of f and h. In our approximations (Sections 3–5), we expand these functions to second order and neglect all but the quadratic nonlinearity. Swerling [27, 28] suggests an approximation to these functions of the form

$$h_i(x, t) = \sum_j \sum_k c_{ijk}\varphi_j{}^i(x)\, \psi_k{}^i(t). \tag{9.3}$$

Then the integrals over x might be computed in advance. This is a laborious process, but needs to be done only once. Real-time implementation of the filter would only involve computing linear combinations of the precomputed integrals. Sunahara [26] proposes to replace the nonlinear functions f and h by quasi-linear functions via stochastic linearization. Neither of these techniques appears to have been tested yet in nonlinear filtering problems.

3. NONLINEAR FILTER APPROXIMATIONS:

CONTINUOUS FILTER

The exact equations of evolution for the conditional mean and covariance matrix are given in Theorem 6.6. We develop here several approximations to these equations. First we assume that third- and higher-order central moments are negligible. This is appropriate if the conditional density is almost symmetric and concentrated near its mean. The resulting approximate filter will be called the *truncated second-order filter*. In the second approximation, we do not neglect fourth central moments, but, assuming the density is Gaussian, we write them as

$$[(x_i - \hat{x}_i)(x_j - \hat{x}_j)(x_k - \hat{x}_k)(x_l - \hat{x}_l)]^\wedge = P_{jk}P_{il} + P_{il}P_{ik} + P_{kl}P_{ij} , \qquad (9.4)$$

where P is the conditional covariance matrix. Odd central moments and higher-order even central moments are neglected. In this approximation, we carry the system nonlinearities to second order only. This approximate filter will be called the *Gaussian second-order filter*.

In order not to confuse the reader with notational details, we first consider the scalar filter ($n = m = r = 1$). In Eq. (6.100) of Theorem 6.6, we are required to compute the conditional expectation of the function f. We first expand f in Taylor series around the conditional mean

$$f(x, t) = f(\hat{x}, t) + f_x(\hat{x}, t)(x - \hat{x}) + \tfrac{1}{2}f_{xx}(\hat{x}, t)(x - \hat{x})^2 + \cdots . \qquad (9.5)$$

Then, taking the conditional expectation in (9.5), we have

$$\hat{f}(x, t) = f(\hat{x}, t) + \tfrac{1}{2}Pf_{xx}(\hat{x}, t) \qquad (9.6)$$

for both approximations. Similarly, for both approximations,

$$\widehat{xf} - \hat{x}\hat{f} = Pf_x(\hat{x}, t). \qquad (9.7)$$

Similar expressions are obtained for $\hat{h}$ and $(\widehat{xh} - \hat{x}\hat{h})$. Now for the *truncated* second-order filter,

$$\widehat{g^2}Q = g^2(\hat{x}, t)Q + g_x^2(\hat{x}, t)PQ + g(\hat{x}, t)g_{xx}(\hat{x}, t)PQ,$$
$$(\widehat{x^2h} - \widehat{x^2}\hat{h} - 2\hat{x}\widehat{xh} + 2\hat{x}^2\hat{h}) = -\tfrac{1}{2}P^2h_{xx}(\hat{x}, t), \qquad (9.8)$$

whereas for the *Gaussian* second-order filter,

$$\widehat{g^2}Q = g^2(\hat{x}, t)Q + g_x^2(\hat{x}, t)PQ + g(\hat{x}, t)g_{xx}(\hat{x}, t)PQ + \tfrac{3}{4}g_{xx}^2(\hat{x}, t)P^2Q,$$
$$(\widehat{x^2h} - \widehat{x^2}\hat{h} - 2\hat{x}\widehat{xh} + 2\hat{x}^2\hat{h}) = P^2h_{xx}(\hat{x}, t). \qquad (9.9)$$

The differences between (9.8) and (9.9) arise because the fourth partials of g^2 and x^2h contain the second partials g_{xx} and h_{xx}, respectively. These extra terms are retained in the Gaussian second-order filter since fourth moments are not neglected but rather approximated as in (9.4).

Then, in view of Theorem 6.6, the *truncated* second-order filter becomes

$$d\hat{x}_t = [f(\hat{x}_t, t) + \tfrac{1}{2}P_t f_{xx}(\hat{x}_t, t)] \, dt + P_t h_x(\hat{x}_t, t) \, R^{-1}(t)$$

$$\times \, [dz_t - (h(\hat{x}_t, t) + \tfrac{1}{2}P_t h_{xx}(\hat{x}_t, t)) \, dt],$$

$$\hspace{9cm}(9.10)$$

$$dP_t = [2P_t f_x(\hat{x}_t, t) + \widehat{g^2 Q} - P_t h_x(\hat{x}_t, t) \, R^{-1}(t) \, h_x(\hat{x}_t, t) \, P_t] \, dt$$

$$- \tfrac{1}{2}P_t^2 h_{xx}(\hat{x}_t, t) \, R^{-1}(t)[dz_t - (h(\hat{x}_t, t) + \tfrac{1}{2}P_t h_{xx}(\hat{x}_t, t)) \, dt],$$

with $\widehat{g^2 Q}$ given in (9.8). The *Gaussian* second-order filter has the same equation for the conditional mean as the truncated second-order filter, but the variance equation becomes

$$dP_t = [2P_t f_x(\hat{x}_t, t) + \widehat{g^2 Q} - P_t h_x(\hat{x}_t, t) \, R^{-1}(t) \, h_x(\hat{x}_t, t) \, P_t] \, dt$$

$$+ P_t^2 h_{xx}(\hat{x}_t, t) \, R^{-1}(t)[dz_t - (h(\hat{x}_t, t) + \tfrac{1}{2}P_t h_{xx}(\hat{x}_t, t)) \, dt], \quad (9.11)$$

with $\widehat{g^2 Q}$ given in (9.9).

The truncated and Gaussian second-order filters for vector systems are given in Appendix 9A.

The truncated second-order filter was developed by Jazwinski [15, 16], and independently by Bass et al. [2]. Schwartz [22], Jazwinski [16], and Fisher [11] independently developed the Gaussian second-order filter. Schwartz and Stear [23] have pointed out that the equations for these filters do not satisfy the (sufficient) conditions for existence and uniqueness of solutions, as given in Theorem 4.5, for example. They consequently modified the approximating expansions for the functions f, g, and h somewhat, and obtained slightly different filter equations which satisfy the conditions of Theorem 4.5. It can be shown, however, that both their modified approximating expansion and the one we use here (Taylor series) produce the same computational results (see Schwartz [22]).

Let us make a few observations about the truncated and Gaussian second-order filters. We first note that both filters reduce to the Kalman–Bucy filter (Theorem 7.3) when the dynamics and observations are linear. Second, we note that, if the terms containing the second partial derivatives f_{xx} and h_{xx} are dropped, and if the noise coefficient matrix G is a function of time only, then both filters reduce to the extended

Kalman filter (see Example 9.1). In both filters, as well as in the extended Kalman filter, the equations for the conditional mean and covariance matrix are coupled; it is no longer possible to precompute the filter gain, as it is in the linear case.

Example 9.1 (Extended Kalman Filter). The extended Kalman filter for continuous-discrete (or discrete) systems is given in Theorem 8.1. This filter is the result of the application of the linear Kalman–Bucy filter to a linearized nonlinear system, where the nonlinear system is relinearized after each observation (see Section 8.3). This has the obvious extension to continuous systems, and we sketch that extension here. Differentiating (8.30) and (8.31) with respect to t_{k+1} (call it t), and using the properties of the state transition matrix, we get

$$[d\hat{x}_t/dt]_d = f(\hat{x}_t, t),$$
$$[dP_t/dt]_d = FP_t + P_tF^T + GQG^T, \tag{9.12}$$

where the matrix F is defined in (8.16) and in Appendix 9A. The subscript d in (9.12) indicates that these are the rates due to the dynamics. Using the limiting procedure described in Example 4.18 and in Eqs. (8.32) and (8.33), we get the following rates of change due to observations:

$$[d\hat{x}_t/dt]_0 = P_tM^TR^{-1}(y_t - h(\hat{x}_t, t)),$$
$$[dP_t/dt]_0 = -P_tM^TR^{-1}MP_t, \tag{9.13}$$

where the matrix M is defined in (8.23) and in Appendix 9A. Combining (9.12) and (9.13), we get the (continuous) *extended Kalman filter*:

$$d\hat{x}_t/dt = f(\hat{x}_t, t) + P_tM^TR^{-1}(y_t - h(\hat{x}_t, t)),$$
$$dP_t/dt = FP_t + P_tF^T + GQG^T - P_tM^TR^{-1}MP_t. \quad \blacktriangle \tag{9.14}$$

A significant feature of both the truncated and Gaussian second-order filters is the presence of the random forcing term (measurement differential dz_t) in the variance equation. This term is, of course, properly present (see Theorem 6.6). It is disquieting, however, that in the one filter that term enters with a minus sign, and in the other filter with a plus sign. Since the equations for the conditional covariance matrix are not exact, it is possible for the random forcing term to produce a negative variance. This possibility is particularly strong in case the noise inputs to the system are small and the filter operates over long times. In that case, we can learn the state very precisely (small P). These considerations lead one to consider a compromise between the truncated and Gaussian second-order filters, which we shall call the *modified second-order filter*:

$$d\hat{x}_t = [f(\hat{x}_t, t) + \tfrac{1}{2}P_t f_{xx}(\hat{x}_t, t)] dt$$
$$+ P_t h_x(\hat{x}_t, t) R^{-1}(t)[dz_t - (h(\hat{x}_t, t) + \tfrac{1}{2}P_t h_{xx}(\hat{x}_t, t)) dt],$$
$$dP_t/dt = 2P_t f_x(\hat{x}_t, t) + \widehat{g^2 Q} - P_t h_x(\hat{x}_t, t) R^{-1}(t) h_x(\hat{x}_t, t) P_t. \tag{9.15}$$

In the modified second-order filter, the random forcing term in the variance equation is neglected.

We remind the reader that the differential equations describing the approximate nonlinear filters are (Itô) stochastic differential equations. The coefficients of dz_t are in general random and, therefore, care must be taken in simulating these filters on a computer. It is tacitly assumed that the Itô equations describing the dynamical system (6.64), (6.65) model the physical system of interest (see Section 4.8). One way of simulating the filters is first to transform the filter equations to the Stratonovich form (see Sections 4.6 and 4.7) and then to use standard integration techniques. Also see Kushner [19] in this connection.

Parallel to the probabilistic results in nonlinear filtering, a number of authors have looked at the problem from the statistical point of view, as described in Chapter 5. They applied the various techniques described in the examples of Chapter 7 (e.g., Examples 7.11–7.13) to nonlinear systems and obtained approximate nonlinear filters by a linearization of one form or another. For example, Cox [5–7] and others have derived the extended Kalman filter in this fashion. Detchmendy and Sridhar [10] derived an approximate nonlinear filter using the invariant imbedding technique described in Example 7.13, plus a linearization. If the arbitrary weighting functions used by Detchmendy and Sridhar are given a probabilistic interpretation, then their filter is of the form

$$d\hat{x}_t/dt = f(\hat{x}_t, t) + P_t h_x(\hat{x}_t, t) R^{-1}(t)(y_t - h(\hat{x}_t, t)),$$
$$dP_t/dt = 2P_t f_x(\hat{x}_t, t) + g^2(\hat{x}_t, t) Q - P_t h_x(\hat{x}_t, t) R^{-1}(t) h_x(\hat{x}_t, t) P_t \tag{9.16}$$
$$+ P_t^2 h_{xx}(\hat{x}_t, t) R^{-1}(t)(y_t - h(\hat{x}_t, t)).$$

There has been, thus far, little simulation experience with the approximate continuous nonlinear filters we have described (with the exception of the extended Kalman filter, which has been used extensively). Schwartz and Stear [24] report simulations of all these filters (truncated second-order, Gaussian second-order, extended Kalman, modified second-order, Detchmendy and Sridhar's statistical filter, and a few others) in scalar systems, but their results are inconclusive. In essence, according to their simulations, the performance of all these filters is similar, so that no filter can be recommended on the basis of superior performance. The extended Kalman filter, which involves no second partial derivatives, is therefore recommended on the basis of simplicity and smallest computational requirements. In their simulations, Schwartz and Stear unfortunately use very large measurement noise (R), so that measurement nonlinearities are masked by the noise. They use rather small initial errors (P_0), so that the effect of system nonlinearities is reduced (see Section 6). One conclusion that can be drawn from their simulations is that the extended Kalman filter is superior in performance to the linear filter with no relinearization. This fact is, of course, well known (see Section 8.7, for example).

Kushner [19] has simulated a Gaussian-type approximate filter on a second-order system (van der Pol equation) with *linear* measurements. Third moments were neglected, and fourth moments were approximated as in (9.4), but all the system nonlinearities (terms in Taylor series expansion) were carried. (This dynamical system has cubic nonlinearities.) In Kushner's simulations, the measurement noise was large and the initial errors (P_0) were large, so that system nonlinearities were very pronounced. Kushner's filter tracked the system trajectory, whereas the Gaussian second-order filter (in this case equivalent to the truncated second-order and the modified second-order filters) was unstable and diverged. No comparison with other estimators was given. Since linear measurements were used, the variance equation was nonrandom. As a result, possible difficulties due to the presence of a random forcing term in the variance equation could not be assessed.

4. NONLINEAR FILTER APPROXIMATIONS:

CONTINUOUS-DISCRETE FILTER

Using the same approximations as were used in Section 3, we develop in this section approximate filters for the continuous-discrete problem. Comparing the exact equations of evolution for the conditional mean and covariance matrix in continuous and continuous-discrete filtering problems (Theorems 6.6 and 6.2, respectively), we see that approximate equations of evolution *between observations* are obtained by simply setting $R^{-1} \equiv 0$ in the approximate continuous filters of Section 3. Restricting ourselves here to the scalar case (the vector case is treated in Appendix 9B), we have from (9.10) that, between observations,

$$d\hat{x}_t/dt = f(\hat{x}_t, t) + \tfrac{1}{2}P_t f_{xx}(\hat{x}_t, t),$$

$$dP_t/dt = 2P_t f_x(\hat{x}_t, t) + \widehat{g^2}Q,$$

(9.17)

where $\widehat{g^2Q}$ is given in (9.8) for the *truncated* second-order filter, and in (9.9) for the *Gaussian* second-order filter.

One can obtain approximate difference equations for the conditional mean and covariance matrix *at an observation* by similarly approximating the expectations on the right-hand side of Eq. (6.26) in Theorem 6.2. This is, in fact, the approach taken by Jazwinski [13] and by Sorenson and Stubberud [25]. In order to approximate these expectations, we must approximate the exponential function $p(y_k \mid x_{t_k})$ by some sort of series expansion. It turns out (see Jazwinski [14]) that this approach is not very satisfactory; the filter is very sensitive to this approximation. We therefore pursue another approach due to Kramer [18] and further developed by Jazwinski [16].

From Theorem 6.1, we have the difference equation for the conditional density at an observation

$$p(x_{t_k} \mid Y_{t_k}) = \frac{p(y_k \mid x_{t_k})\,p(x_{t_k} \mid Y_{t_{k-1}})}{p(y_k \mid Y_{t_{k-1}})}, \qquad (9.18)$$

which is essentially Bayes' rule. Now

$$p(y_k \mid x_{t_k}) = p(y_k \mid x_{t_k}, Y_{t_{k-1}}),$$

and

$$p(y_k \mid x_{t_k}, Y_{t_{k-1}})\,p(x_{t_k} \mid Y_{t_{k-1}}) = p(x_{t_k}, y_{t_k} \mid Y_{t_{k-1}}).$$

Therefore, (9.18) becomes

$$p(x_{t_k}, y_{t_k} \mid Y_{t_{k-1}}) = p(x_{t_k} \mid Y_{t_k})\,p(y_{t_k} \mid Y_{t_{k-1}}). \qquad (9.19)$$

To determine how moments propagate across an observation, we multiply (9.19) by the product of the two functions $s(x_{t_k})\,u(y_{t_k})$ and integrate over x and y:

$$\iint s(x)\,u(y)\,p(x, y \mid Y_{t_{k-1}})\,dx\,dy = \iint s(x)\,u(y)\,p(x \mid Y_{t_k})\,p(y \mid Y_{t_{k-1}})\,dx\,dy. \qquad (9.20)$$

The right-hand side of (9.20) can be written

$$\iint s(x)\,u(y)\,p(x \mid Y_{t_k})\,p(y \mid Y_{t_{k-1}})\,dx\,dy$$

$$= \int \left[\int s(x)\,p(x \mid Y_{t_k})\,dx\right] u(y)\,p(y \mid Y_{t_{k-1}})\,dy,$$

and, combining this with (9.20), we have

$$\mathscr{E}\{s(x_{t_k})\, u(y_{t_k}) \mid Y_{t_{k-1}}\} = \mathscr{E}\{\mathscr{E}\{s(x_{t_k}) \mid Y_{t_k}\}\, u(y_{t_k}) \mid Y_{t_{k-1}}\}. \qquad (9.21)$$

Equation (9.21) provides the fundamental relationship we are after. The moments we want to compute are of the form

$$\mathscr{E}\{s(x_{t_k}) \mid Y_{t_k}\},$$

so that (9.21) gives us a relationship between (any) moments prior to an observation and after an observation. We use (9.21) in the following way. Suppose we want to compute the conditional mean

$$\hat{x}(t_k \mid t_k) \equiv \mathscr{E}\{x_{t_k} \mid Y_{t_k}\}.$$

We observe that $\hat{x}(t_k \mid t_k)$ is a function of y_{t_k}. If we assume that $\hat{x}(t_k \mid t_k)$ is a power series in y_{t_k}, then (9.21) can be used to determine the coefficients in that power series. In this way, we can obtain approximate difference equations for any moment at an observation. The use of (9.21) is demonstrated in the following example.

Example 9.2 (Kalman Filter Rederived). Suppose observations are linear,

$$y_{t_k} = M(k)\, x_{t_k} + v_{t_k},$$

and assume that the conditional mean and covariance matrix are of the form

$$\hat{x}(t_k \mid t_k) = a + B(y_{t_k} - \hat{y}_{t_k}^{t_{k-1}}),$$
$$P(t_k \mid t_k) = C, \qquad (9.22)$$

where

$$\hat{y}_{t_k}^{t_{k-1}} \equiv \mathscr{E}\{y_{t_k} \mid Y_{t_{k-1}}\},$$

and a, B, C are, respectively, $n \times 1$, $n \times m$, $n \times n$ matrices to be determined. We shall rederive the Kalman filter [(7.29) of Theorem 7.2] using (9.21). That is, we shall use (9.21) to determine the coefficients a, B, and C.

Setting $s(x_{t_k}) = x_{t_k}$ and $u(y_{t_k}) = 1$ in (9.21), we have

$$\hat{x}(t_k \mid t_{k-1}) = \mathscr{E}\{\hat{x}(t_k \mid t_k) \mid Y_{t_{k-1}}\}. \qquad (9.23)$$

With

$$s = x_{t_k}, \qquad u = (y_{t_k} - \hat{y}_{t_k}^{t_{k-1}})^{\mathrm{T}},$$

(9.21) becomes

$$\mathscr{E}\{x_{t_k}(y_{t_k} - \hat{y}_{t_k}^{t_{k-1}})^{\mathrm{T}} \mid Y_{t_{k-1}}\} = \mathscr{E}\{\hat{x}(t_k \mid t_k)(y_{t_k} - \hat{y}_{t_k}^{t_{k-1}})^{\mathrm{T}} \mid Y_{t_{k-1}}\}, \qquad (9.24)$$

and, with $s = [x_{t_k} - \hat{x}(t_k \mid t_k)][x_{t_k} - \hat{x}(t_k \mid t_k)]^{\mathrm{T}}$, $u = 1$, (9.21) is

$$\mathscr{E}\{[x_{t_k} - \hat{x}(t_k \mid t_k)][x_{t_k} - \hat{x}(t_k \mid t_k)]^{\mathrm{T}} \mid Y_{t_{k-1}}\} = \mathscr{E}\{P(t_k \mid t_k) \mid Y_{t_{k-1}}\}. \qquad (9.25)$$

Now, substituting (9.22) in (9.23)–(9.25), we get

$$\hat{x}(t_k \mid t_{k-1}) = a, \qquad (9.26)$$

$$\mathscr{E}\{x_{t_k}(y_{t_k} - \hat{y}_{t_k}^{t_{k-1}})^{\mathrm{T}} \mid Y_{t_{k-1}}\} = \mathscr{E}\{[a + B(y_{t_k} - \hat{y}_{t_k}^{t_{k-1}})][y_{t_k} - \hat{y}_{t_k}^{t_{k-1}}]^{\mathrm{T}} \mid Y_{t_{k-1}}\},$$
$$(9.27)$$

$$\mathscr{E}\{[x_{t_k} - a - B(y_{t_k} - \hat{y}_{t_k}^{t_{k-1}})][x_{t_k} - a - B(y_{t_k} - \hat{y}_{t_k}^{t_{k-1}})]^{\mathrm{T}} \mid Y_{t_{k-1}}\} = C.$$
$$(9.28)$$

Thus we have already determined the vector a [in (9.26)]. Since

$$y_{t_k} - \hat{y}_{t_k}^{t_{k-1}} = M(k)[x_{t_k} - \hat{x}(t_k \mid t_{k-1})] + v_{t_k},$$

(9.27) becomes

$$P(t_k \mid t_{k-1})\, M^{\mathrm{T}}(k) = B[M(k)\, P(t_k \mid t_{k-1})\, M^{\mathrm{T}}(k) + R(k)]. \qquad (9.29)$$

Using (9.26) in (9.28), we get

$$P(t_k \mid t_{k-1}) - P(t_k \mid t_{k-1})\, M^{\mathrm{T}}(k)\, B^{\mathrm{T}} - BM(k)\, P(t_k \mid t_{k-1})$$
$$+ B[M(k)\, P(t_k \mid t_{k-1})\, M^{\mathrm{T}}(k) + R(k)]\, B^{\mathrm{T}} = C. \qquad (9.30)$$

Finally, substituting for a, B, and C from (9.26), (9.29), and (9.30), our filter (9.22) becomes

$$\hat{x}(t_k \mid t_k) = \hat{x}(t_k \mid t_{k-1}) + P(t_k \mid t_{k-1})\, M^{\mathrm{T}}(k)$$
$$\times [M(k)\, P(t_k \mid t_{k-1})\, M^{\mathrm{T}}(k) + R(k)]^{-1}$$
$$\times [y_{t_k} - M(k)\, \hat{x}(t_k \mid t_{k-1})],$$
$$P(t_k \mid t_k) = P(t_k \mid t_{k-1}) - P(t_k \mid t_{k-1})\, M^{\mathrm{T}}(k) \qquad (9.31)$$
$$\times [M(k)P(t_k \mid t_{k-1})\, M^{\mathrm{T}}(k) + R(k)]^{-1}$$
$$\times M(k)\, P(t_k \mid t_{k-1}),$$

which is the Kalman filter.   ▲

In developing approximate equations for the conditional mean and covariance at an observation, we assume the power series

$$\hat{x}(t_k \mid t_k) = \sum_{i=0}^{N} a_i (y_{t_k} - \hat{y}_{t_k}^{t_{k-1}})^i,$$

$$P(t_k \mid t_k) = \sum_{i=0}^{M} b_i (y_{t_k} - \hat{y}_{t_k}^{t_{k-1}})^i,$$

(9.32)

and use (9.21) to evaluate the coefficients in these series. (We restrict ourselves to the scalar case in this section; the vector case is treated in Appendix 9B.) Using Example 9.2 as a guide, it is easy to see that the following choice of the functions s and u in (9.21) will determine the coefficients:

$$s = x_{t_k}, \qquad u = (y_{t_k} - \hat{y}_{t_k}^{t_{k-1}})^i$$

will determine a_i, $i = 0, 1, ..., N$;

$$s = (x_{t_k} - \hat{x}(t_k \mid t_k))^2, \qquad u = (y_{t_k} - \hat{y}_{t_k}^{t_{k-1}})^i$$

will determine b_i, $i = 0, 1, ..., M$. If we were interested in the coefficients of the expansion for the third central moment, we would use

$$s = (x_{t_k} - \hat{x}(t_k \mid t_k))^3, \qquad u = (y_{t_k} - \hat{y}_{t_k}^{t_{k-1}})^i,$$

and so on. The extension to the vector case is straightforward, although the notation becomes rather messy.

For practical (computational) reasons, we carry the series in (9.32) to first order only and assume the forms

$$\hat{x}(t_k \mid t_k) = a + B(y_{t_k} - \hat{y}_{t_k}^{t_{k-1}}),$$

$$P(t_k \mid t_k) = C + D(y_{t_k} - \hat{y}_{t_k}^{t_{k-1}}).$$

(9.33)

Then from (9.21) we have for the coefficients

$a = \hat{x},$

$B = (\widehat{xh} - \hat{x}\hat{h})[[(h - \hat{h})^2]^\wedge + R]^{-1},$

$C = P(t_k \mid t_{k-1}) - (\widehat{xh} - \hat{x}\hat{h})[[(h - \hat{h})^2]^\wedge + R]^{-1}(\widehat{hx} - \hat{h}\hat{x}),$ (9.34)

$D = \{[(x - \hat{x})^2 h]^\wedge - 2B[(h - \hat{h})(x - \hat{x})h]^\wedge + B^2[[(h - \hat{h})^2 h]^\wedge + R\hat{h}] - C\hat{h}\}$

$\qquad \times \{[(h - \hat{h})^2]^\wedge + R\}^{-1},$

where, to save space, we dropped the subscript t_k on x. The measurement function h is evaluated at x_{t_k}, R is $R(k)$, and the caret denotes the expectation using $p(x_{t_k} \mid Y_{t_{k-1}})$, that is,

$$\hat{\varphi} = [\varphi]^{\wedge} = \mathscr{E}\{\varphi(x_{t_k}) \mid Y_{t_{k-1}}\}.$$

We now approximate the expectations appearing in (9.34) as we did for the continuous filter in Section 3. For both the truncated and Gaussian second-order filters [see (9.7)],

$$\widehat{xh} - \hat{x}\hat{h} = P(t_k \mid t_{k-1}) h_x(\hat{x}, t_k). \tag{9.35}$$

For the *truncated* second-order filter,

$$[(h - \hat{h})^2]^{\wedge} = h_x(\hat{x}, t_k) P(t_k \mid t_{k-1}) h_x(\hat{x}, t_k) - \tfrac{1}{4}P^2(t_k \mid t_{k-1}) h_{xx}^2(\hat{x}, t_k), \tag{9.36}$$

$$\begin{aligned} D = -\tfrac{1}{2}P^2(t_k \mid t_{k-1}) h_{xx}[[(h - \hat{h})^2]^{\wedge} + R]^{-1} \\ \times \{1 - 3P(t_k \mid t_{k-1}) h_x{}^2[[(h - \hat{h})^2]^{\wedge} + R]^{-1} \\ - P(t_k \mid t_{k-1}) h_x{}^2[[(h - \hat{h})^2]^{\wedge} + R]^{-2}[R + \tfrac{1}{4}P^2(t_k \mid t_{k-1}) h_{xx}^2] \\ + 2P^2(t_k \mid t_{k-1}) h_x{}^4[[(h - \hat{h})^2]^{\wedge} + R]^{-2}\}, \end{aligned} \tag{9.37}$$

whereas for the *Gaussian* second-order filter,

$$[(h - \hat{h})^2]^{\wedge} = h_x(\hat{x}, t_k) P(t_k \mid t_{k-1}) h_x(\hat{x}, t_k) + \tfrac{1}{2}P^2(t_k \mid t_{k-1}) h_{xx}^2(\hat{x}, t_k), \tag{9.38}$$

$$\begin{aligned} D = P^2(t_k \mid t_{k-1}) h_{xx}[[(h - \hat{h})^2]^{\wedge} + R]^{-1} \\ \times \{1 - \tfrac{9}{2}P(t_k \mid t_{k-1}) h_x{}^2[[(h - \hat{h})^2]^{\wedge} + R]^{-1} \\ + \tfrac{1}{2}P(t_k \mid t_{k-1}) h_x{}^2[[(h - \hat{h})^2]^{\wedge} + R]^{-2}[R + 7P(t_k \mid t_{k-1}) h_x{}^2]\}. \end{aligned} \tag{9.39}$$

Of course, in both filters,

$$\hat{y}_{t_k}^{t_{k-1}} = h(\hat{x}, t_k) + \tfrac{1}{2}P(t_k \mid t_{k-1}) h_{xx}(\hat{x}, t_k). \tag{9.40}$$

Let

$$Y^t \triangleq h_x P(t_k \mid t_{k-1}) h_x + R(k) - \tfrac{1}{4}P^2(t_k \mid t_{k-1}) h_{xx}^2, \tag{9.41}$$

$$Y^G \triangleq h_x P(t_k \mid t_{k-1}) h_x + R(k) + \tfrac{1}{2}P^2(t_k \mid t_{k-1}) h_{xx}^2, \tag{9.42}$$

where h, h_x, and h_{xx} are evaluated at $(\hat{x}(t_k \mid t_{k-1}), t_k)$. Then the *truncated* second-order filter (at an observation) is given by

$$\begin{aligned} \hat{x}(t_k \mid t_k) = \hat{x}(t_k \mid t_{k-1}) + P(t_k \mid t_{k-1}) h_x[Y^t]^{-1}[y_{t_k} - h - \tfrac{1}{2}P(t_k \mid t_{k-1}) h_{xx}], \\ P(t_k \mid t_k) = P(t_k \mid t_{k-1}) - P(t_k \mid t_{k-1}) h_x[Y^t]^{-1} h_x P(t_k \mid t_{k-1}) \\ + D[y_{t_k} - h - \tfrac{1}{2}P(t_k \mid t_{k-1}) h_{xx}], \end{aligned} \tag{9.43}$$

where D is given in (9.37). The *Gaussian* second-order filter is given by

$$\hat{x}(t_k \mid t_k) = \hat{x}(t_k \mid t_{k-1}) + P(t_k \mid t_{k-1})\, h_x[Y^G]^{-1}\, [y_{t_k} - h - \tfrac{1}{2}P(t_k \mid t_{k-1})\, h_{xx}],$$

$$P(t_k \mid t_k) = P(t_k \mid t_{k-1}) - P(t_k \mid t_{k-1})\, h_x[Y^G]^{-1}\, h_x P(t_k \mid t_{k-1}) \qquad (9.44)$$

$$+ D[y_{t_k} - h - \tfrac{1}{2}P(t_k \mid t_{k-1})\, h_{xx}],$$

with D given in (9.39). We also define the *modified truncated* and *modified Gaussian* second-order filters, which consist of (9.43) and (9.44), respectively, but with $D \equiv 0$. All these filters of course use (9.17) between observations.

We observe that all the approximate nonlinear filters reduce to the Kalman filter (Theorem 7.1) when the dynamics and observations are linear. When the second partial derivatives f_{xx} and h_{xx} are dropped, they reduce to the extended Kalman filter. Using the limiting procedure introduced in Example 4.18, the reader may verify that these approximate continuous-discrete filters limit to their continuous counterparts developed in Section 3.

The remarks made in Section 3 in regard to continuous approximate filters apply here as well, with the following exception. Because observations are discrete, these filters do not involve stochastic differential equations, and there is therefore no problem in simulating them. Some simulations are described in Sections 8–10. Since the Gaussian approximation has a wider range of validity than the truncated approximation (see the first paragraph of Section 3), we do not use the truncated second-order filters in the simulations, but simulate only the Gaussian second-order filters.

We note that the modified Gaussian second-order filter was developed independently by Athans *et al.* [1].

5. NONLINEAR FILTER APPROXIMATIONS:

DISCRETE FILTER

We do not develop the details of approximate discrete filters. However, all the ingredients have essentially been derived, and the reader can easily complete the details himself. Equations of evolution for the mean and variance (between observations) of scalar discrete systems are given in Section 3.9. Approximate difference equations at an observation, given in the preceding section, are of course directly applicable to the discrete filter.

Approximations for the nonlinear discrete filtering problem have been developed by Friedland and Bernstein [12] using statistical methods (Section 5.3). We have already mentioned in Section 4 the work of Sorenson and Stubberud [25]. However, we do not pursue those methods in this book.

6. DISCUSSION OF NONLINEAR EFFECTS

Inspection of the approximate nonlinear filters reveals that the size of nonlinearities depends not only on the size of the second partial derivatives (f_{xx} and h_{xx}), but also on the estimation error variance. The second-order terms appear as

$$Ph_{xx}$$

in the case of measurement nonlinearity, and as

$$Pf_{xx}$$

in the case of system nonlinearity. These are the *expected values* of the second-order terms in the Taylor series expansions. As a consequence, these nonlinear terms can be large because the second partial derivatives are large, or because the estimation error variance is large, or both. We cannot be definitive as to what we mean here by "large" in an absolute sense, but we shall subsequently have something to say about the *significance* of nonlinear effects.

It is important to distinguish between the several ways in which these nonlinear terms can be large. If the nonlinearity is large because the second partial derivatives are large, we call it a *real* nonlinearity. If it is large because the estimation error variance is large, we call it an *induced* nonlinearity. If it is large because the second partials and the error variance are both large, we call it a *mixed* nonlinearity. We note that a nonlinearity can always be induced by setting the initial estimation error variance P_0 sufficiently large.

Now, we expect our approximate filters to be useful and effective when nonlinearities are of the *real* type, that is, when nonlinear effects are significant because the second partials are large while error variances are small. For if the error variance P is small, then P^2, P^4, etc., are smaller still, and the terms that we dropped in our approximations are small relative to the terms we have retained.[1] In case the nonlinearity is induced or mixed, the terms we have neglected can be significant relative to the terms we have retained in these approximations.

---

[1] To be rigorous, we should be talking here about the magnitude of the remainder in a Taylor series. See footnote 17 in Chapter 3.

Regardless of whether nonlinearities are real, induced, or mixed, it is important to determine when they are *significant*. Denham and Pines [9], Jazwinski [14], Carney and Goldwyn [4], and Sorenson and Stubberud [25] have observed experimentally that measurement non-linearities become significant when the measurement noise variance R is small while the estimation error variance P is relatively large. Sorenson and Stubberud [25] have also observed that system nonlinearities become significant when the system noise variance Q is small. In general, nonlinear effects appear to be significant when noise inputs are small while the estimation error variance is relatively large. This appears to be rather intuitively obvious, perhaps because of hindsight. Large noise inputs can effectively "cover" neglected nonlinearities.

To assess qualitatively the significance of system nonlinearities, consider Eqs. (9.17), which describe the approximate evolution of the conditional mean and variance between observations. If Q is sufficiently large, then P cannot become very small as a result of information contained in the observations, since we are constantly adding noise to the system. As long as P is large, a small error or bias in the mean $\hat{x}$ is not very significant. [Such a bias in $\hat{x}$ would be produced by neglecting the nonlinear term in the first of Eqs. (9.17).] Now if Q is small and P is initially large but subsequently becomes small, even a small bias in $\hat{x}$ can be significant. The approximate mean can easily fall outside the 3σ limits. Of course if P remains large, significant errors can be made in prediction over long time intervals because of the integrated effect of neglected nonlinearities. Neglected system nonlinearities have a biasing effect on the estimate.

Let us now consider the measurement nonlinearities. We shall look specifically at approximate continuous-discrete filters.

It is difficult to assess the importance of the random forcing term in the variance equation [term with coefficient D in (9.43) or (9.44)], except to observe that its presence can lead to serious difficulties. As was noted in Section 3, that term can produce a negative error variance P. It is easy to see that this random forcing term is significant if $Ph_{xx}/Y^{1/2}$ is comparable to, or larger than, Ph_x^2/Y. Since simulations (Section 8) indeed show that this term can produce negative error variances, we shall restrict our attention to the modified second-order filters.

The forcing term $\frac{1}{2}Ph_{xx}$ in the equation for the conditional mean [in (9.43) or (9.44)] is clearly a (nonlinearity) bias correction term. It ensures that

$$\mathscr{E}\{y_{t_k} - \hat{y}_{t_k}^{t_{k-1}}\} = 0$$

to second order. The significance of this bias correction can be assessed by looking at the variance of

$$(y_{t_k} - \hat{y}_{t_k}^{t_{k-1}}),$$

which is Y^l or Y^G, depending on the approximation used. This will also enable us to assess the significance of nonlinearities in the filter gain itself, since the gain is inversely proportional to Y.

Looking at Eq. (9.41) or (9.42), we see that measurement nonlinearities are significant if

$$P^2 h_{xx}^2 \gtrsim R. \tag{9.45}$$

This observation was also made independently by Denham and Pines [9] and substantiates the experimental findings cited earlier. Measurement nonlinearities are significant if they are comparable to, or larger than, the measurement noise. Significant neglected measurement nonlinearities bias the estimate and result in an incorrect weighting of observations.

Some of the qualitative effects of nonlinearities described in this section will be demonstrated in Section 8 via simulations.

7. LOCAL ITERATIONS

In Section 8.3, we derived two local iteration schemes which, we claim, reduce the (adverse) effect of nonlinearities on the performance of the linear filter applied to a linearized nonlinear system. The iterated extended Kalman filter (Theorem 8.2) reduces the effect of measurement nonlinearities, whereas the iterated linear filter-smoother (Theorem 8.3) reduces the effect of system nonlinearities as well. These iterators were developed as relatively straightforward extensions of the linear theory.

That these iterators are indeed effective in nonlinear problems has been demonstrated experimentally by Denham and Pines [9] and by Wishner *et al.* [29]. We report some simulation experience with these iterators in Sections 8 and 10.

In the present section, we show that these iterators also have a probabilistic interpretation. They are related to maximum likelihood (Bayesian) estimation (see Section 5.3 for definitions of maximum likelihood estimates).

In Example 6.2, we derived the equations for the conditional mode at an observation, assuming, for simplicity of notation, that observations are scalar ($m = 1$). By the conditional mode, we mean the mode of

$$p(x_{t_k} \mid Y_{t_k}).$$

Assuming that $p(x_{t_k} \mid Y_{t_{k-1}})$ is Gaussian with mean $\hat{x}(t_k \mid t_{k-1})$ and covariance matrix $\hat{P}(t_k \mid t_{k-1})$, the mode of $p(x_{t_k} \mid Y_{t_k})$, call it m_{t_k}, satisfies [from (6.52)]

$$\varphi_x(m_{t_k}) \triangleq -M^{\mathrm{T}}(m_{t_k})\, R^{-1}(k)[y_{t_k} - h(m_{t_k}, t_k)] + P^{-1}(t_k \mid t_{k-1})$$
$$\times\, [m_{t_k} - \hat{x}(t_k \mid t_{k-1})] = 0. \tag{9.46}$$

The matrix of second partial derivatives φ_{xx} is [from (6.53)]

$$\varphi_{xx}(m_{t_k}) = P^{-1}(t_k \mid t_{k-1}) + M^{\mathrm{T}}(m_{t_k})\, R^{-1}(k)\, M(m_{t_k})$$
$$- h_{xx}(m_{t_k})\, R^{-1}(k)[y_{t_k} - h(m_{t_k}, t_k)], \tag{9.47}$$

where

$$h_{xx} = \left[\frac{\partial^2 h}{\partial x_i\, \partial x_j}\right] \qquad (n \times n).$$

Suppose we solve (9.46) for the mode using Newton–Raphson iteration.[2] Then our iterator would be

$$\bar{\eta}_{i+1} = \bar{\eta}_i - \varphi_{xx}^{-1}(\bar{\eta}_i)\, \varphi_x(\bar{\eta}_i). \tag{9.48}$$

This iteration involves the computation of the matrix of second partials h_{xx}. Kelley and Denham [17] recommend the Davidon optimization method for computing the mode (maximum likelihood estimate). The Davidon method does not require the second partial derivatives φ_{xx}, yet possesses desirable convergence properties similar to Newton–Raphson iteration.

Let us modify the Newton–Raphson iteration (9.48) by setting $h_{xx} \equiv 0$ in φ_{xx} of (9.47). Then iteration (9.48) becomes

$$\eta_{i+1} = \eta_i + [P^{-1}(t_k \mid t_{k-1}) + M^{\mathrm{T}}(\eta_i)\, R^{-1}(k)\, M(\eta_i)]^{-1}$$
$$\times\, [M^{\mathrm{T}}(\eta_i)\, R^{-1}(k)\, (y_{t_k} - h(\eta_i, t_k))$$
$$+ P^{-1}(t_k \mid t_{k-1})(\hat{x}(t_k \mid t_{k-1}) - \eta_i)]. \tag{9.49}$$

Applying the matrix equalities given in Appendix 7B to the right-hand side of (9.49), we get

$$\eta_{i+1} = \hat{x}(t_k \mid t_{k-1}) + P(t_k \mid t_{k-1})\, M^{\mathrm{T}}(\eta_i)$$
$$\times\, [M(\eta_i)\, P(t_k \mid t_{k-1})\, M^{\mathrm{T}}(\eta_i) + R(k)]^{-1}$$
$$\times\, [y_{t_k} - h(\eta_i, t_k) + M(\eta_i)(\eta_i - \hat{x}(t_k \mid t_{k-1}))]. \tag{9.50}$$

---

[2] The reader unfamiliar with this iteration can consult any textbook on numerical analysis.

Comparing this with Theorem 8.2, we see that we have derived the iterator of the iterated extended Kalman filter. Iteration (9.50) [or (9.49)] will clearly converge (locally) to the mode as long as

$$P^{-1}(t_k \mid t_{k-1}) + M^T(\eta_i) R^{-1}(k) M(\eta_i) > 0.$$

Therefore, the iterated extended Kalman filter consists of the following. Between observations, the conditional mean and covariance matrix propagate according to first-order (not linear) theory [(8.30) and (8.31)]. At an observation, assuming the prior density is Gaussian, this filter solves for the conditional mode of the posterior density. The conditional covariance matrix is then computed according to first-order theory [(8.33)]. The conditional mode is then used for the conditional mean.

Let the reader show that the iterated linear filter-smoother (Theorem 8.3) can be similarly derived by solving for the mode of the conditional density

$$p(x_{t_k}, x_{t_{k-1}} \mid Y_{t_k}).$$

In the two iterated filters just described, the mode of the posterior density is used for the conditional mean. Rauch [21] describes an approximate method of obtaining the mean from the mode by a local expansion in the vicinity of the mode. This expansion is valid when the mean is close to the mode, and the correction is of second order. Breakwell [3] has pointed out that these local iterations produce a *biased* estimate. This is clearly so, since the mode is not equivalent to the mean in nonlinear problems. However, as the error variance becomes small, so does the bias in the estimate. Recall (Section 6) that nonlinearities are proportional to P.

8. A NONLINEAR EXAMPLE

We describe here the results of numerical simulations of some of the approximate nonlinear filters we have developed. Our continuous dynamical system is scalar, but nonlinear, and nonlinear observations are taken at discrete time instants. The system is sufficiently simple so that the conditional probability density function can be computed, thus making it possible to obtain a true evaluation of approximate filter performance. Similar simulations have been previously reported by this author [14].

Our dynamical system consists of the scalar Ricatti equation

$$dx_t/dt = ax_t + bx_t^2 \qquad (a, b \text{ const}), \qquad t \geqslant t_0, \qquad (9.51)$$

and the discrete observations

$$y_{t_k} = x_{t_k}^2 + x_{t_k}^3 + v_k \qquad [h(x) = x^2 + x^3]. \tag{9.52}$$

Since (9.51) is noise-free, we will be able to map the conditional density between observations with the aid of Theorem 2.7. The Ricatti equation (9.51) has the solution

$$\varphi(t, t_0, x_{t_0}) = \frac{a x_{t_0} \, e^{a(t-t_0)}}{a + b x_{t_0}[1 - e^{a(t-t_0)}]}. \tag{9.53}$$

Consider the function $\varphi(t_{k+1}, t_k, \cdot)$, and let $\psi(t_{k+1}, t_k, \cdot)$ be the inverse function. Clearly,

$$\psi(t_{k+1}, t_k, x) = \frac{a x \, e^{a(t_k - t_{k+1})}}{a + b x[1 - e^{a(t_k - t_{k+1})}]}. \tag{9.54}$$

Then, according to Theorem 2.7,

$$p(x, t_{k+1} ; Y_{t_k}) = p(\psi(t_{k+1}, t_k, x), t_k ; Y_{t_k})(\partial \psi / \partial x), \tag{9.55}$$

where we used the notation defined in Sect. 3.2,

$$p(x, t_l ; Y_{t_k}) = p(x_{t_l} \mid Y_{t_k}).$$

The difference equation for the conditional density at an observation is of course Eq. (6.14) of Theorem 6.1.

Having described the evolution of the conditional density for our filtering problem, we define the *exact* nonlinear filter. In the exact filter, the conditional mean and variance are computed from the conditional density. That is, the expectations in (6.26) of Theorem 6.2 are computed by numerical quadratures. The result is the *exact* conditional mean and *exact* conditional variance. We note that it is not feasible to compute the exact mean and variance in all but scalar problems, or perhaps two-dimensional problems. We of course have in mind real-time filtering applications when we speak of computational feasibility.

In order to assess the effect of the Gaussian approximation, we introduce what we shall call the *Gaussian* filter. We simulate this filter in static problems only ($\dot{x}_t \equiv 0$). Assuming that the prior density $p(x_{t_k} \mid Y_{t_{k-1}})$ is Gaussian with mean $\hat{x}(t_k \mid t_{k-1})$ and variance $P(t_k \mid t_{k-1})$, this filter computes $\hat{x}(t_k \mid t_k)$ and $P(t_k \mid t_k)$ via Eq. (6.26) of Theorem 6.2.

Simulations of the Gaussian and truncated second-order filters quickly showed that our concern about the presence of a random forcing term in the variance equation was indeed justified. These filters frequently produced negative variances, and we were forced to abandon them.

We present here some simulations of the exact, Gaussian, iterated extended Kalman, modified Gaussian second-order, and extended Kalman filters. Since the exact conditional mean and variance are available (exact filter), we are able to give a true assessment of approximate filter performance. We emphasize that hundreds of simulations were made in the course of this study, with different (random) initial conditions and noise samples. The simulations presented here are representative.

To assess the effect of measurement nonlinearities, we first present simulations for the static problem [$a = b = 0$ in (9.51)]. With $x_{t_0} = 2.0$, $\hat{x}_{t_0} = 2.3$, $P_{t_0} = 0.01$, and observations every 0.1 units of time, starting at 0.1, the various filters are simulated for three values of measurement noise variance R (0.09, 0.01, 0.0001). The results, presented in Figs. 9.1–9.3, enable us to study the significance of measurement nonlinearities.

We first observe that differences between the exact, Gaussian, and iterated extended Kalman filters are negligible, and do not even show on the scale used for plotting in these figures. This means that for our problem the Gaussian approximation is an excellent one. If the conditional density were precisely Gaussian, then the Gaussian and iterated extended Kalman filters would give the exact conditional mean. In

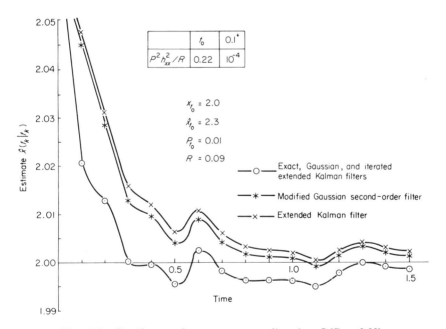

FIG. 9.1. Significance of measurement nonlinearity—I ($R = 0.09$).

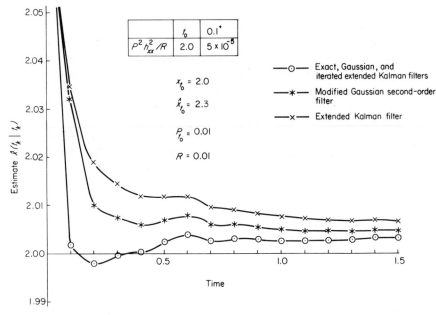

FIG. 9.2.   Significance of measurement nonlinearity—II ($R = 0.01$).

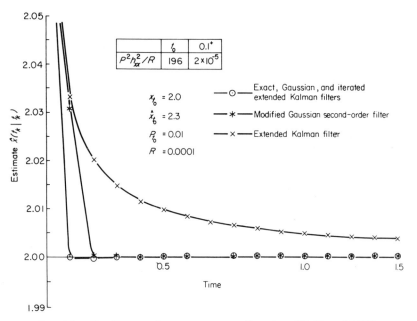

FIG. 9.3.   Significance of measurement nonlinearity—III ($R = 0.0001$).

Fig. 9.1, where the measurement nonlinearity is relatively small ($P^2 h_{xx}^2/R = 0.22$ initially), nonlinear effects are not very significant. Although the exact filter is initially best, subsequent performance of the various filters does not differ substantially. The extended Kalman filter would be chosen here as "best" on the basis of simplicity. As we go to larger measurement nonlinearities in Figs. 9.2 and 9.3, the picture changes significantly. The performance of the modified Gaussian second-order filter approaches that of the exact filter, whereas the extended Kalman filter estimate is biased. We note, however, that in all simulations the extended Kalman filter eventually converges to the true state. The larger the nonlinearity, the slower is the convergence of the extended Kalman filter. Simulations indicate that the modified Gaussian second-order filter adequately accounts for the biasing effect of measurement nonlinearities, although some bias still remains in cases where the measurement nonlinearity is marginally significant (see Fig. 9.2).

As we observed in Section 6, measurement nonlinearities are affected by h_{xx} and P_{t_0}. These effects are shown in Figs. 9.4 and 9.5. In Fig. 9.4, measurement nonlinearity is decreased relative to that in Fig. 9.3 by decreasing h_{xx}. In Fig. 9.5, the measurement nonlinearity is increased

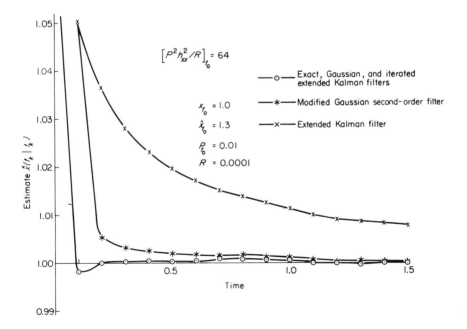

FIG. 9.4. Effect of h_{xx}

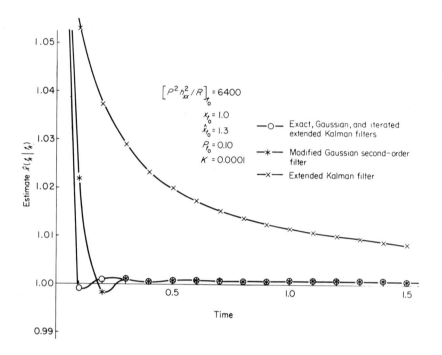

FIG. 9.5.   Effect of P_{t_0}.

TABLE 9.1

CONDITIONAL VARIANCES ASSOCIATED WITH FIG. 9.4

	$P(t_k \mid t_k) \times 10^6$				
Time	Exact	Gaussian	IE–Kalman[a]	MGSO[b]	E–Kalman[c]
0.1	4.019	4.019	4.020	81.851	1.700
0.2	2.003	2.007	2.004	3.295	1.132
0.4	0.998	1.002	1.001	1.234	0.696
1.0	0.399	0.400	0.400	0.431	0.331
2.0	0.200	0.200	0.200	0.207	0.179

[a] Iterated extended Kalman filter.
[b] Modified Gaussian second-order filter.
[c] Extended Kalman filter.

relative to that in Fig. 9.4 by increasing P_{t_0}. The behavior of the non-linear filters is as predicted, and as just described.

The conditional variances associated with the various filters, for the run in Fig. 9.4, are given in Table 9.1. Here we see again that the exact, Gaussian, and iterated extended Kalman filters are in excellent agree-

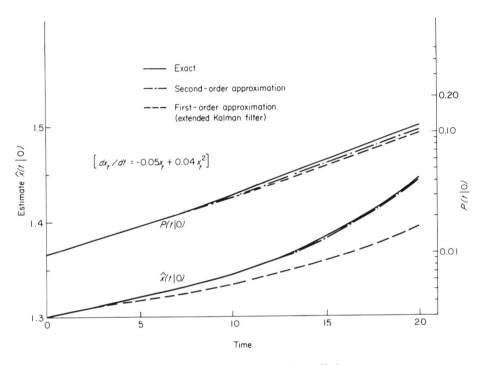

FIG. 9.6. System nonlinearities in prediction.

ment. The modified Gaussian second-order filter overestimates the variance, which is generally desirable. The extended Kalman filter usually underestimates the variance, which has a detrimental effect. The extended Kalman estimate is biased, but optimistic. This filter thinks it is performing very well, whereas, in reality, it is not.

When dynamics are introduced into the simulations we have described, the results remain essentially unchanged. The effect of system non-linearities is not felt in the short time intervals between observations. The effect of measurements dominates.

The effect of system nonlinearities in (long-time) prediction is shown in Fig. 9.6. The effect on the conditional variance is not very significant, whereas the effect on the estimate itself is. First-order theory, which neglects second-order nonlinearities, produces a significantly biased prediction. The second-order approximation accounts for most of this bias.

It should be noted that in our simulations the measurement non-linearity decreased sharply after the first observation was processed. This is, of course, due to the sharp reduction in P. As a consequence, it is important to account for measurement nonlinearities early in the estimation. Subsequently, nonlinearities become insignificant, and the extended Kalman filter can then be used effectively. This can be seen in the simulation of the iterated extended Kalman filter. Generally, four iterations are required in processing the first observation. Sub-sequently, one iteration is usually sufficient. One iteration of the iterated extended Kalman filter is equivalent to the extended Kalman filter itself. This conclusion should not be generalized however. Non-linearities can occur later in the estimation, as in the reentry filtering problem described in Section 10.

Simulations in our simple nonlinear example indicate that the iterated extended Kalman filter is extremely effective in accounting for measurement nonlinearities. The modified Gaussian second-order filter is also very effective, although it does not remove the biasing effect of non-linearities completely. The first filter involves iteration, whereas the second one requires the computation of second partial derivatives of the measurement function. As a consequence, it is difficult to choose between the two on the basis of this example. We saw (Fig. 9.6) that the second-order approximation is very effective in removing the biasing effect of system nonlinearities. These nonlinear filters give significantly better performance than the extended Kalman filter.

9. APPLICATIONS IN ORBIT DETERMINATION

Extensive orbit determination experience at Goddard (see Wolf [30]) appears to indicate that the extended Kalman filter is adequate in this application. Second-order nonlinear effects seem to be negligible. This conclusion appears correct in view of this author's experiments with the modified Gaussian second-order filter in simple orbit determination problems.

We have in mind here orbit determination situations in which many data are available. Nonlinear effects in long-time prediction may still be significant. In that case, however, the effect of dynamic model errors is very important (see Chapter 8, particularly Section 8.8). The problem of model accuracy must clearly come first. It makes little sense to take into account second derivatives of a function, when the function itself is not accurately known.

10. APPLICATIONS IN REENTRY

The first application of approximate nonlinear filters in a problem of practical significance containing significant nonlinearities was made by Athans, *et al.* [1], and Wishner *et al.* [29] in reentry filtering. The reader can refer to Section 8.7 and Appendix 8B for a description of reentry dynamics and filtering. We saw in Section 8.7 that nonlinearities are significant in reentry. They are particularly severe during high decelerations which occur midway in the reentry.

These authors simulated the extended Kalman filter, the iterated linear filter-smoother (Theorem 8.3), and the modified Gaussian second-order filter. Their problem is too complex to permit the computation of the exact conditional mean and covariance matrix, so that filter performance must be assessed on the basis of average properties. For this purpose, the average bias

$$b_j(t_k) = N^{-1} \sum_{i=1}^{N} [x_{t_k}^{(i)} - \hat{x}^{(i)}(t_k \mid t_k)]_j \, ,$$

and the *RMS* estimation error

$$RMS_j(t_k) = \left\{ N^{-1} \sum_{i=1}^{N} [(x_{t_k}^{(i)})_j - (\hat{x}^{(i)}(t_k \mid t_k))_j]^2 \right\}^{1/2}$$

for each filter are used. Here subscript j denotes the jth component of the vector, and superscript (i) denotes the ith simulation run.

These authors found that the iterated linear filter-smoother and the modified Gaussian second-order filter gave substantially better average performance than the extended Kalman filter. The improved performance was particularly striking in the high deceleration region and in the remainder of the reentry trajectory. In general, system nonlinearities appear much more significant than measurement nonlinearities in their examples. The modified Gaussian second-order filter successfully removes the biasing effect of nonlinearities and has the smallest average bias of the filters tested. On the other hand, the iterated linear filter-smoother shows somewhat better *RMS* error performance than the modified Gaussian second-order filter. This suggests the possibility of using the second-order filter (propagation of the conditional mean between observations) in the iterator.

In their simulations, the authors found that the modified Gaussian second-order filter consumed about 20% more computing time than the extended Kalman filter. Two iterations of the iterated filter of course consume twice as much time as the extended Kalman filter. Most of the performance improvement is fortunately achieved with two iterations.

11. SUMMARY

The extended Kalman filter has been successfully applied to numerous nonlinear filtering problems. If nonlinearities are significant, however, filter performance can be substantially improved by local iteration or inclusion of second-order effects. The modified Gaussian second-order filter has been applied in several problems with considerable success. The improvement in performance realized with the modified Gaussian second-order filter can be particularly significant in real-time applications where decisions must be made on the basis of the estimate and estimation error statistics. Such is the case in guidance and control applications.

It must be emphasized, however, that, in at least one highly nonlinear problem (see Kushner [19]), the modified Gaussian second-order filter diverged! The problems in which the modified Gaussian second-order filter performed well are problems in which the extended Kalman filter performed also, although perhaps not as well.

It is of course possible to develop higher-order approximations and to consider higher-order moments and even skewness (third moments). The practical usefulness of higher-order approximations is, however, questionable, especially in problems of high dimension (>3, say). Furthermore, it is questionable whether higher-order approximations would improve performance in cases where the extended Kalman filter does not work at all (diverges). The crucial defect in our approximations is that we replace global properties of a function (its average) by local properties (its derivatives). If the first-order approximation is completely inadequate, then finite higher-order approximation will tend to be useless as well.

In the final analysis, our approximations are *ad hoc* and must be tested by simulation. Thus, much more simulation experience with approximate nonlinear filters is highly desirable. Simulations could give considerable insight into the stability properties of these filters, as well as determine their statistical performance. Such simulations might also suggest different types of approximations.

Along theoretical lines, research is needed in techniques of parametrizing probability density functions.

Let us conclude by observing that, although approximate nonlinear filters have been found useful in some applications, their value in a particular problem cannot be assessed *a priori*, but must be determined by simulations.

APPENDIX 9A

APPROXIMATE CONTINUOUS FILTERS: VECTOR CASE

We develop in this appendix the truncated and Gaussian second-order filters for continuous vector systems. The approximations on which these filters are based are given in Section 3. They involve approximations to the expectations appearing in the equations for the conditional mean and covariance matrix in Theorem 6.6.

For both filters, we have

$$\hat{f}_i(x, t) = f_i(\hat{x}, t) + \frac{1}{2} \sum_{j,k=1}^{n} P_{jk} \frac{\partial^2 f_i(\hat{x}, t)}{\partial x_j\, \partial x_k}, \tag{9A.1}$$

$$\widehat{x_i h_j} - \hat{x}_i \hat{h}_j = \sum_{k=1}^{n} P_{ik} \frac{\partial h_j(\hat{x}, t)}{\partial x_k}, \tag{9A.2}$$

where we have used the symmetry of the covariance matrix P. Similar expressions hold for $\hat{h}_i$ and $(\widehat{x_i f_j} - \hat{x}_i \hat{f}_j)$. For the *truncated* second-order filter, we get

$$[\widehat{x_i x_j h_k} - \widehat{\hat{x}_i x_j h_k} - \widehat{\hat{x}_i x_i h_k} - \widehat{\hat{x}_j x_i h_k} + 2\hat{x}_i \hat{x}_j \hat{h}_k]$$

$$= -\tfrac{1}{2} P_{ij} \sum_{q,r=1}^{n} \frac{\partial^2 h_k(\hat{x}, t)}{\partial x_q\, \partial x_r} P_{qr}, \tag{9.A3}$$

whereas for the *Gaussian* second-order filter this quantity (in brackets) is

$$[\cdot] = \sum_{q,r=1}^{n} \frac{\partial^2 h_k(\hat{x}, t)}{\partial x_q\, \partial x_r} P_{iq} P_{jr}. \tag{9A.4}$$

Let us define the quantities

$$(P\, \partial^2 f)_i \triangleq \sum_{j,k=1}^{n} P_{jk} \frac{\partial^2 f_i(\hat{x}, t)}{\partial x_j\, \partial x_k}, \tag{9A.5}$$

$$(P^2\, \partial^2 h)_{ijk} \triangleq \sum_{q,r=1}^{n} \frac{\partial^2 h_k(\hat{x}, t)}{\partial x_q\, \partial x_r} P_{iq} P_{jr}, \tag{9A.6}$$

and the matrices

$$F \triangleq \left[\frac{\partial f_i(\hat{x}, t)}{\partial x_j} \right] \quad (n \times n),$$

$$M \triangleq \left[\frac{\partial h_i(\hat{x}, t)}{\partial x_j} \right] \quad (m \times n),$$

$$(P \, \partial^2 f) \triangleq [(P \, \partial^2 f)_1 ,..., (P \, \partial^2 f)_n]^{\mathrm{T}} \quad (n \times 1),$$

$$(P \, \partial^2 h) \triangleq [(P \, \partial^2 h)_1 ,..., (P \, \partial^2 h)_m]^{\mathrm{T}} \quad (m \times 1).$$

(9A.7)

Furthermore, if ξ is an m-vector, we define the operation

$$(P^2 \, \partial^2 h) : \xi \triangleq \left[\sum_{k=1}^{m} (P^2 \, \partial^2 h)_{ijk} \, \xi_k \right], \tag{9A.8}$$

which is an $n \times n$ symmetric matrix.

Then, from Theorem 6.6, we have the *truncated* second-order filter

$$d\hat{x}_t = [f(\hat{x}_t , t) + \tfrac{1}{2}(P_t \, \partial^2 f)] \, dt$$
$$+ P_t M^{\mathrm{T}} R^{-1}[dz_t - (h(\hat{x}_t , t) + \tfrac{1}{2}(P_t \, \partial^2 h)) \, dt], \tag{9A.9}$$

$$dP_t = [FP_t + P_t F^{\mathrm{T}} + \widehat{GQG^{\mathrm{T}}} - P_t M^{\mathrm{T}} R^{-1} M P_t] \, dt$$
$$- \tfrac{1}{2} P_t \{ (P_t \, \partial^2 h)^{\mathrm{T}} R^{-1}[dz_t - (h(\hat{x}_t , t) + \tfrac{1}{2}(P_t \, \partial^2 h)) \, dt] \}. \tag{9A.10}$$

Note that the quantity in braces in (9A.10) is a scalar. The *Gaussian* second-order filter has Eq. (9A.9) for the conditional mean, but has the variance equation

$$dP_t = [FP_t + P_t F^{\mathrm{T}} + \widehat{GQG^{\mathrm{T}}} - P_t M^{\mathrm{T}} R^{-1} M P_t] \, dt$$
$$+ (P_t^2 \, \partial^2 h) : \{ R^{-1}[dz_t - (h(\hat{x}_t , t) + \tfrac{1}{2}(P_t \, \partial^2 h)) \, dt] \}. \tag{9A.11}$$

We leave it for the reader to work out $[GQG^{\mathrm{T}}]^\wedge$, which is different for the two filters.

APPENDIX 9B

APPROXIMATE CONTINUOUS-DISCRETE FILTERS:

VECTOR CASE

In this appendix, we derive the truncated, Gaussian, and modified second-order filters for continuous-discrete vector systems. The approximations involved in these filters are described in Section 3.

The approximate equations of evolution for the conditional mean and covariance matrix *between observations* are the vector analogs of (9.17), obtained from (9A.9) and (9A.10) by setting $R^{-1} \equiv 0$:

$$d\hat{x}_t/dt = f(\hat{x}_t, t) + \tfrac{1}{2}(P_t \, \partial^2 f),$$

$$dP_t/dt = FP_t + P_t F^{\mathrm{T}} + \widehat{GQG^{\mathrm{T}}}. \tag{9B.1}$$

The n-vector $P_t \, \partial^2 f$ and the $n \times n$ matrix F of partial derivatives are defined in Appendix 9A. The evaluation of $[GQG^{\mathrm{T}}]\hat{}$ for the truncated and Gaussian approximations is left to the reader [see (9.8) and (9.9) for the scalar case].

At an observation, we assume the form

$$\hat{x}(t_k \mid t_k) = a + B(y_{t_k} - \hat{y}_{t_k}^{t_{k-1}}),$$

$$[P(t_k \mid t_k)]_{ij} = C_{ij} + \sum_{l=1}^{m} D_{ijl}(y_{t_k} - \hat{y}_{t_k}^{t_{k-1}})_l. \tag{9B.2}$$

Here a is an n-vector, B is $n \times m$, and C is $n \times n$ with elements C_{ij}. Equations (9.23)–(9.25), which result from the application of (9.21), hold in the nonlinear case, except that now

$$y_{t_k} - \hat{y}_{t_k}^{t_{k-1}} = h - \hat{h} + v_k. \tag{9B.3}$$

Using the notation described below Eq. (9.34), Eqs. (9.23)–(9.25) give

$$a = \hat{x}(t_k \mid t_{k-1}), \tag{9B.4}$$

$$B = (\widehat{xh^{\mathrm{T}}} - \hat{x}\hat{h}^{\mathrm{T}})[[(h - \hat{h})(h - \hat{h})^{\mathrm{T}}]\hat{} + R]^{-1}, \tag{9B.5}$$

$$C = P(t_k \mid t_{k-1}) - (\widehat{xh^{\mathrm{T}}} - \hat{x}\hat{h}^{\mathrm{T}})[[(h - \hat{h})(h - \hat{h})^{\mathrm{T}}]\hat{} + R]^{-1}(\widehat{hx^{\mathrm{T}}} - \hat{h}\hat{x}^{\mathrm{T}}). \tag{9B.6}$$

The coefficients D_{ijl} are evaluated by setting

$$s = [x_{t_k} - \hat{x}(t_k \mid t_k)]_i \, [x_{t_k} - \hat{x}(t_k \mid t_k)]_j,$$

$$u = (y_{t_k} - \hat{y}_{t_k}^{t_{k-1}})_l$$

in (9.21), whereby we obtain

$$[(x - \hat{x})_i \, (x - \hat{x})_j \, (h - \hat{h})_l]\hat{} - [(x - \hat{x})_i \, (B(h - \hat{h}))_j \, (h - \hat{h})_l]\hat{}$$

$$- [(B(h - \hat{h}))_i \, (x - \hat{x})_j \, (h - \hat{h})_l]\hat{} + [(B(h - \hat{h}))_i \, (B(h - \hat{h}))_j \, (h - \hat{h})_l]\hat{}$$

$$= \sum_{q=1}^{m} D_{ijq}[[(h - \hat{h})_q \, (h - \hat{h})_l]\hat{} + R_{ql}],$$

which, after some manipulation, reduces to

$$D_{ijl} = \sum_{q=1}^{m} \left\{ [(x - \hat{x})_i \, (x - \hat{x})_j \, h_q]\hat{} - \sum_{r=1}^{m} B_{jr}[(h - \hat{h})_r \, (x - \hat{x})_i \, h_q]\hat{} \right.$$

$$- \sum_{r=1}^{m} B_{ir}[(h - \hat{h})_r (x - \hat{x})_j \, h_q]\hat{}$$

$$\left. + \sum_{r,p=1}^{m} B_{ir} B_{jp}[[(h - \hat{h})_r \, (h - \hat{h})_p \, h_q]\hat{} + R_{rp}\hat{h}_q] - C_{ij}\hat{h}_q \right\}$$

$$\times \{[(h - \hat{h})(h - \hat{h})^{\mathsf{T}}]\hat{} + R\}_{ql}^{-1}. \tag{9B.7}$$

Since the coefficients D_{ijl} are extremely cumbersome, although straightforward, to evaluate, and because of the numerical results of Section 8, we shall not evaluate them here. The interested reader may evaluate them himself, or refer to Jazwinski [16]. We shall develop here the modified second-order filters of the form

$$\hat{x}(t_k \mid t_k) = a + B(y_{t_k} - \hat{y}_{t_k}^{t_{k-1}}),$$
$$P(t_k \mid t_k) = C. \tag{9B.8}$$

For both the truncated and Gaussian approximations,

$$\widehat{xh^{\mathsf{T}}} - \hat{x}\hat{h}^{\mathsf{T}} = P(t_k \mid t_{k-1}) \, M^{\mathsf{T}}(t_k), \tag{9B.9}$$

where the matrix of partial derivatives M is defined in (9A.7). For the *truncated* approximation,

$$[(h - \hat{h})(h - \hat{h})^{\mathsf{T}}]\hat{} = M(t_k) \, P(t_k \mid t_{k-1}) \, M^{\mathsf{T}}(t_k) - \tfrac{1}{4}(P \, \partial^2 h)(P \, \partial^2 h)^{\mathsf{T}}, \tag{9B.10}$$

where the vector $P \, \partial^2 h$ is defined in (9A.7); whereas for the *Gaussian* approximation,

$$[(h - \hat{h})(h - \hat{h})^{\mathsf{T}}]\hat{} = M(t_k) \, P(t_k \mid t_{k-1}) \, M^{\mathsf{T}}(t_k) + \tfrac{1}{2}(\partial^2 h P^2 \, \partial^2 h), \tag{9B.11}$$

where $\partial^2 h P^2 \, \partial^2 h$ is the symmetric $m \times m$ matrix with elements (i, j)

$$\sum_{k,l,p,q=1}^{n} \frac{\partial^2 h_i}{\partial x_k \, \partial x_l} P_{lp} P_{kq} \frac{\partial^2 h_j}{\partial x_p \, \partial x_q} \qquad [P = P(t_k \mid t_{k-1})]. \tag{9B.12}$$

Now, defining the $m \times m$ matrices,

$$Y^t \triangleq M(t_k) \, P(t_k \mid t_{k-1}) \, M^{\mathsf{T}}(t_k) + R(k) - \tfrac{1}{4}(P \, \partial^2 h)(P \, \partial^2 h)^{\mathsf{T}}, \tag{9B.13}$$

$$Y^G \triangleq M(t_k) \, P(t_k \mid t_{k-1}) \, M^{\mathsf{T}}(t_k) + R(k) + \tfrac{1}{2}(\partial^2 h P^2 \, \partial^2 h), \tag{9B.14}$$

the *modified truncated second-order filter* becomes (at an observation)

$$\hat{x}(t_k \mid t_k) = \hat{x}(t_k \mid t_{k-1}) + P(t_k \mid t_{k-1}) \, M^\mathrm{T}(t_k)[Y^t]^{-1}$$
$$\times \, [y_{t_k} - h(\hat{x}(t_k \mid t_{k-1}), t_k) - \tfrac{1}{2}(P(t_k \mid t_{k-1}) \, \partial^2 h)], \qquad (9\mathrm{B}.15)$$
$$P(t_k \mid t_k) = P(t_k \mid t_{k-1}) - P(t_k \mid (t_{k-1}) \, M^\mathrm{T}(t_k)[Y^t]^{-1} M(t_k) \, P(t_k \mid t_{k-1}).$$

The *modified Gaussian second-order filter* is given by

$$\hat{x}(t_k \mid t_k) = \hat{x}(t_k \mid t_{k-1}) + P(t_k \mid t_{k-1}) \, M^\mathrm{T}(t_k)[Y^G]^{-1}$$
$$\times \, [y_{t_k} - h(\hat{x}(t_k \mid t_{k-1}), t_k) - \tfrac{1}{2}(P(t_k \mid t_{k-1}) \, \partial^2 h)], \qquad (9\mathrm{B}.16)$$
$$P(t_k \mid t_k) = P(t_k \mid t_{k-1}) - P(t_k \mid t_{k-1}) \, M^\mathrm{T}(t_k)[Y^G]^{-1} M(t_k) \, P(t_k \mid t_{k-1}).$$

REFERENCES

1. M. Athans, R. P. Wishner, and A. Bertolini, Suboptimal State Estimation for Continuous-Time Nonlinear Systems from Discrete Noisy Measurements, *1968 Joint Automatic Control Conf.*, Ann Arbor, Michigan, pp. 364–382, June 1968.
2. R. W. Bass, V. D. Norum, and L. Schwartz, Optimal Multichannel Nonlinear Filtering, *J. Math. Anal. Appl.* 16, 152–164 (1966).
3. J. V. Breakwell, Estimation with Slight Non-Linearity. Unpublished communication, TRW Systems, Redondo Beach, California, November 29, 1967.
4. T. M. Carney and R. M. Goldwyn, Numerical Experiments with Various Optimal Estimators, *J. Optimization Theory Appl.* 1, 113–130 (1967).
5. H. Cox, On Estimation of State Variables and Parameters, *Joint AIAA-IMS-SIAM-ONR Symp. Control and System Optimization*, Monterey, California, 1964.
6. H. Cox, On the Estimation of State Variables and Parameters for Noisy Dynamic Systems, *IEEE Trans. Automatic Control* 9, 5–12 (1964).
7. H. Cox, Estimation of State Variables via Dynamic Programming, *Proc. 1964 Joint Automatic Control Conf.*, pp. 376–381, Stanford, California, 1964.
8. H. Cramer, "Mathematical Methods of Statistics." Princeton Univ. Press, Princeton, New Jersey, 1961.
9. W. F. Denham and S. Pines, Sequential Estimation When Measurement Function Nonlinearity is Comparable to Measurement Error, *AIAA J.* 4, 1071–1076 (1966).
10. D. M. Detchmendy and R. Sridhar, Sequential Estimation of States and Parameters in Noisy Non-Linear Dynamical Systems, *Proc. 1965 Joint Automatic Control Conf.*, Troy, New York, pp. 56–63, 1965.
11. J. R. Fisher, Optimal Nonlinear Filtering, *Advan. Control Systems* 5, 198–301 (1967).
12. B. Friedland and I. Bernstein, Estimation of the State of a Nonlinear Process in the Presence of Nongaussian Noise and Disturbances, *J. Franklin Inst.* 281, 455–480 (1966).
13. A. H. Jazwinski, Nonlinear Filtering with Discrete Observations, *AIAA 3rd Aerospace Sci. Meeting, New York* Paper No. 66-38 (January 1966).
14. A. H. Jazwinski, Nonlinear Filtering—Numerical Experiments, *NASA-Goddard Astrodynamics Conf.*, Greenbelt, Maryland (April 1966).

15. A. H. Jazwinski, Filtering for Nonlinear Dynamical Systems, *IEEE Trans. Automatic Control* 11, 765–766 (1966).
16. A. H. Jazwinski, Stochastic Processes with Application to Filtering Theory, Analytical Mechanics Associates, Inc., Seabrook, Maryland, Rept. No. 66-6, 1966.
17. H. J. Kelley and W. F. Denham, Orbit Determination with the Davidon Method, *Proc. 1966 Joint Automatic Control Conf.*, *Seattle, Washington*, pp. 251–254, August 1966.
18. J. D. R. Kramer, Partially Observable Markov Processes, Ph.D. Thesis, MIT, Operations Research Center, Cambridge, Massachusetts, April 1964.
19. H. J. Kushner, Approximations to Optimal Nonlinear Filters, *IEEE Trans. Automatic Control* 12, 546–556 (1967).
20. P. I. Kuznetsov, R. L. Stratonovich, and V. I. Tikhonov, "Non-Linear Transformations of Stochastic Processes." Pergamon Press, New York, 1965.
21. H. E. Rauch, Advanced Orbit Mechanics Studies: Differential Correction and Least Squares Estimation, Lockheed Missiles and Space Company, Palo Alto, California. Rept. C-64-67-1, March 1967.
22. L. Schwartz, Approximate Continuous Nonlinear Minimal-Variance Filtering, Hughes Aircraft Company, Space Systems Division, El Segundo, California. SSD 60472R, Report Number 18, December 1966.
23. L. Schwartz and E. B. Stear, A Valid Mathematical Model for Approximate Nonlinear Minimal-Variance Filtering, *J. Math. Anal. Appl.* 21, 1–6 (1968).
24. L. Schwartz and E. B. Stear, A Computational Comparison of Several Nonlinear Filters, *IEEE Trans. Automatic Control* 13, 83–86 (1968).
25. H. W. Sorenson and A. R. Stubberud, Non-Linear Filtering by Approximation of the A Posteriori Density, *Intern. J. Control* 8, 33–51 (1968).
26. Y. Sunahara, An Approximation Method of State Estimation for Nonlinear Dynamical Systems, Brown University, Center for Dynamical Systems, Providence, Rhode Island, Tech. Rept. 67-8, December 1967.
27. P. Swerling, Classes of Signal Processing Procedures Suggested by Exact Minimum Mean Square Error Procedures, *SIAM J. Appl. Math.* 14, 1199–1224 (1966).
28. P. Swerling, Note On a New Computational Data Smoothing Procedure Suggested by Minimum Mean Square Error Estimation, *IEEE Trans. Information Theory* 12, 9–12 (1966).
29. R. P. Wishner, J. A. Tabaczynski, and M. Athans, On the Estimation of the State of Noisy Nonlinear Multivariable Systems, *IFAC Symp. Multivariable Control Systems, Dusseldorf, West Germany* (October 1968).
30. H. Wolf, An Assessment of the Work in Minimum Variance Orbit Determination, Analytical Mechanics Associates, Inc., Westbury, New York. Rept. No. 68-3, January 1968.

Author Index

Numbers in parentheses are reference numbers and indicate that an author's work is referred to, although his name is not cited in the text. Numbers in italics show the page on which the complete reference is listed.

A

Albert, A., 157 (1), *159*
Anderson, T. W., 46 (1), *46*, 87 (1), *92*
Andrews, A., 324 (1), *329*
Aoki, M., 233 (1), *262*, 270 (2), 324 (2), *329*
Astrom, K. J., 126 (1), *140*
Athans, M., 280 (40), *331*, 346 (1), 349 (29), 359 (1, 29), *365*, *366*

B

Bailie, A. E., 267 (24), 268 (24), 293 (24), 307 (24), 311 (15), 312 (15), 313 (15), 314 (15), *330*
Baker, R. M. L., 324 (3), *329*
Bartlett, M. S., 50 (2), 57 (2), *92*
Bass, R. W., 337 (2), *365*

Battin, R. H., 145 (24), 157 (2), *159*, *160*, 208 (2, 3), 247 (3), *262*, 267 (4), 268 (5), *329*
Bellantoni, J. F., 324 (6), *329*
Bellman, R. E., 43 (2), *46*, 158 (3–5), *159*, 225 (4), 227 (5), *262*
Bernstein, I., 145 (17), 153 (17), 157 (17), 158 (17), *160*, 347 (12), *365*
Bertolini, A., 346 (1), 359 (1), *365*
Bertram, J. E., 239 (30, 31), 240 (30, 31), *264*
Bharucha-Reid, A. T., 81 (3), *92*, 131 (2), *140*
Bodewig, E., 262 (6), *262*
Bogdanoff, J. L., 136 (3), *140*
Breakwell, J. V., 351 (3), *365*
Brown, J. L., 148 (6), *159*
Bryson, A. E., Jr., 145 (7), 153 (7), 157 (7), 159 (8, 9, 27), *159*, *160*, 205 (10), 213 (9), 218 (9), 223 (7), 224 (7), 230 (8, 39), 231 (7), *263*, *264*

367

Subject Index